# Lecture Notes in Physics

## New Series m: Monographs

## The Editorial Policy for Monographs

The series Lecture Notes in Physics reports new developments in physical research and teaching - quickly, informally, and at a high level. The type of material considered for publication in the New Series m includes monographs presenting original research or new angles in a classical field. The timeliness of a manuscript is more important than its form, which may be preliminary or tentative. Manuscripts should be reasonably self-contained. They will often present not only results of the author(s) but also related work by other people and will provide sufficient motivation, examples, and applications.

The manuscripts or a detailed description thereof should be submitted either to one of the series editors or to the managing editor. The proposal is then carefully refereed. A final decision concerning publication can often only be made on the basis of the complete manuscript, but otherwise the editors will try to make a preliminary decision as definite as they can on the basis of the available information.

Manuscripts should be no less than 100 and preferably no more than 400 pages in length. Final manuscripts should preferably be in English, or possibly in French or German. They should include a table of contents and an informative introduction accessible also to readers not particularly familiar with the topic treated. Authors are free to use the material in other publications. However, if extensive use is made elsewhere, the publisher should be informed.

Authors receive jointly 50 complimentary copies of their book. They are entitled to purchase further copies of their book at a reduced rate. As a rule no reprints of individual contributions can be supplied. No royalty is paid on Lecture Notes in Physics volumes. Commitment to publish is made by letter of interest rather than by signing a formal contract. Springer-Verlag secures the copyright for each volume.

## The Production Process

The books are hardbound, and quality paper appropriate to the needs of the author(s) is used. Publication time is about ten weeks. More than twenty years of experience guarantee authors the best possible service. To reach the goal of rapid publication at a low price the technique of photographic reproduction from a camera-ready manuscript was chosen. This process shifts the main responsibility for the technical quality considerably from the publisher to the author. We therefore urge all authors to observe very carefully our guidelines for the preparation of camera-ready manuscripts, which we will supply on request. This applies especially to the quality of figures and halftones submitted for publication. Figures should be submitted as originals or glossy prints, as very often Xerox copies are not suitable for reproduction. In addition, it might be useful to look at some of the volumes already published or, especially if some atypical text is planned, to write to the Physics Editorial Department of Springer-Verlag direct. This avoids mistakes and time-consuming correspondence during the production period.

As a special service, we offer free of charge LaTeX and TeX macro packages to format the text according to Springer-Verlag's quality requirements. We strongly recommend authors to make use of this offer, as the result will be a book of considerably improved technical quality. The typescript will be reduced in size (75% of the original). Therefore, for example, any writing within figures should not be smaller than 2.5 mm.

Manuscripts not meeting the technical standard of the series will have to be returned for improvement.

For further information please contact Springer-Verlag, Physics Editorial Department II, Tiergartenstrasse 17, W-6900 Heidelberg, FRG.

R. K. Zeytounian

# Meteorological Fluid Dynamics

Asymptotic Modelling, Stability and
Chaotic Atmospheric Motion

Springer-Verlag Berlin Heidelberg GmbH

**Author**

Radyadour K. Zeytounian
Université de Lille I, Laboratoire de Mécanique de Lille
F-59655 Villeneuve d'Ascq Cedex, France

ISBN 978-3-662-13842-7    ISBN 978-3-540-38386-4 (eBook)
DOI 10.1007/978-3-540-38386-4

© Springer-Verlag Berlin Heidelberg 1991

Originally published by Springer-Verlag Berlin Heidelberg New York in 1991
Softcover reprint of the hardcover 1st edition 1991

2153/3140-543210 - Printed on acid-free paper

## PREFACE

This short course on Meteorological Fluid Dynamics (MFD) is strongly influenced by the author's own conception of meteorology as a fluid mechanics discipline, which is a privileged area for applied mathematics techniques.

One of the key features of MFD is the need to combine model equations of the basic "exact" Navier-Stokes (N-S) equations for atmospheric motions with a careful and rational fluid dynamics analysis. Therefore, much of the discussion of this course is directed towards the subsequent derivation and analysis of systematic approximations and consistent model equations to the exact N-S equations for atmospheric phenomena.

Obviously, the process of research towards developing appropriate fluid dynamics models, rooted in a rational use of modeling, is very important for a basic approach to the difficult problem of inserting atmospheric flows in a meteorological context. A complete consistent rational modeling of atmospheric phenomena is a long way in the future, but fairly sophisticated fluid dynamics models of various aspects of the individual motions of the atmosphere are available today. Unfortunately, at the present time a considerable gap still exists between fluid dynamics modeling of various atmospheric motions and the application to the problem of numerical weather prediction.

Naturally, the development of atmospherical-meteorological models from the point of view of fluid dynamics proceeds by considering models of submotions, which, when they prove to be successful, can be linked together. It may well be that in the next ten years it will be this aspect of MFD which makes the greatest advances. I feel that, parallel to a "practical" meteorology, whose goal is mainly to (numerically) predict the weather, we should develop a fluid dynamics meteorology, which would be considered one of the branches of theoretical fluid dynamics. In my opinion, this return of meteorology to the family of fluid mechanics will be of value to both meteorologists and fluid mechanics specialists.

It is important to understand that, in the majority of cases, the establishment of models is an intuitive, heuristic matter and so it is not clear how to insert the model under consideration into a hierarchy of rational approximations which in turn result from the general equations chosen at the beginning (either the N-S or Euler equations). It seems obvious that an improvement in weather forecasting depends largely on the obtaining of more efficient models and not only on the development of numerical techniques of analysis and calculation as is thought by certain specialists in the field of numerical weather forecasting.

The science of meteorology and, more particularly, numerical weather prediction is seen to be suffering today from an excess of "experimentation". Thus the realistic modeling of atmospheric phenomena is lagging behind. I am of the opinion, however, that only conceptually coherent theoretical modeling can bring to light the time problems to be solved in order to achieve a significant improvement in the reliability of predictions. Of course, it must not be forgotten that such modeling must be a mathematical expression of real atmospheric phenomena that permits their interpretation. Thus it is necessary from the start to choose sufficiently realistic equations and conditions which reflect the essential characteristics of atmospheric phenomena such as gravity, compressibility, stratification, viscosity, rotation and baroclinity. The fluid mechanics theorist now has avaible conceptual tools which permit the modeling of atmospheric phenomena - above all. I naturally have in mind the asymptotic techniques which have proven so decisive in fluid mechanics. I believe that these asymptotic techniques should find new applications in the special field of meteorology - a meaningful illustration of this tendency can be found in my recent book *Asymptotic Modeling of Atmospheric Flows* (Springer-Verlag, Heidelberg 1990). The present "short course" is a good preparation for the reading of this latter book, which presents various rational asymptotic models for application in meteorology and, especially, for short-term and local weather predictions.

Meteorological fluid dynamics is a relatively young science and I hope that the present course will aid the development of fluid dynamical studies in meteorology. In this course I have been highly selective in my choice of topics and in many cases the choice of topics for analysis is based on my own interest and judgement. In fact, the purpose of this short course is only to give a fluid mechanics description of a certain class of atmospheric phenomena. To that extent the text is a personal expression of my view of the subject and is constituted by ten chapters and two appendices.

Note that this course presupposes familiarity with the basic notions of fluid dynamics; nevertheless, they are briefly summarized, primarily to introduce suitable notation.

I am most grateful to Springer-Verlag for the publication of this book. I ask for the indulgence of English-speaking readers, thinking that they might prefer a text in not quite perfect English rather than in "perfect" French. Finally I thank Prof.Dr.W.Beiglböck for offering me the possibility of presenting these ideas on meteorological fluid dynamics.

Villeneuve d'Ascq                                Radyadour Kh. ZEYTOUNIAN
February 1991

CONTENTS

# CHAPTER I

## THE ROTATING EARTH
## AND ITS ATMOSPHERE

### 1 THE GRAVITATIONAL ACCELERATION

The earth revolves about its axis once in every 23 h 56 min and 4 s or a total of 86164s. The frequency of rotation or the angular velocity of the earth is:

$$(1,1) \qquad \Omega_0 = \frac{2 \, \Pi}{86164} = 7.292 \times 10^{-5} \frac{rad}{s} \, .$$

The radius of the earth at the geographic latitude of $\varphi = 45°$ is $a_0 = 6370.1$ km and the *true* gravitational acceleration owing to the pull of the earth, on the surface and at a geographic latitude of $\varphi = 45°$, is :

$$(1,2) \qquad |\vec{f}| = 9.82357 \ m/s^2,$$

and therefore, it will be assumed here that the body force, $\rho\vec{f}$ , in the momentum equation, is the true gravitational force, where $\rho$ is the atmosphere density.

To distinguish the experiences of a fixed and a rotating observer, let a subscript "a" denote quantities referred to an absolute, inertial frame of reference, and a subscript "r", quantities referred to a frame rotating whith angular velocity $\vec{\Omega} = \Omega_0 \vec{e}$ with respect to the absolute frame.

Let $\vec{i}$, $\vec{j}$ and $\vec{k}$ denote respectively the unit vectors pointing east, north and vertically upward, then :

$$(1,3) \qquad \vec{e} = \vec{k} \sin \varphi \ + \ \vec{j} \cos \varphi \, .$$

The absolute velocity is

$$\vec{V}_a \equiv \frac{\partial_a}{\partial t} \left[ \vec{X}(\vec{a}, t) \right] = (\frac{\partial_r}{\partial t} + \vec{\Omega} \wedge)\vec{X}(\vec{a}, t) \, ,$$

and since the rotating observer sees only the change $\frac{\partial_r}{\partial t}\left[\vec{X}(\vec{a},\ t)\right] \equiv \vec{V}_r$ in the position vector $\vec{X}$ of a point moving with the atmosphere, the respective velocities for the two observers are related by[†] :

(1,4) $\qquad \vec{V}_a = \vec{V}_r + \vec{\Omega} \wedge \vec{X}\ (\vec{a},t)$ .

Thus we obtain for the absolute acceleration the following relation:

$$\vec{\gamma}_a = \vec{\gamma}_r + 2\vec{\Omega} \wedge \vec{V}_r + \vec{\Omega} \wedge (\vec{\Omega} \wedge \vec{X})\ ,$$

and since $\vec{\Omega} \wedge \vec{X} \equiv \vec{\Omega} \wedge \vec{X}_\perp$ , where subscript $\perp$ denotes the equatorial component, $\vec{\Omega} \wedge (\vec{\Omega} \wedge \vec{X}) = -\Omega_0^2\vec{X}_\perp$ and

(1,5) $\qquad \vec{\gamma}_a = \vec{\gamma}_r + 2\vec{\Omega} \wedge \vec{V}_r - \Omega_0^2\vec{X}_\perp$ .

Then to the gravitational pull we should add vectorially the centrifugical force per unit mass and obtain a modified gravitational acceleration $\vec{g}$, such that :

(1,6) $\qquad \vec{g} = \vec{f} + \Omega_0^2\vec{X}_\perp$ ,

and resultant vector is slightly inclined away from the radius of the earth because the order of magnitude of the centrifugical acceleration is smaller; thus

$$|\vec{g}| \equiv g = |\vec{f}| - \Omega_0^2|\vec{X}_\perp| = 9.82357\ \frac{m}{s^2} - \Omega_0^2 a_0 \cos^2\varphi\ ,$$

or

(1,7) $\qquad g = 9.8066\ m/s^2$, since $\Omega_0^2 a_0 \cos^2\varphi = 0.0169\ m/s^2$ at $\varphi = 45°$.

The *Froude number* ,Fr, is a measure of the significance of the gravitational acceleration (the force of gravity). It is defined by

(1,8) $\qquad Fr = U_0/\sqrt{gL_0}$

where $|\vec{U}| = U_0$, and $\vec{U}$ is a characteristic velocity whereas $L_0$ is a characteric length.

---

[†] We may forgo the subscipt on $\vec{X}$, since we are at liberty to assume that the two frames coincide at the particular time t under consideration. At this point $\vec{a}$ denote the position vector at some chosen time, which may then be called t=0.

## 2 THE CORIOLIS ACCELERATION

In the equation (1,5) the terme $2\vec{\Omega} \wedge \vec{u}$ , where $\vec{u}$ denotes the relative velocity $\vec{V}_r$, is an apparent acceleration known as the *Coriolis acceleration* which exist only if there is motion with reference to a moving frame such as the earth. For the Coriolis acceleration we have :

$$(2,1) \qquad 2\vec{\Omega} \wedge \vec{u} = 2\Omega_0 \sin\varphi \ (\vec{k} \wedge \vec{u}) + 2\vec{\Omega}_0 \cos\varphi \ (\vec{j} \wedge \vec{u}) \ .$$

If u, v and w are the components of the relative velocity $\vec{u}$ :
$\vec{u} = u\,\vec{i} + v\,\vec{j} + w\,\vec{k}$ , then (2,1) becomes :

$$(2,2) \qquad \begin{aligned} 2\vec{\Omega} \wedge \vec{u} = \ &(2\Omega_0 w \cos\varphi - 2\Omega_0 v \sin\varphi)\ \vec{i} \\ &+ 2\Omega_0 u \sin\varphi\ \vec{j} \\ &- 2\Omega_0 u \cos\varphi\ \vec{k} \ . \end{aligned}$$

If we let the symbol

$$(2,3) \qquad f = 2\Omega_0 \sin\varphi$$

be called the *local Coriolis parameter*, the final expression for the Coriolis acceleration in terms of its components is

$$(2,4) \qquad 2\vec{\Omega} \wedge \vec{u} = f(u\vec{j} - v\vec{i}) + \frac{df}{d\varphi} (w\vec{i} - u\vec{k}) \ .$$

The importance of the Coriolis acceleration in relation to the inertial forces is given by the *Rossby number*, Ro, which is defined as

$$(2,5) \qquad Ro = \frac{U_0/L_0}{f_0} \ ,$$

where $f_0 \equiv 2\Omega_0 \sin\varphi_0$, with $\varphi_0 =$ Constant .
When Ro $\gg$ 1[†] , Coriolis forces are likely to cause only a slight modification of the flow pattern, but when Ro $\ll$ 1[††] the effects of the Coriolis force are likely to be dominant . When Ro $\sim$ 1[†††] we have the in-between situation . For the *synoptic - scale* atmospheric motions, we have :

---

[†]   That is for the *high* Rossby numbers.
[††]   That is for the *small* Rossby numbers.
[†††]   That is for Ro = O(1).

$L_0 \cong 10^6$m and Ro $\cong 10^{-1}$ , but for the *meso* (or *regional*) - *scale* atmospheric motions, we have : $L_0 \cong 10^5$m and Ro $\cong 1$. Finally, for the case of *local-scale* atmospheric process, we have rather : $L_0 \cong 10^4$m and Ro $\cong 10$ .

Here we prefer to use, instead of the Rossby number Ro, the parameter

$$(2,6) \qquad \frac{1/f_0}{t_0} = Ki \ ,$$

which is called the *Kibel number*. In (2,6), $t_0$ is a characteristic time scale and if $t_0$ is the advective time scale, $L_0/U_0$, then : $Ki \equiv Ro$ .

## BJERKNES' THEOREM

It is natural for the rotating observer to define the circulation of a circuit $\mathscr{C}$ as

$$\int_\mathscr{C} \vec{u} . d\vec{X} = \Gamma_r(\mathscr{C})$$

and the absolute circulation is then

$$(2,7) \qquad \Gamma_a(\mathscr{C}) \equiv \int_\mathscr{C} \vec{V}_a . d\vec{X} = \Gamma_r(\mathscr{C}) + \int_\mathscr{C} (\vec{\Omega} \wedge \vec{X}_\perp) . d\vec{X} \ .$$

But $(\vec{\Omega} \wedge \vec{X}) . d\vec{X} = (\vec{\Omega} \wedge \vec{X}_\perp) . d\vec{X}_\perp$ , and therefore

$$(2,8) \qquad \Gamma_a - \Gamma_r = \vec{\Omega} . \int_\mathscr{C} (\vec{X}_\perp \wedge d\vec{X}_\perp) = 2\Omega_0 \textstyle\sum(\mathscr{C}_\perp) \ ,$$

where $\sum(\mathscr{C}_\perp)$ is the area enclosed by the normal projection $\mathscr{C}_\perp$ of $\mathscr{C}$ onto the equatorial plane, provided that the orientation of $\mathscr{C}_\perp$ corresponding to that of $\mathscr{C}$ is related to $\vec{\Omega}$ by the right-hand screw rule.

This is the part of Bjerknes' theorem concerning the effect of rotation. As an illustration, consider a circuit $\mathscr{C}_0$ which is horizontal at latitude $\varphi > 0$, has enclosed area $A_0$, and the greatest diameter of which is small compared with the earth's radius $a_0$.

Then

$$\textstyle\sum(\mathscr{C}_\perp) = A_0 \sin \varphi$$

approximately, and hence, a meridional translation, without any deformation, of the atmosphere mass containing $\mathscr{C}_0$ is sufficient to create relative circulation $\Gamma_r$.

# 3 THE ATMOSPHERE AS A CONTINUUM

The general laws of mechanics and thermodynamics establish the basic working scientific principles of the atmosphere. The meaning that properties have in this atmosphere depends on an understandings of a continuum. For instance the meaning of pressure in the atmosphere is often defined as the total force per unit area imposed on each of the unit areas of the atmosphere and can be thought of as the force per unit area on a solid unit surface immersed at any point in the atmosphere owing to the continuous impingement and bouncing off of molecules at the surface. A given mass of atmospheric air in a constant volume and at a constant temperature is always under the same pressure. This is true for the thermodynamically pure substances. In particular this statement may be verified from Boyle's and Charles's laws pertaining to perfect gases. Boyle's law states that during an isothermal process the ratio of the pressure to the density is constant. Charles's law states that an isobaric (constant pressure) process the product of the density and the absolute temperature is constant. From a mathematical point of view, a continuum let us postulate that the properties at any one point can be expressed in terms of the properties at a neighboring point. This is because the property and its derivatives are continuous in their variations with space.

From the mechanics point of view, the atmosphere is a thin layer of gaseous mixture surrounding the surface of the earth which remains attached to the earth by the pull of the gravity.

The atmosphere is made up of a number of layers, each characterized by a distincly different temperature distribution.

The layer nearest the surface of the earth, characterized by a linear decrease of temperature with altitude, is called the *troposphere* or the *mixed* layer where the average rate of temperature decrease with altitude is approximately 6.5°K/km . We shall see that the temperature gradient in the lowest part of the troposphere varies a great deal, whereas in the upper layers it remains essentially unchanged. The troposphere contains about 80 percent of the total atmospheric mass. It is the layer in contact with the earth's surface and therefore it is most influenced by energy transfer through radiation, evaporation, condensation and convection. This layer is approximately 15 km in thickness and represents the limit within which conventional air flights take place. The troposphere is also the layer in which man-made pollution from industrial wastes in principally confined, and where most cloud formations are found.

Dynamically speaking, the troposphere is stable, but those portions of the layer nearest the surface of the earth are often unstable.

The atmosphere constitute, from a mechanical point of view, a category of continua called fluids- the atmosphere is a *Newtonian fluid*. Concerning the thermodynamics we assume that the atmosphere consists of *dry air* rulled by the law of perfect gases with constant specific $C_p$ and $C_v$ heats, namely :

$$(3,1) \qquad p = R \rho T \text{ and } e = C_v T \text{ ,}$$

$R = C_v (\gamma - 1)$ , $\gamma = C_p / C_v$, where p is the atmosphere pressure , $\rho$ is the atmosphere density, T is the absolute temperature of the atmosphere and e its internal energy per unit mass.

For dry air : $C_p = 1.015$ J°K/g and g = 9.8066 m/s$^2$, then the adiabatic lapse rate

$$(3,2) \qquad \frac{g}{C_p} \equiv \frac{\gamma-1}{\gamma} \frac{g}{R} = 9.66°K/km \quad ^\dagger \text{ .}$$

If the relative velocities are small the pressure will be only slightly disturbed from the value it would have in the absence of motion, $p_\infty(z_\infty)$, defined by the relations :

$$(3,3) \qquad \frac{dp_\infty}{dz_\infty} + g \, \rho_\infty = 0 \quad \text{and} \quad \rho_\infty(z_\infty) = \frac{p_\infty(z_\infty)}{RT_\infty(z_\infty)} \text{ ,}$$

and we can think of $p_\infty(z_\infty)$ , $\rho_\infty(z_\infty)$ and $T_\infty(z_\infty)$, functions of but one variable, namely the "*standard*" altitude $z_\infty$, as defining a standard atmosphere, i.e ,a basic state upon which fluctuations due to the motion occur. The basic standard state is assumed known, although in fact its determination from first principles requires the consideration of mechanism such a radiative transfer in the atmosphere.

In simple case we have from the first law thermodynamics, for the standard temperature $T_\infty(z_\infty)$, the following equation

$$(3,4) \qquad k(T_\infty) \frac{dT_\infty}{dz_\infty} + \hat{R}_\infty(T_\infty) = 0 \quad \text{with} \quad \frac{d\hat{R}_\infty(T_\infty)}{dz_\infty} \equiv \rho_\infty \hat{Q}_\infty(T_\infty),$$

---

† Note that for dry air $\gamma=1.4$ and the gas constant R=287 J°K/kg.

where $k(T_\infty)$ is the coefficient of thermal conductivity and $\hat{Q}_\infty(T_\infty)$ is the rate of heat supply per unit mass by radiative heat transfer. For our purpose, in this Course, it is sufficient to assume that $\hat{Q}_\infty$ is a known function of $T_\infty(z_\infty)$ and we suppose also that the influence of the rate of heat by radiative heat transfer on the atmospheric motions is essential even with the standard atmosphere. Doing this we consider only a mean standard heat source and ignore variation therefrom. For our purpose this modelling of the radiative heat transfer will be sufficient and as a result the thermodynamics reference quantities are $p_\infty(0)$, $\rho_\infty(0)$ and $T_\infty(0)$ (e.g. the values of standard state at the ground level).

In this case the hydrostatic equation (3,3) involve the following non dimensional ratio of reference quantities :

$$(3,5) \qquad Bo = \frac{gH_0}{\dfrac{p_\infty(0)}{\rho_\infty(0)}} = \frac{H_0}{\dfrac{R\,T_\infty(0)}{g}}$$

and this ratio Bo is called the *Boussinesq number*. In (3,5), $H_0$ is a characteristic length scale for the vertical motion and it follows from (3,5) that

$$(3,6) \qquad H_\infty \equiv R\,T_\infty(0)\,/g$$

is a characteristic length scale for the standard altitude $z_\infty$. It follows also from (1,8) and (3,5) that

$$(3,7) \qquad Fr^2 = \gamma\,\frac{M_0^2\,\varepsilon_0}{Bo}\,,$$

where

$$(3,8) \qquad M_0 = U_0\,/\,\sqrt{\gamma\,R\,T_\infty(0)}$$

is called the characteristic *Mach number* and

$$(3,9) \qquad \varepsilon_0 = H_0\,/\,L_0$$

is called the *hydrostatic parameter*.
The $N_\infty(z_\infty)$ defined by

$$(3,10) \qquad N_\infty^2(z_\infty) = \frac{g}{T_\infty(z_\infty)} \left[ \frac{\gamma - 1}{\gamma} \frac{g}{R} + \frac{dT_\infty}{dz_\infty} \right]$$

is called the *Brunt-Väisälä frequency* or the natural frequency of oscillations of a vertical column of "standard" atmospheric mass given a small displacement from its equilibrium position. *The standard atmosphere is statically stable when $N_\infty$ is real*

The existence of characteristic scales is exploited by the introduction of nondimensional quantities denoted by primes, i.e,

$$z \equiv H_0 z' \quad , \quad z_\infty = H_\infty z'_\infty \quad , \quad T_\infty = T_\infty(0) \ T'_\infty \ ,$$

and then

$$(3,11) \qquad \alpha_\infty^0 \ N_\infty'^2(z'_\infty) \equiv \frac{Bo}{T'_\infty} \left\{ \frac{\gamma - 1}{\gamma} + \frac{dT'_\infty}{dz'_\infty} \right\} \ ,$$

where $\alpha_\infty^0$ is a dimensionless *measure of the standard stability* and $N_\infty'(z'_\infty)$ is the dimensionless Väisälä internal frequency.

Finally, we obtain the following relation :

$$(3,12) \qquad z'_\infty = Bo \ z' \ ,$$

between the dimensionless $z'_\infty$ and $z'$ (local altitude).

*MODELLING OF THE TURBULENCE*

If $\hat{F}(\vec{u})$ is the frictional force in the atmosphere, then for Newtonian fluid like dry air

$$(3,13) \qquad \hat{F}(\vec{u}) = \mu \ \vec{\nabla}^2 \ \vec{u} + \frac{\mu}{3} \ \vec{\nabla} \ (\vec{\nabla}.\vec{u}),$$

where $\vec{\nabla}$ is the gradient operator, $\mu$ is the coefficient of viscosity, but the representation (3,13) based on laminar friction is valid only if $\mu$ is constant ($\mu \equiv \mu_0$) and if the so-called Stokes hypothesis (which amounts the neglecting the bulk viscosity) is adopted. These approximations will be adequate for our purpose.

According to our view, in this Course, the problem of modelling turbulence should not be considered at the same level as the one of building theoretical models for atmospheric motions.

The coefficient $\nu_0 \equiv \mu_0/\rho_\infty(0)$ is the kinematic viscosity and for dry air the value is $\nu_0 = 5$ m$^2$/s.

Strictly speaking $\nu_0$ is the kinematic eddy (turbulent) coefficient of viscosity and it is only from this eddy coefficient of viscosity that we can take into account the effect of "turbulence" in the boundary layer close to the earth's surface. Of course, such a procedure is much over-simplified but it will be sufficient for our purpose.

In the lowest few kilometres of the atmosphere, the eddy viscosity depends on both the topography and the background winds.

The simplified form (3,13) for the frictional force (viscous force density) is a bastardized representation, valid only in Cartesian coordinates, which represents the three-component relations.

A form of the viscous force dencity that is invariant under a change of coordinate system is:

$$(3,14) \qquad \hat{F}(\vec{u}) = \mu \left\{ \frac{4}{3} \vec{\nabla} (\vec{\nabla}.\vec{u}) - \vec{\nabla} \wedge (\vec{\nabla} \wedge \vec{u}) \right\} .$$

But if $\nu_0$ is constant, as it is in the lowest 40 km of the atmosphere, then it is not possible to use (3,14), because if $\nu_0$ is constant in the troposphere, then $\mu$ varies, so that a more complicated form must be used for the viscous force density

$$(3,15) \qquad \hat{F}(\vec{u}) = (\vec{\nabla}.\mu\vec{\nabla}) \, \vec{u} + \frac{1}{3} \vec{\nabla}(\mu\vec{\nabla}.\vec{u}) + \left\{ (\vec{\nabla}\mu \wedge \vec{\nabla}) \wedge \vec{u} \right\} .$$

The complete hydrodynamic equations for a viscous atmosphere involve not only the addition of a term like (3,13) to the momentum equation. To be completely accurate, viscosity must be included in the energy equation as well. That involve the addition of a term $\hat{\chi}(\vec{u})$, which is a quadratic functional of $\vec{u}$ and it is the production of thermal energy by the viscous dissipation of mechanical energy, in the first law of thermodynamics.

The *Reynolds number* is

$$(3,16) \qquad Re = U_0 L_0 / \nu_0$$

and this show relative importance of the inertia to the viscosity.

The measure of the ratio between the frictional and Coriolis forces is the *Ekman number*

$$(3,17) \qquad \text{Ek} \equiv \frac{\text{Ro}}{\text{Re}} = \frac{\nu_0}{f_0 L_0^2} \ .$$

An important feature of large synoptic scale motions in the atmosphere is that both the Kibel and Ekman numbers are small. A typical value of Ek in the earth's troposphere is $10^{-3}$, using value of eddy viscosity $\nu_0 = 5$ $m^2/s$. Except in the immediate vicinity of the equator, we have Ki<<1, if the characteristic time scale $t_0 >> 10^4$ s (for the synoptic motions we have to $t_0 \sim 10^5$ s).

In any realistic situation, in the atmosphere, $M_0 << 1$ and usually a meteorological synoptic situation corresponds to : $\varepsilon_0 \sim 10^{-2}$ and Bo $\sim$ 1. But, frequently, for the prediction of atmospheric phenomena at regional and local scales, one may assume that Bo<<1 (but $\varepsilon_0 \sim 1$).

*BACKGROUND READING*

For further details concerning the physical nature of the atmosphere the reader is referred to :

ESKINASI, S. (1975) _ *Fluid Mechanics and thermodynamics of our environment*,
                    Academic Press Inc., New York.

HOUGHTON, J.T. (1977) _ *The Physics of Atmospheres*,
                    Cambridge University Press.

SCORER, R.S. (1978) _ *Environmental Aerodynamics*,
                    Ellis Horwood limited,
                    Publisher, Chichester (England).

# CHAPTER II

# DYNAMICAL AND THERMODYNAMICAL EQUATIONS
# FOR ATMOSPHERIC MOTIONS

## 4 . THE BASIC EQUATIONS

In a coordinate frame rotating with the earth the momentum equations is

$$(4,1) \qquad \rho \, \frac{D\vec{u}}{Dt} = - \, \vec{\nabla} \, p + \rho \, \vec{g} - \rho \, (2 \, \vec{\Omega} \wedge \vec{u}) + \mu_0 \, \vec{\nabla}^2 \, \vec{u} + \frac{\mu_0}{3} \, \vec{\nabla} \, (\vec{\nabla}.\vec{u}) \, ,$$

where, $\vec{u}$ is the velocity vector as observed in the earth frame and

$$(4,2) \qquad \frac{D}{Dt} = \frac{\partial}{\partial t} + \vec{u}.\vec{\nabla} \, ,$$

is the material (or convective) derivative .

The conservation law of mass is expressed by the equation of continuity :

$$(4,3) \qquad \frac{D}{Dt}(\text{Log } \rho) + \vec{\nabla}.\vec{u} = 0$$

and the first law of thermodynamics by the energy equation :

$$(4,4) \qquad \rho \, \frac{De}{Dt} + p \, \rho \, \frac{D}{Dt}(1/\rho) = k_0 \vec{\nabla}^2 \, T + \hat{\chi}(\vec{u}) + \rho_\infty \, \hat{Q}_\infty(T_\infty) \, ,$$

where $k_0$ is the constant coefficient of thermal eddy conductivity and $\hat{\chi}(\vec{u})$ is expressible as :

$$\hat{\chi}(\vec{u}) = (\vec{\nabla} \, \vec{u}).\hat{\tau} \, ,$$

where [†]

---

[†] The transpose $\hat{A}^* = (a^*_{ij})$ of a tensor $\hat{A}=(a_{ij})$ is defined by $a^*_{ji}=a_{ij}$. The tensor $\hat{I}=(\delta_{ij})$, where $\delta_{ij}$ is the Kronecker delta, is the unit tensor. The gradient of the velocity: $\vec{\nabla}\vec{u}$ is a tensor, but $\vec{\nabla}.\vec{u}$ is a scalar and $\vec{\nabla}\wedge\vec{u}$ is a vector.

(4,5)
$$\hat{\tau} = \mu \left[ \vec{\nabla}\,\vec{u} + (\vec{\nabla}\,\vec{u})^* \right] - \frac{2}{3}\,\mu\,(\vec{\nabla}.\vec{u})\,\hat{I}$$

is called the viscous stress tensor.

Thus, as $\mu \equiv \mu_0$ ,

(4,6)
$$\hat{\chi}(\vec{u}) = \mu_0 \left\{ (\vec{\nabla}\,\vec{u}) . \left[ \vec{\nabla}\,\vec{u} + (\vec{\nabla}\,\vec{u})^* \right] - \frac{2}{3}\,(\vec{\nabla}.\vec{u})^2 \right\}$$

$$= \mu_0 \left\{ -\frac{2}{3}\,(\vec{\nabla}.\vec{u})^2 + \frac{1}{2}\left[(\text{def}\,\vec{u}) . (\text{def}\,\vec{u})\right] \right\} ,$$

where def $\vec{u} \equiv \vec{\nabla}\,\vec{u} + (\vec{\nabla}\,\vec{u})^*$ is the velocity deformation symmetric tensor .

For the acceleration $D\vec{u} / Dt$ a small calculation shows that

(4,7)
$$\frac{D\vec{u}}{Dt} = \frac{\partial\vec{u}}{\partial t} + (\vec{u}.\vec{\nabla})\,\vec{u}$$

$$= \frac{\partial\vec{u}}{\partial t} + \vec{\nabla}\,(\frac{q^2}{2}) + 2\,\vec{\omega} \wedge \vec{u} ,$$

where $q = |\vec{u}|$ and $\vec{\omega} = \frac{1}{2}(\vec{\nabla} \wedge \vec{u})$, and from this form, the components of $D\vec{u} / Dt$ are easily found for any system of curvilinear orthogonal coordinates. With the equation of state

(4,8)
$$p = R\,\rho\,T$$

and the thermodynamic relation

(4,9)
$$e = C_v T ,$$

the equations (4,1) , (4,3) and (4,4) provide a closed system for $\vec{u}$, $\rho$ and T .

*SPHERICAL COORDINATES*

It is helpful to employ now spherical coordinates : $\lambda$, $\varphi$, r and let u, v, w denote the corresponding velocity components in the direction of increasing azimuth ($\lambda$), latitude ($\varphi$) and radius (r) respectively .

In terms of these coordinates the line elements is given by :

(4,10)
$$d\sigma^2 = r^2 \cos^2\varphi\,d\lambda^2 + r^2\,d\varphi^2 + dr^2 ;$$

hence the metric coefficient are :

$$h_1 = r \cos \varphi \quad , \quad h_2 = r \quad \text{and} \quad h_3 = 1 \ .$$

Thus

(4,11)
$$\vec{\nabla} = \frac{1}{r \cos \varphi} \frac{\partial}{\partial \lambda} \vec{i} + \frac{1}{r} \frac{\partial}{\partial \varphi} \vec{j} + \frac{\partial}{\partial r} \vec{k}$$

and

(4,12)
$$\vec{\nabla} . \vec{u} = \frac{1}{r \cos \varphi} \frac{\partial u}{\partial \lambda} + \frac{1}{r \cos \varphi} \frac{\partial}{\partial \varphi} (\cos\varphi \ v) + \frac{\partial w}{\partial r} + \frac{2w}{r} \ ,$$

(4,13)
$$\vec{u} . \vec{\nabla} = \frac{u}{r \cos \varphi} \frac{\partial}{\partial \lambda} + \frac{v}{r} \frac{\partial}{\partial \varphi} + w \frac{\partial}{\partial r} \ .$$

Finally, for the Laplacian operator we have :

(4,14)
$$\vec{\nabla}^2 = \frac{1}{r^2 \cos^2\varphi} \frac{\partial^2}{\partial \lambda^2} + \frac{1}{r^2} \frac{\partial^2}{\partial \varphi^2} + \frac{\partial^2}{\partial r^2} - \frac{1}{r^2} \tan \varphi \frac{\partial}{\partial \varphi} + \frac{2}{r} \frac{\partial}{\partial r} \equiv \Delta \ .$$

The changes in unit vectors during the differentiation are :

(4,15a)
$$\frac{\partial \vec{i}}{\partial \varphi} = 0 \quad , \quad \frac{\partial \vec{j}}{\partial \varphi} = -\vec{k} \quad , \quad \frac{\partial \vec{k}}{\partial \varphi} = \vec{j} \ ;$$

(4,15b)
$$\frac{\partial \vec{i}}{\partial r} = \frac{\partial \vec{j}}{\partial r} = \frac{\partial \vec{k}}{\partial r} \equiv 0 \ ;$$

(4,15c)
$$\frac{\partial \vec{i}}{\partial \lambda} = -k \cos \varphi + \vec{j} \sin \varphi \ , \quad \frac{\partial \vec{j}}{\partial \lambda} = -\vec{i} \sin \varphi \ , \quad \frac{\partial \vec{k}}{\partial \lambda} = \vec{i} \cos \varphi \ .$$

From (4,13) and (4,15) it follows that

(4,16)
$$\frac{D\vec{u}}{Dt} = \left\{ \frac{\partial u}{\partial t} + \frac{u}{r \cos \varphi} \frac{\partial u}{\partial \lambda} + \frac{v}{r} \frac{\partial u}{\partial \varphi} + w \frac{\partial u}{\partial r} + \frac{uw}{r} - \frac{uv}{r} \tan \varphi \right\} \vec{i}$$
$$+ \left\{ \frac{\partial v}{\partial t} + \frac{u}{r \cos \varphi} \frac{\partial v}{\partial \lambda} + \frac{v}{r} \frac{\partial v}{\partial \varphi} + w \frac{\partial v}{\partial r} + \frac{vw}{r} + \frac{u^2}{r} \tan \varphi \right\} \vec{j}$$
$$+ \left\{ \frac{\partial w}{\partial t} + \frac{u}{r \cos \varphi} \frac{\partial w}{\partial \lambda} + \frac{v}{r} \frac{\partial w}{\partial \varphi} + w \frac{\partial w}{\partial r} - \frac{u^2 + v^2}{r} \right\} \vec{k} \ ,$$

and from (4,14), with (4,15), the Laplacian operator of $\vec{u}$ is :

$$\vec{\nabla}^2\vec{u} = \left\{ \Delta u - \frac{u}{r^2\cos^2\varphi} + \frac{2}{r^2\cos\varphi}\frac{\partial w}{\partial\lambda} - 2\frac{\tan\varphi}{r^2\cos\varphi}\frac{\partial v}{\partial\lambda} \right\}\vec{i}$$

(4,17)

$$+ \left\{ \Delta v + \frac{2}{r^2}\frac{\partial w}{\partial\varphi} - \frac{v}{r^2\cos^2\varphi} + 2\frac{\tan\varphi}{r^2\cos\varphi}\frac{\partial u}{\partial\lambda} \right\}\vec{j}$$

$$+ \left\{ \Delta w - \frac{2w}{r^2} - \frac{2}{r^2}\frac{\partial v}{\partial\varphi} + 2\frac{\tan\varphi}{r^2}v - \frac{2}{r^2\cos\varphi}\frac{\partial u}{\partial\lambda} \right\}\vec{k} \; .$$

In the equation of energy (4,4), the expression for the dissipation function becomes :

$$\hat{\chi}(\vec{u}) = \frac{\mu_0}{6}\left\{ (d_{11} - d_{22})^2 + (d_{22} - d_{33})^2 + (d_{33} - d_{11})^2 \right\}$$

(4,18)

$$+ \mu_0\left[ d_{12}^2 + d_{23}^2 + d_{31}^2 \right] \; ,$$

where $d_{ij}$ ( $i, j = 1,2$ and $3$ ) are the components of the rate-of-strain tensor $\hat{D} \equiv \operatorname{def}\vec{u} = \vec{\nabla}\vec{u} + (\vec{\nabla}\vec{u})^*$, and for the compenents of the viscous stress tensor $\hat{\tau}$ we have the expression :

$$\tau_{ij} = \mu_0\left\{ d_{ij} - 2\,(\vec{\nabla}.\vec{u})\,\delta_{ij} \right\} \; .$$

The expression (4,18) for $\hat{\chi}(\vec{u})$ in terms of $d_{ij}$ is valid in general orthogonal coordinates and in spherical coordinates,

(4,19)

$$d_{11} = 2\left[ \frac{1}{r\cos\varphi}\frac{\partial u}{\partial\lambda} + \frac{w}{r} - \frac{\tan\varphi}{r}v \right] \; ;$$

$$d_{22} = 2\left[ \frac{1}{r}\frac{\partial v}{\partial\varphi} + \frac{w}{r} \right] \; ;$$

$$d_{33} = 2\frac{\partial w}{\partial r} \; ;$$

$$d_{12} = \frac{\cos\varphi}{r}\frac{\partial}{\partial\varphi}\left(\frac{u}{\cos\varphi}\right) + \frac{1}{r\cos\varphi}\frac{\partial v}{\partial\lambda} ;$$

$$d_{23} = r\frac{\partial}{\partial r}\left(\frac{v}{r}\right) + \frac{1}{r}\frac{\partial w}{\partial\varphi} \; ;$$

$$d_{31} = \frac{1}{r\cos\varphi}\frac{\partial w}{\partial\lambda} + r\frac{\partial}{\partial r}\left(\frac{u}{r}\right) \; .$$

*NON DIMENSIONAL FORM*

We introduce the following transformation :

(4,20) $\qquad x = a_0 \cos\varphi_0 \, \lambda \quad , \quad y = a_0(\varphi - \varphi_0) \quad , \quad z = r - a_0 \, ,$

where $\varphi_0$ is a reference latitude. It follows immediately that

(4,21) $\qquad \dfrac{\partial}{\partial\lambda} = a_0 \cos\varphi_0 \, \dfrac{\partial}{\partial x} \quad , \quad \dfrac{\partial}{\partial\varphi} = a_0 \dfrac{\partial}{\partial y} \quad , \quad \dfrac{\partial}{\partial r} = \dfrac{\partial}{\partial z} \; .$

The origin of this right handed curvilinear coordinates system lies on the earth surface (for a flat ground we have $r = a_0$) at latitude $\varphi_0$ and longitude $\lambda = 0$ .

We suppose therefore that the atmospheric motion occurs in a mid-latitude region, distant from the equator, around some central latitude $\varphi_0$ and therefore, $\sin\varphi_0$ , $\cos\varphi_0$ and $\tan\varphi_0$ are all of order unity. Although x and y are in principle new longitude and latitude coordinates in terms of which the basic equations may be rewritten without approximation, they are obviously introduced in the expectation that for small

(4,22) $\qquad \delta_0 = L_0 \, / \, a_0 \, ,$

the ratio between the characteristic horizontal length of atmospheric motion ($L_0$) and the earth radius ($a_0$), they will be the Cartesian coordinates of the so-called $\beta$-plane approximation (see, the section 5).

At first it is convenient to introduce the nondimensional quantities and to put the basic equations in nondimensional form .

The nondimensional variables (marked by primes) are introduced as follows :

(4,23) $\qquad \begin{cases} t' = t \, / \, t_0 \; , \quad x' = x \, / \, L_0 \; , \quad y' = y \, / \, L_0 \; , \quad z' = z \, / \, H_0 \; , \\[2ex] u' = u \, / \, U_0 \; , \quad v' = v \, / \, U_0 \; , \quad w' = w \, / \, W_0 \; , \quad p' = p \, / \, p_\infty(0), \\[2ex] \rho' = \rho \, / \, \rho_\infty(0) \; , \quad T' = T \, / \, T_\infty(0) \; , \\[2ex] \hat{Q}'_\infty = \hat{Q}_\infty \, / \, \hat{Q}_\infty(1) \; , \end{cases}$

where $W_0$ is a characteristic vertical velocity scale and $\hat{Q}_\infty(1)$ is a characteristic value of mean radiative transfer .

Using (4,23), (4,20) and (4,21) in (4,12) and (4,13) we readily obtain, after some rearrangement of the terms, the *continuity equation* in the nondimensional form :

$$(4,24) \quad S \frac{D'\rho'}{Dt'} + \rho' \left\{ \frac{\eta_0}{\varepsilon_0} \frac{\partial w'}{\partial z'} + 2 \, \eta_0 \delta_0 \, \frac{w'}{1 + \varepsilon_0 \delta_0 z'} \right. $$

$$ + \frac{1 \, / \, \cos \varphi}{1 + \varepsilon_0 \delta_0 z'} \, \frac{\partial}{\partial y'} (\cos \varphi \, v') $$

$$ \left. + \frac{(\cos \varphi_0)/(\cos \varphi)}{1 + \varepsilon_0 \delta_0 z'} \, \frac{\partial u'}{\partial x'} \right\} = 0 \; , $$

where

$$(4,25) \quad S \frac{D'}{Dt'} \equiv S \frac{\partial}{\partial t'} + \frac{\dfrac{\cos \varphi_0}{\cos \varphi}}{1 + \varepsilon_0 \delta_0 z'} \, u' \frac{\partial}{\partial x'} + \frac{1}{1 + \varepsilon_0 \delta_0 z'} \, v' \frac{\partial}{\partial y'} + \frac{\eta_0}{\varepsilon_0} \, w' \frac{\partial}{\partial z'} \; , $$

$$ \eta_0 \equiv W_0 \, / \, U_0 \quad , \quad S = \frac{L_0}{U_0 \, t_0} \; , $$

and S is the *Strouhal number* ($SR_0 \equiv K_1$).

Then, a convenient scale for the vertical velocity is:

$$(4,26) \quad \eta_0 \, / \, \varepsilon_0 \equiv 1 \quad \Longrightarrow \quad W_0 \equiv \varepsilon_0 U_0 \; . $$

Using this choice of scale together, with the relations (4,20)-(4,23) and (4,11), (4,16), (4,17), the *azimutal component of the momentum equation* (4,1) now becomes :

$$(4,27) \quad \rho' \left\{ S \frac{D'u'}{Dt'} + \frac{\varepsilon_0 \delta_0}{1 + \varepsilon_0 \delta_0 z'} \, u'w' - \delta_0 \frac{\tan \varphi}{1 + \varepsilon_0 \delta_0 z'} \, u'v' \right. $$

$$ \left. - \frac{1}{Ro} \frac{\sin \varphi}{\sin \varphi_0} \, v' + \frac{\varepsilon_0}{Ro} \frac{\cos \varphi}{\sin \varphi_0} \, w' \right\} + \frac{\cos \varphi_0 / \cos \varphi}{1 + \varepsilon_0 \delta_0 z'} \, \frac{1}{\gamma M_0^2} \frac{\partial p'}{\partial x'} $$

$$ = \frac{1}{Re} \left\{ \Delta' u' + \frac{1}{3} \frac{\cos \varphi_0 / \cos \varphi}{1 + \varepsilon_0 \delta_0 z'} \, \frac{\partial}{\partial x'} (\vec{\nabla}' . \vec{u}') - \delta_0^2 \frac{1/\cos^2 \varphi}{(1 + \varepsilon_0 \delta_0 z')^2} \, u' \right. $$

$$ \left. + 2 \, \delta_0 \varepsilon_0 \frac{\cos \varphi_0 / \cos \varphi}{(1 + \varepsilon_0 \delta_0 z')^2} \, \frac{\partial w'}{\partial x'} - 2 \, \delta_0 \tan \varphi \, \frac{\cos \varphi_0 / \cos \varphi}{(1 + \varepsilon_0 \delta_0 z')^2} \, \frac{\partial v'}{\partial x'} \right\} \; , $$

where

$$\Delta' \equiv \frac{\cos^2\varphi_0/\cos^2\varphi}{(1 + \varepsilon_0\delta_0 z')^2} \frac{\partial^2}{\partial x'^2} + \frac{1}{(1 + \varepsilon_0\delta_0 z')^2} \frac{\partial^2}{\partial y'^2}$$

(4,28)

$$+ \frac{1}{\varepsilon_0^2} \frac{\partial^2}{\partial z'^2} - \delta_0 \frac{\tan\varphi}{(1 + \varepsilon_0\delta_0 z')^2} \frac{\partial}{\partial y'} + \frac{2}{1 + \varepsilon_0\delta_0 z'} \frac{\delta_0}{\varepsilon_0} \frac{\partial}{\partial z'} \,,$$

and

$$\vec{\nabla}' \cdot \vec{u}' \equiv \frac{\partial w'}{\partial z'} + 2\,\varepsilon_0\delta_0 \frac{w'}{1 + \varepsilon_0\delta_0 z'} + \frac{1/\cos\varphi}{1 + \varepsilon_0\delta_0 z'} \frac{\partial}{\partial y'}(\cos\varphi\, v')$$

(4,29)

$$+ \frac{\cos\varphi_0/\cos\varphi}{1 + \varepsilon_0\delta_0 z'} \frac{\partial u'}{\partial x'} \,.$$

Analogous *momentum equations* can be written down *for* v' *and* w':

$$\rho'\left\{ S \frac{D'v'}{Dt'} + \frac{\varepsilon_0\delta_0}{1 + \varepsilon_0\delta_0 z'} v'w' + \delta_0 \frac{\tan\varphi}{1 + \varepsilon_0\delta_0 z'} u'^2 \right.$$

$$\left. + \frac{1}{Ro} \frac{\sin\varphi}{\sin\varphi_0} u' \right\} + \frac{1}{1 + \varepsilon_0\delta_0 z'} \frac{1}{\gamma M_0^2} \frac{\partial p'}{\partial y'}$$

(4,30)

$$= \frac{1}{Re}\left\{ \Delta'v' + \frac{1}{3} \frac{1}{1 + \varepsilon_0\delta_0 z'} \frac{\partial}{\partial y'}(\vec{\nabla}' \cdot \vec{u}') - \delta_0^2 \frac{1/\cos^2\varphi}{(1 + \varepsilon_0\delta_0 z')^2} v' \right.$$

$$\left. + 2\,\delta_0\varepsilon_0 \frac{1}{(1 + \varepsilon_0\delta_0 z')} \frac{\partial w'}{\partial y'} - 2\,\delta_0\tan\varphi \frac{\cos\varphi_0/\cos\varphi}{(1 + \varepsilon_0\delta_0 z')^2} \frac{\partial u'}{\partial x'} \right\} \,,$$

$$\rho'\left\{ S\varepsilon_0^2 \frac{D'w'}{Dt'} + \varepsilon_0\delta_0 \frac{u'^2 + v'^2}{1 + \varepsilon_0\delta_0 z'} - \frac{\varepsilon_0}{Ro} \frac{\cos\varphi}{\cos\varphi_0} u' \right\}$$

$$+ \frac{1}{\gamma M_0^2}\left( \frac{\partial p'}{\partial z'} + Bo\,\rho' \right)$$

(4,31)

$$= \frac{1}{Re}\left\{ \varepsilon_0^2 \Delta'w' - 2\,\delta_0^2\varepsilon_0^2 \frac{1}{(1 + \varepsilon_0\delta_0 z')^2} w' \right.$$

$$- 2\,\delta_0\varepsilon_0 \frac{1}{(1 + \varepsilon_0\delta_0 z')^2} \frac{\partial v'}{\partial y'} + 2\delta_0^2\varepsilon_0\tan\varphi \frac{1}{(1 + \varepsilon_0\delta_0 z')^2} v'$$

$$\left. - 2\,\delta_0\varepsilon_0 \frac{\cos\varphi_0/\cos\varphi}{(1 + \varepsilon_0\delta_0 z')^2} \frac{\partial u'}{\partial x'} \right\} \,.$$

The nondimensional form of the equation of state (4,8) is

(4,32) $\qquad p' = \rho'\, T'$.

Finally, using (4,20)-(4,23) and (4,9), (4,18), (4,19) we readily obtain the equation of energy (4,4) in the nondimensional form :

(4,33)

$$\rho' S \frac{D'T'}{Dt'} - \frac{\gamma-1}{\gamma} S \frac{D'p'}{Dt'} = \frac{1}{Pr}\frac{1}{Re}\Delta'T' + \frac{\gamma-1}{\varepsilon_0^2\,Re}M_0^2\,\hat{\chi}'(\vec{u}';\varepsilon_0,\delta_0)$$

$$+ \frac{1}{Pr}\frac{Bo^2\sigma_0}{\varepsilon_0^2\,Re}\frac{d\hat{R}'_\infty(T'_\infty)}{dz'_\infty} ,$$

where$^\dagger$ $z'_\infty = Bo\ z'$ , $\sigma_0 = \frac{R}{gk_0}\hat{R}_\infty(1)$ and

(4,34)
$$Pr = \frac{\mu_0 C_p}{k_0}$$

is the *Prandtl number* .
For $\hat{\chi}'$ we have the following expression :

(4,35)

$$\hat{\chi}' = \frac{\varepsilon_0^2}{(1+\varepsilon_0\delta_0 z')^2}\Bigg\{\Bigg\{\frac{2}{3}\Bigg\{\Bigg[\frac{\cos\varphi_0}{\cos\varphi}\frac{\partial u'}{\partial x'} - \delta_0\tan\varphi\,v' - \frac{\partial v'}{\partial y'}\Bigg]^2$$

$$+ \Bigg[\frac{\partial v'}{\partial y'} + \delta_0\varepsilon_0 w' - (1+\varepsilon_0\delta_0 z')\frac{\partial w'}{\partial z'}\Bigg]^2 + \Bigg[(1+\varepsilon_0\delta_0 z')\frac{\partial w'}{\partial z'}$$

$$- \frac{\cos\varphi_0}{\cos\varphi}\frac{\partial u'}{\partial x'} + \delta_0\tan\varphi\,v' - \delta_0\varepsilon_0 w'\Bigg]^2\Bigg\}$$

$$+ \Bigg[\cos\varphi\frac{\partial}{\partial y'}(\frac{u'}{\cos\varphi}) + \frac{\cos\varphi_0}{\cos\varphi}\frac{\partial v'}{\partial x'}\Bigg]^2$$

$$+ \Bigg[\frac{(1+\varepsilon_0\delta_0 z')^2}{\varepsilon_0}\frac{\partial}{\partial z'}(\frac{v'}{1+\varepsilon_0\delta_0 z'}) + \varepsilon_0\frac{\partial w'}{\partial y'}\Bigg]^2$$

$$+ \Bigg[\varepsilon_0\frac{\cos\varphi_0}{\cos\varphi}\frac{\partial w'}{\partial x'} + \frac{(1+\varepsilon_0\delta_0 z')^2}{\varepsilon_0}\frac{\partial}{\partial z'}(\frac{u'}{1+\varepsilon_0\delta_0 z'})\Bigg]^2\Bigg\}\Bigg\}.$$

---

$^\dagger$ $\rho'(z'_\infty)\hat{Q}_\infty(T'_\infty) \equiv \frac{d\hat{R}'_\infty(T'_\infty)}{dz'_\infty}$ , and $\hat{R}_\infty(1)$ is the characteristic value of $\hat{R}_\infty(T_\infty)$;

$\rho'_\infty=\rho_\infty/\rho_\infty(0)$ and $T'_\infty=T_\infty/T_\infty(0)$.

# 5 . THE $f_0$-PLANE AND $\beta$-PLANE APPROXIMATIONS

If the horizontal scale of the atmospheric motion considered is "small" with respect to the earth radius, i.e. if $\delta_0 \ll 1$, this can be exploited in the nondimensional equations (4,24), (4,27), (4,30), (4,31) and (4,33) by performing the following limiting process :

(5,1)          $\delta_0 \to 0$ , with t', x' ,y' and z' fixed.

When $\delta_0 \to 0$ , in Cartesian metric centered at some reference latitude $\varphi_0$ and longitude $\lambda=0$, that locally approximates the spherical metric in this chosen neighborhood of the earth's surface, the effects of sphericity are retained by approximating $f=2\Omega_0 \sin \varphi$ ,the local vertical (or radial) component of $2\vec{\Omega}$, with a linear function of y, a latitudinal coordinate which is measured positive northward from the reference latitude. This approximation is known as the $\beta$-plane approximation and was first introduced by Rossby. The symbol $\beta$ has been traditionally used to denote $df/dy \equiv (1/a_0)/(df/d\varphi)$.

If is important to mention here that the $\beta$-plane approximation consists solely of a set of geometric approximations and it does not involve any assumptions regarding the relative magnitudes of the various dynamic terms in the basic equations. Since this point has not been emphasized in the standard references on the $\beta$-plane approximation and because of the fundamental importance of the so-called $\beta$-plane equations, we feel that it is important to present a thorough account of this approximation.

Let us now assume that we have chosen our scales correctly in the normalization (4,23), and that the quantities in (4,24), (4,27), (4,30), (4,31) and (4,33) are all of order unity. Since we shall be dealing with motions whose lateral scale $L_0$ will generally be smaller than $a_0$, the quantity $\delta_0 y'$, in : $\varphi=\varphi_0+\delta_0 y'$, will be small compared to unity and $\cos\varphi$, $\sin\varphi$, $\tan\varphi$ can be expanded in a convergent Taylor series about the reference latitude $\varphi_0$ (y'=0) :

(5,2)
$$\begin{cases} \cos \varphi = \cos \varphi_0 \left\{ 1-\delta_0 \tan \varphi_0 y' + O(\delta_0^2) \right\} ; \\[2mm] \sin \varphi = \sin \varphi_0 \left\{ 1+\delta_0 \frac{1}{\tan \varphi_0} y' + O(\delta_0^2) \right\} ; \\[2mm] \tan \varphi = \tan \varphi_0 \left\{ 1+\delta_0 \frac{1}{\cos \varphi_0 \sin \varphi_0} y' + O(\delta_0^2) \right\} . \end{cases}$$

From (4,25) and (5,2) it is clear that :

$$S \frac{D'}{Dt} = S \frac{\partial}{\partial t}, + u' \frac{\partial}{\partial x}, + v' \frac{\partial}{\partial y}, + w' \frac{\partial}{\partial z},$$

$$- \delta_0 \left\{ \tan \varphi_0 \; y'u' \; \frac{\partial}{\partial x}, + \varepsilon_0 z' (u' \frac{\partial}{\partial x}, + v' \frac{\partial}{\partial y}, ) \right\} + O(\delta_0^2) \; ,$$

and to zeroth order in $\delta_0$ we have, obviously,

(5,3)     $$S \frac{D'}{Dt}, \sim S \frac{\partial}{\partial t}, + u' \frac{\partial}{\partial x}, + v' \frac{\partial}{\partial y}, + w' \frac{\partial}{\partial z}, \; .$$

Thus, although according to

$$x' = \frac{\cos \varphi_0}{\delta_0} \lambda \quad , \quad y' = \frac{\varphi - \varphi_0}{\delta_0} \; ,$$

x' and y' are orthogonal curvilinear coordinates lying on the spherical surface z'=0, under the approximation $\delta_0 \ll 1$, x' and y' behave as ordinary Cartesian coordinates that can be thought of as spanning a plane (the so-called $\beta$-plane) that is tangent to the surface z'=0 at the reference latitude $\varphi_0$ and longitude $\lambda$=0.

For scales $L_0 < 10^6$m, we have $\delta_0 < 10^{-1}$ and hence (5,3) will then be valid provide $\varphi_0$ corresponds to mid-or low latitudes, so that $\tan \varphi_0 \leq 1$.

Indeed approximation

(5,4)     $$\delta_0 \tan \varphi_0 \ll 1$$

rules out the application of the resulting $f_0$-plane equations (see, eqs.(5,6)-(5,10) below) to be study of motions at high latitudes or in polar area. For these cases, a different approach involving cylindrical coordinates must be used.

If now we expand the metric factors in the nondimensional equations (4,24), (4,27), 4,30), (4,31) and (4,33), according to (5,2) and if in these equations $\delta_0$ is set equal to zero, the so-called $f_0$-plane approximation emerges. Dropping the primes, we readily obtain the following *set of $f_0$-plane equations*:

(5,5)     $$S \frac{D\rho}{Dt} + \rho \vec{\nabla} . \vec{u} = 0 \; ;$$

(5,6)

$$\rho \left\{ S \frac{D\vec{v}}{Dt} + \frac{1}{Ro} (\vec{k} \wedge \vec{v}) + \frac{\varepsilon_0}{\tan \varphi_0} \frac{1}{Ro} w\vec{i} \right\} + \frac{Bo}{\gamma M_0^2} \vec{\nabla} p$$

$$= \frac{1}{Re} \left\{ \vec{\nabla}^2 \vec{v} + \frac{1}{\varepsilon_0^2} \frac{\partial^2 \vec{v}}{\partial z^2} + \frac{1}{3} \vec{\nabla} (\vec{\nabla} . \vec{u}) \right\} \; ;$$

$$\rho \left\{ S\varepsilon_0^2 \frac{Dw}{Dt} - \frac{\varepsilon_0}{\tan \varphi_0} \frac{1}{Ro} \vec{v}.\vec{\imath} \right\} + \frac{1}{\gamma M_0^2} \frac{\partial p}{\partial z} + \frac{Bo}{\gamma M_0^2} \rho$$

(5,7)

$$= \frac{1}{Re} \left\{ \varepsilon_0^2 \vec{D}^2 w + \frac{1}{\varepsilon_0^2} \frac{\partial^2 w}{\partial z^2} + \frac{1}{3} \frac{\partial}{\partial z} (\vec{\nabla}.\vec{u}) \right\} ;$$

(5,8)     $p = \rho T$ ;

$$\rho S \frac{DT}{Dt} - \frac{\gamma-1}{\gamma} S \frac{Dp}{Dt} = \frac{1}{Pr} \frac{1}{Re} \left\{ \vec{D}^2 T + \frac{1}{\varepsilon_0^2} \frac{\partial^2 T}{\partial z^2} \right\}$$

$$+ \frac{\gamma-1}{\varepsilon_0^2 Re} M_0^2 \left\{ \left\{ (\frac{\partial u}{\partial z} + \varepsilon_0^2 \frac{\partial w}{\partial x})^2 + (\frac{\partial v}{\partial z} + \varepsilon_0^2 \frac{\partial w}{\partial y})^2 \right. \right.$$

(5,9)

$$+ \varepsilon_0^2 \left\{ (\frac{\partial u}{\partial y} + \frac{\partial v}{\partial x})^2 + 2 \left[ (\frac{\partial u}{\partial x})^2 + (\frac{\partial v}{\partial y})^2 + (\frac{\partial w}{\partial z})^2 \right] \right.$$

$$\left. \left. - \frac{2}{3} (\vec{\nabla}.\vec{u})^2 \right\} \right\} \right\} + \frac{1}{Pr} \frac{Bo^2 \sigma_0}{\varepsilon_0^2 Re} \frac{d\hat{R}_\infty (T_\infty)}{dz_\infty} .$$

In these $f_0$-plane nondimensional equations (5,5)-(5,9) we have

(5,10)
$$\begin{cases} \vec{u} = \vec{v} + \varepsilon_0 w \vec{k} \quad , \quad \vec{\nabla} = \vec{D} + \frac{1}{\varepsilon_0} \frac{\partial}{\partial z} \vec{k} \quad , \\[2mm] \vec{D} = \frac{\partial}{\partial x} \vec{\imath} + \frac{\partial}{\partial y} \vec{\jmath} \quad , \quad \vec{v} = u \vec{\imath} + v \vec{\jmath} \quad , \\[2mm] S \frac{D}{Dt} = S \frac{\partial}{\partial t} + \vec{v}.\vec{D} + w \frac{\partial}{\partial z} \quad . \end{cases}$$

In many studies of atmospheric motions on synoptic scale ($\delta_0 \sim 10^{-1}$) the system of "exact" nondimensional equations (4,24), (4,27), (4,30), (4,31) and (4,33), is reduced to describe the so-called $\beta$-plane approximation. In this approximation the variability of the Coriolis parameter with latitude is retained while the earth curvature is neglected. This is accomplished (from an intuitive and heuristic argument) by neglecting all $O(\delta_0)$-terms in the nondimensional equations in spherical coordinates *except* for the second term in the Taylor expansion of the Coriolis parameter.

The resulting equations are of the same form as (5,5)-(5,9). However, in this case the constant Coriolis parameter $f_0$ is substituted by :

(5,11a)     $f \cong f_0 + \beta_0 y$

with

(5,11b)     $\beta_0 = (\frac{1}{a_0} \frac{df}{d\varphi})_{\varphi=\varphi_0}$ .

In this case in the equation (5,6), for $\vec{v}$, we substitute the term $\frac{1}{Ro}(\vec{k} \wedge \vec{v})$ by the term :

(5,12)     $(\frac{1}{Ro} + \beta y)(\vec{k} \wedge \vec{v})$,

where $\beta$ is the following nondimensional parameter

(5,13)     $\beta = \dfrac{\delta_0}{\tan \varphi_0 \; Ro} \equiv O(1)$.

Synoptic-scale atmospheric motions are characterized by $U_0 \sim 10$ m/s and $L_0 \sim 10^6$m.

But in mid-latitude : $f_0 \sim 10^{-4}$ 1/s, so that

$$\delta_0 \sim 10^{-1} \text{ and } Ro \sim 10^{-1}$$

and if these scales are used we have : $\beta \sim 1$ .

For these synoptic-scale atmospheric motions, we have likewise

$$\varepsilon_0 = \frac{H_0}{L_0} \ll 1 \implies \varepsilon_0 \sim Ro^2 \sim 10^{-2},$$

since the characteristic vertical scale of motions, $H_0$, is of the order of $10^4$m in all cases of interest, in the troposphere.

Thus, it is necessary to examine, in the nondimensional basic equations (4,24), (4,27), (4,30), (4,31) and (4,33) the following limiting process (for the case of Re$\equiv\infty$) :

(5,14)     $\begin{cases} \delta_0 \to 0 \; , \; Ro \to 0 \; , \; \varepsilon_0 \to 0 \\ \text{with } \beta, \; \alpha = \dfrac{\varepsilon_0}{Ro^2} \; \text{ and } t', \; x', \; y', \; z' \text{ fixed} . \end{cases}$

The limiting process (5,14), with $\beta = O(\frac{\delta_0}{Ro})$ chosen to be of order one, has special significance because it examines *geostrophic dynamics*, when the planetary vorticity gradient contributes equally with the relative vorticity gradient to the overall vorticity balance. This recognition of the particular meaning of this ordering relation then gives confidence that the observed numerical relations between the parameters are not merely fortuitous.

The detailed derivation and justification of the geostrophic approximation and the quasi-geostrophic main equation is deferred to section 9 and Chapter V .

## 6 . THE EQUATIONS FOR THE LARGE-SYNOPTIC SCALE
   ATMOSPHERIC PROCESSES

In any realistic situation, in the atmosphere, Re>>1 and now we consider the following hydrostatic main limiting process:

$$(6,1) \quad \begin{cases} \varepsilon_0 \to 0 \text{ , and Re} \to \infty \text{ ,} \\[2mm] \text{with} \quad \varepsilon_0^2 \text{Re} \equiv \text{Re}_\perp = O(1) \\[2mm] \text{and t', x', y', z' fixed .} \end{cases}$$

*If we suppose that* $\delta_0 = O(1)$, then we obtain, from the full equations (4,24), (4,27), (4,30), (4,31) and (4,33), the equations for the *large-synoptic scale atmospheric processes.*

If we drop the primes we can write the following set of nondimensional *hydrostatic equations* for the *non tangent*, viscous and non adiabatic atmospheric motions :[†]

$$(6,2) \quad S \frac{D\rho}{Dt} + \rho\left(\frac{\partial w}{\partial z} + \frac{1}{\cos \varphi} \frac{\partial}{\partial y}(\cos\varphi\; v) + \frac{\cos \varphi_0}{\cos \varphi} \frac{\partial u}{\partial x}\right) = 0 \; ;$$

$$(6,3) \quad \rho\left\{S \frac{Du}{Dt} - \delta_0 \tan\varphi\; uv - \frac{1}{Ro} \frac{\sin\varphi}{\sin\varphi_0} v\right\} + \frac{\cos\varphi_0}{\cos\varphi} \frac{1}{\gamma M_0^2} \frac{\partial p}{\partial x} = \frac{1}{Re_\perp} \frac{\partial^2 u}{\partial z^2} \; ;$$

$$(6,4) \quad \rho\left\{S \frac{Dv}{Dt} + \delta_0 \tan\varphi\; u^2 + \frac{1}{Ro} \frac{\sin\varphi}{\sin\varphi_0} u\right\} + \frac{1}{\gamma M_0^2} \frac{\partial p}{\partial y} = \frac{1}{Re_\perp} \frac{\partial^2 v}{\partial z^2} \; ;$$

$$(6,5) \quad \frac{\partial p}{\partial z} + B_0 \rho = 0 \; ; \quad p = \rho T;$$

$$(6,6) \quad \rho\, S \frac{DT}{Dt} - \frac{\gamma-1}{\gamma} S \frac{Dp}{Dt} = \frac{1}{Pr} \frac{1}{Re_\perp} \frac{\partial^2 T}{\partial z^2} + \frac{\gamma-1}{Re_\perp} M_0^2 \left[(\frac{\partial u}{\partial z})^2 + (\frac{\partial v}{\partial z})^2\right]$$

$$+ \frac{1}{Pr} \frac{1}{Re_\perp} Bo^2 \sigma_0 \frac{d\hat{R}_\infty}{dz_\infty} \; ,$$

---

[†] Where $\varphi = \varphi_0 + \delta_0 y$ , but $\delta_0 \cong 1$.

where

$$(6,7) \qquad S \frac{D}{Dt} = S \frac{\partial}{\partial t} + \frac{\cos\varphi_0}{\cos\varphi} u \frac{\partial}{\partial x} + v \frac{\partial}{\partial y} + w \frac{\partial}{\partial z} .$$

These hydrostatic model equations (6,2)-(6,6) constitute a very significant system for the large-synoptic scale atmospheric motions in a thin layer as the troposphere around the earth sphere.

## 7 . THE CLASSICAL PRIMITIVE EQUATIONS

When $Re_\perp \equiv \infty$ we obtain instead, of hydrostatic equations (6,2)-(6,6) for non tangent, viscous and non adiabatic large-synoptic scale atmospheric motions, the set of *generalized primitive equations* for the *non tangent* but non viscous and adiabatic atmospheric motions.

In particular case, if $\delta_0 \to 0$, with the relation (5,13) ($\beta$-effect), in these last generalized primitive equations, then we obtain the classical, *Kibel*, *primitive equations for the tangent hydrostatic non viscous and adiabatic atmospheric motions:*

$$(7,1) \qquad \begin{cases} S \dfrac{D\rho}{Dt} + \rho(\dfrac{\partial u}{\partial x} + \dfrac{\partial v}{\partial y} + \dfrac{\partial w}{\partial z}) = 0 ; \\[2mm] \rho\left[ S \dfrac{Du}{Dt} - (\dfrac{1}{Ro} + \beta y)v \right] + \dfrac{1}{\gamma M_0^2} \dfrac{\partial p}{\partial x} = 0 ; \\[2mm] \rho\left[ S \dfrac{Dv}{Dt} - (\dfrac{1}{Ro} + \beta y)u \right] + \dfrac{1}{\gamma M_0^2} \dfrac{\partial p}{\partial y} = 0 ; \\[2mm] \dfrac{1}{\rho} \dfrac{\partial p}{\partial z} + Bo = 0 ; \\[2mm] p = \rho T ; \\[2mm] S \rho \dfrac{DT}{Dt} - \dfrac{\gamma-1}{\gamma} S \dfrac{Dp}{Dt} = 0 ; \end{cases}$$

where

$$S \frac{D}{Dt} = S \frac{\partial}{\partial t} + u \frac{\partial}{\partial x} + v \frac{\partial}{\partial y} + w \frac{\partial}{\partial z} .$$

These equations (7,1) are usually rewritten in pressure coordinates developed by Eliassen .

## THE PRIMITIVE EQS. IN PRESSURE COORDINATES

In the system of pressure coordinates, the usual $(x,y)$ coordinates denote a point's position projected onto a horizontal plane, but the pressure p denotes its location along the vertical axis. The "horizontal" derivatives of a variable are its differences from one point to another in the some isobaric (p=constant) surface with respect to corresponding differences of position projected onto a horizontal plane. The "vertical" derivative of a variable is its derivative with respect to pressure, but is directed along the vertical axis.

The dependent variables are unaffected by the coordinate transformation except that p becomes one of the independent variables, the height z of a particular isobaric surface becomes a dependent variable, and the role of the vertical air speed dz/dt is taken over by $\omega$, the total derivative of pressure

(7,2) $$\omega = \frac{Dp}{Dt} \ .$$

A useful form of the continuity equation in which the pressure p is the vertical coordinate is found by considering an elemental column of atmosphere confined between the isobaric surfaces p and p$-\delta$p.

The mass of the column $\delta$m is equal to $\rho \delta x . \delta y . \delta z$ which by the hydrostatic equation ( $\frac{\delta p}{\delta z} + \rho Bo = 0$) gives :

$$\frac{1}{\delta m} \frac{D}{Dt}(\delta m) \equiv \frac{Bo}{\delta x . \delta y . \delta p} \frac{D}{Dt} \left(\frac{\delta x . \delta y . \delta p}{Bo}\right) = 0 \ ,$$

and carrying out the differentiation and taking the limite we find :

(7,3) $$\frac{\partial u}{\partial x} + \frac{\partial v}{\partial y} + \frac{\partial \omega}{\partial p} = 0 \ .$$

Note that, as with many other primitive equations of tangent atmospheric motions written in the isobaric system, it does not involve density

(7,4) $$\rho = \frac{p}{T} \ .$$

Naturally, the main advantage of the pressure coordinates steem to the fact that according to the hydrostatic approximation the atmosphere, for the synoptic scale tangent motions, is to leading order in hydrostatic equilibrium. This guarantee that the pressure is a monotonic function of altitude z at x and y fixed and consequently the change is mathematically sound.

The procedure for transforming derivatives in the (x, y, z) coordinates into derivatives in the (x, y, p) coordinates is very simple. In dimensionless form we have :

$$(7,5) \quad \begin{cases} \dfrac{\partial}{\partial x} = \dfrac{\partial}{\partial x} + Bo\rho \dfrac{\partial \mathcal{H}}{\partial x}\dfrac{\partial}{\partial p} \; ; \quad \dfrac{\partial}{\partial y} = \dfrac{\partial}{\partial y} + Bo\rho \dfrac{\partial \mathcal{H}}{\partial y}\dfrac{\partial}{\partial p} \; ; \\[3mm] \dfrac{\partial}{\partial z} = -Bo\rho \dfrac{\partial}{\partial p}; \quad S\dfrac{\partial}{\partial t} = S\dfrac{\partial}{\partial t} + S\,Bo\rho\dfrac{\partial \mathcal{H}}{\partial t}\dfrac{\partial}{\partial p} \; ; \end{cases}$$

where

$$(7,6) \quad z = \mathcal{H}(t,x,y,p)$$

is a dependent variable with $\mathcal{H}$ as the local height above the flat ground of an isobaric surface.

As consequence of (7,5) we have :

$$S\frac{D}{Dt} = S\frac{\partial}{\partial t} + u\frac{\partial}{\partial x} + v\frac{\partial}{\partial y} + Bo\rho(S\frac{\partial \mathcal{H}}{\partial t} + u\frac{\partial \mathcal{H}}{\partial x} + v\frac{\partial \mathcal{H}}{\partial y} - w)\frac{\partial}{\partial p} \; ;$$

but in nondimensional form

$$(7,7) \quad \omega \equiv S\frac{\partial p}{\partial t} + u\frac{\partial p}{\partial x} + v\frac{\partial p}{\partial y} + w\frac{\partial p}{\partial z} = Bo\rho(S\frac{\partial \mathcal{H}}{\partial t} + u\frac{\partial \mathcal{H}}{\partial x} + v\frac{\partial \mathcal{H}}{\partial y} - w) \; .$$

Therefore

$$(7,8) \quad \begin{aligned} S\frac{D}{Dt} &= S\frac{\partial}{\partial t} + u\frac{\partial}{\partial x} + v\frac{\partial}{\partial y} + \omega\frac{\partial}{\partial p} \\[2mm] &= S\frac{\partial}{\partial t} + \vec{v}.\vec{D} + \omega\frac{\partial}{\partial p} \; . \end{aligned}$$

Summarize the simplifications attached to the use of pressure coordinates, we rewrite the complete set of resulting tangent primitive equations in the following form :

$$(7,9) \quad \begin{cases} S\dfrac{\partial \vec{v}}{\partial t} + \vec{v}.\vec{D}\vec{v} + \omega\dfrac{\partial \vec{v}}{\partial p} + (\dfrac{1}{Ro} + \beta y)(\vec{k} \wedge \vec{v}) + \dfrac{Bo}{\gamma M_0^2}\vec{D}\mathcal{H} = 0 \; ; \\[4mm] \vec{D}.\vec{v} + \dfrac{\partial \omega}{\partial p} = 0 \; ; \\[4mm] T = -Bo\rho \dfrac{\partial \mathcal{H}}{\partial p} \; ; \\[4mm] S\dfrac{\partial T}{\partial t} + \vec{v}.\vec{D}T + \omega(\dfrac{\partial T}{\partial p} - \dfrac{\gamma-1}{\gamma}\dfrac{T}{p}) = 0 \; . \end{cases}$$

For the primitive equations (7,9) we have only the possibility to impose *two initial* conditions :

(7,10)                 $t=0$ : $\vec{v}=\vec{v}^\circ$ and $T=T^\circ$ ,

as the system of primitive equations (7,9) is of the *third* order relative to time .

For these primitive equations (7,9) we have the following *slip condition* on the *flat* ground :

(7,11)              $\omega = \rho Bo(S \dfrac{\partial \mathcal{H}}{\partial t} + \vec{v}.\vec{D}\mathcal{H})$, on $\mathcal{H}=0$ ,

according to (7,7).


## 8 . THE BOUSSINESQ MODEL EQUATIONS

We consider a nonviscous adiabatic atmospheric motion (Re≡∞) and we suppose in the $f_0$-plane equations (5,5)-(5,9) that

$$\varepsilon_0 \equiv 1 \text{ , but Ro} \equiv \infty \text{ and } \beta \equiv 0 .$$

To discuss the Boussinesq approximation, first introduced by Boussinesq in 1903, it is convenient to introduce the hydrostatic equilibrium standard state described by :

(8,1)             $\vec{v}=0$ , $w=0$ , $p=p_\infty(z_\infty)$ , $\rho=\rho_\infty(z_\infty)$ , $T=T_\infty(z_\infty)$,

where $z_\infty = Boz$ (see,(3,12)) .

Equations (5,5), (5,6), (5,9), with Re≡∞ , $\varepsilon_0 \equiv 1$, Ro≡∞ and $\beta \equiv 0$, are then indentically satisfied and (5,7), (5,8) reduces to

(8,2)              $\dfrac{dp_\infty}{dz_\infty} + \rho_\infty = 0$ , $p_\infty = \rho_\infty T_\infty$ , since $z_\infty = Bo\ z$.

Thus for any given temperature distribution $T_\infty(z_\infty)$, (8,2) can be integrated to find $p_\infty(z_\infty)$ and also $\rho_\infty(z_\infty)$.

To describe the atmospheric motions which represent departures from the static standard state (8,1), (8,2), we introduce the perturbation pressure $\pi$, the perturbation density $\omega$ and the perturbation temperature $\theta$ defined by the relations :

$$
(8,3) \quad
\begin{cases}
p = p_\infty(z_\infty)(1+\pi) \; ; \\[2ex]
\rho = \rho_\infty(z_\infty)(1+\omega) \; ; \\[2ex]
T = T_\infty(z_\infty)(1+\theta) \; .
\end{cases}
$$

Substituting (8,3) into (5,5)-(5,9), where $Re \equiv \infty$, $\varepsilon_0 \equiv 1$, $Ro \equiv \infty$ and $\beta \equiv 0$, we find the following *exact* set of equations, for $\vec{u} = \vec{v} + w\vec{k}$, $\pi$, $\omega$ and $\theta$ :

$$
(8,4) \quad
\begin{cases}
(1+\omega)S \dfrac{D\vec{u}}{Dt} + \dfrac{T_\infty(z_\infty)}{\gamma M_0^2} \vec{\nabla}\pi = (1+\omega) \dfrac{Bo}{\gamma M_0^2} \theta \vec{k} \; ; \\[3ex]
S \dfrac{D\omega}{Dt} + (1+\omega)(\vec{\nabla}.\vec{u}) = (1+\omega) \dfrac{Bo}{T_\infty(z_\infty)} \left[ 1 + \dfrac{dT_\infty(z_\infty)}{dz_\infty} \right] \vec{u}.\vec{k} \; ; \\[3ex]
(1+\omega)S \dfrac{D\theta}{Dt} - \dfrac{\gamma-1}{\gamma} S \dfrac{D\pi}{Dt} + (1+\pi) \dfrac{Bo}{T_\infty(z_\infty)} \left[ \dfrac{\gamma-1}{\gamma} + \dfrac{dT_\infty}{dz_\infty} \right] \vec{u}.\vec{k} = 0 \; ; \\[3ex]
\pi = \omega + \theta + \omega\theta \; .
\end{cases}
$$

Let us consider the "*Boussinesq case*", first introduced by Zeytounian in 1974, when :

$$
(8,5) \qquad Bo \to 0 \quad \text{and} \quad M_0 \to 0, \text{ for } t, x, y, z \text{ fixed,}
$$

with the *similarity condition*

$$
(8,6) \qquad \frac{Bo}{M_0} \equiv \hat{B} = O(1) \; .
$$

In this case : $z_\infty = Boz \equiv \hat{B}M_0 z \to 0$, as $M_0 \to 0$ .

If we represent the solution of equations (8,4) by asymptotic expansions of the form[†] :

$$
(8,7) \quad
\begin{cases}
\vec{v} = \vec{v}_B + \ldots \; , \\[1ex]
w = w_B + \ldots \; , \\[1ex]
\pi = M_0^2 \pi_B + \ldots \; , \\[1ex]
\omega = M_0 \omega_B + \ldots \; , \\[1ex]
\theta = M_0 \theta_B + \ldots \; ,
\end{cases}
$$

---

[†] In the Appendix 1 we give a brief introduction to boundary layer techniques as they are used to study of singular perturbation problems .

we can easily show that the limiting functions, $\vec{v}_B$, $w_B$, $\pi_B$, $\omega_B$ and $\theta_B$, satisfy the *nonviscous, adiabatic Boussinesq equations*, namely

(8,8)
$$
\begin{cases}
S \dfrac{D\vec{v}_B}{Dt} + \dfrac{1}{\gamma} \vec{D}\pi_B = 0 \; ; \\[2mm]
S \dfrac{Dw_B}{Dt} + \dfrac{1}{\gamma} \dfrac{\partial \pi_B}{\partial z} = \dfrac{\hat{B}}{\gamma}\, \theta_B \; ; \\[2mm]
\vec{D}.\vec{v}_B + \dfrac{\partial w_B}{\partial z} = 0 \; ; \\[2mm]
S \dfrac{D\theta_B}{Dt} + \hat{B}\left[ \dfrac{\gamma-1}{\gamma} + \left(\dfrac{dT_\infty}{dz_\infty}\right)_{z_\infty=0} \right] w_B = 0 \; ; \\[2mm]
\omega_B = -\theta_B \; ,
\end{cases}
$$

where

$$
S \dfrac{D}{Dt} = S \dfrac{\partial}{\partial t} + \vec{v}_B.\vec{D} + w_B \dfrac{\partial}{\partial z} \; .
$$

We note that in the limiting Boussinesq process (8,5), (8,6), we have (in dimensionless form) that : $T_\infty(z_\infty) \longrightarrow T_\infty(0) \equiv 1$.

The asymptotic theory, allows us to obtain not only the classical Boussinesq equations but also to define the limits of the approximation through which these equations are obtained. Namely, for the characteristic vertical height $H_0$ we have, from the (8,6), the following limitation :

(8,9)
$$
\hat{B} = \dfrac{Bo}{M_0} \sim 1 \implies H_0 \sim \dfrac{U_0}{g} \left[ \dfrac{RT_\infty(0)}{\gamma} \right]^{1/2} .
$$

Then $H_0 \sim 10^3$m, for the usual values of $U_0$ and $T_\infty(0)$.

Otherwise if we consider (3,7), with (8,6), we see that:

$$
Fr^2 \equiv \dfrac{U_0^{\,2}}{gH_0} \sim \gamma M_0 \longrightarrow 0, \quad \text{as } M_0 \longrightarrow 0,
$$

when $\hat{B} = \dfrac{Bo}{M_0} \sim 1$ and $\varepsilon_0 \equiv 1$ ($\gamma$ is fixed).

# 9 . THE QUASI-GEOSTROPHIC MODEL EQUATION

We start from the tangent primitive equations (7,9), for the nonviscous, adiabatic atmospheric motions and we introduce, in the place of Ro, the Kibel number Ki=SRo. In this case we rewrite the equations (7,9) in the following form :

$$\begin{cases} Ki\left\{\dfrac{\partial \vec{v}}{\partial t} + \dfrac{1}{S}\,(\vec{v}.\vec{D}\vec{v} + \omega\,\dfrac{\partial \vec{v}}{\partial p})\right\} + (1 + \dfrac{\beta}{S}\,Ki\ y)(\vec{k}\wedge\vec{v}) + \dfrac{\lambda_0 Bo}{Ki}\,\vec{D}\mathcal{H} = 0; \\[4mm] \vec{D}.\vec{v} + \dfrac{\partial \omega}{\partial p} = 0 \ ; \\[4mm] T = -\ Bop\,\dfrac{\partial \mathcal{H}}{\partial p} \ ; \\[4mm] \dfrac{\partial T}{\partial t} + \dfrac{1}{S}\,(\vec{v}.\vec{D}T + \omega\left[\dfrac{\partial T}{\partial p} - \dfrac{\gamma-1}{\gamma}\,\dfrac{T}{p}\right]) = 0 \ , \end{cases}$$

(9,1)

where

(9,2)    $\lambda_0 \equiv \dfrac{1}{\gamma S}(\dfrac{Ki}{M_0})^2.$

We consider now the *quasi-geostrophic limiting process* :

(9,3)    $Ki \rightarrow 0$ and $M_0 \rightarrow 0$, for t, x, y, p fixed,

with the *similarity condition*

(9,4)    $\dfrac{1}{\gamma S}\,(\dfrac{Ki}{M_0})^2 \equiv \lambda_0 = 0(1)$ .

We represent the solution of equations (9,1) by asymptotic expansions of the form :

$$\begin{cases} \vec{v} = \vec{v}_{qg} + Ki\ \vec{v}_{ag} + \dots\ , \\[3mm] \omega = Ki\ \omega_{qg} + \dots\ , \\[3mm] \mathcal{H} = \mathcal{H}_0(p) + Ki\ \mathcal{H}_{qg} + Ki^2\mathcal{H}_{ag} + \dots\ , \\[3mm] T = T_0(p) + Ki\ T_{qg} + \dots\ , \end{cases}$$

(9,5)

where the standard temperature $T_0(p)$ is related to the heat balance equation and $\mathcal{H}_0(p)$ is deduced from the hydrostatic equilibrium, namely

(9,6)    $\mathcal{H}_0(p) = Bo \displaystyle\int_1^P \dfrac{T_0(q)}{q}\ dq$

if use is made of the condition $\mathcal{H}_0(1) = 0$, on the ground[†] .

From the first equation of (9,1) we find :

---

[†] When $Ki \rightarrow 0$, $\mathcal{H}=\mathcal{H}_0(p)+Ki\mathcal{H}_{qg}+\dots =0$ (see,(7,11)) leads : $\mathcal{H}_0(p)=0$ and we suppose that p=0 (in dimensionless form) is the solution of the equation $\mathcal{H}_0(p)=0$.

$$\vec{k} \wedge \vec{v}_{qg} + \lambda_0 Bo \vec{D} \mathcal{H}_{qg} = 0 \; ,$$

and that is the well known *geostrophic relation* :

(9,7) $$\vec{v}_{qg} = \lambda_0 Bo(\vec{k} \wedge \vec{D} \mathcal{H}_{qg}).$$

From the second equation of (9,1) we get :

(9,8) $$\vec{D}.\vec{v}_{qg} = 0 \; ,$$

and

(9,9) $$\vec{D}.\vec{v}_{ag} + \frac{\partial \omega_{qg}}{\partial p} = 0 \; .$$

From the first equation of (9,1) we find afterwards, for $\vec{v}_{ag}$, the following relation :

(9,10) $$\vec{v}_{ag} = \vec{k} \wedge \left\{ Bo\lambda_0 \vec{D}\mathcal{H}_{ag} + \frac{\partial \vec{v}_{qg}}{\partial t} + \frac{1}{S} \vec{v}_{qg}.\vec{D}\vec{v}_{qg} \right\} - \frac{\beta}{S}y\vec{v}_{qg}.$$

From the hydrostatic equation we have

(9,11) $$T_{qg} = - Bop \frac{\partial \mathcal{H}_{qg}}{\partial p}$$

and equation for T leads :

(9,12) $$\omega_{qg} = \frac{p}{K_0(p)} \left[ S \frac{\partial T_{qg}}{\partial t} + \vec{v}_{qg}.\vec{D}T_{qg} \right]$$

$$= -Bo \frac{p^2}{K_0(p)} \left[ S \frac{\partial}{\partial t} + \vec{v}_{qg}.\vec{D} \right] \frac{\partial \mathcal{H}_{qg}}{\partial p} \; ,$$

where

(9,13) $$K_0(p) \equiv T_0(p) \left[ \frac{\gamma-1}{\gamma} - p \frac{d \, Log \, T_0}{dp} \right] \neq 0 \; .$$

Using (9,9), with the (9,10) and (9,12), we derive the *main equation of the quasi-geostrophic model* :

(9,14) $$\left\{ \frac{\partial}{\partial t} + \frac{Bo\lambda_0}{S} (\frac{\partial \mathcal{H}_{qg}}{\partial x} \frac{\partial}{\partial y} - \frac{\partial \mathcal{H}_{qg}}{\partial y} \frac{\partial}{\partial x}) \right\} \hat{\Lambda} \, \mathcal{H}_{qg} + \frac{\beta}{S} \frac{\partial \mathcal{H}_{qg}}{\partial x} = 0 \; ,$$

where

(9,15) $$\hat{\Lambda} \equiv \lambda_0 \vec{D}^2 + S \frac{\partial}{\partial p} \left[ \frac{p^2}{K_0(p)} \frac{\partial}{\partial p} \right] \; ,$$

with

$$\vec{D}^2 \equiv \frac{\partial^2}{\partial x^2} + \frac{\partial^2}{\partial y^2} \; .$$

We observe that (9,14) contains only one derivation with respect to t and, as consequence, only one initial condition must be supplied ?

We observe that this loss of initial conditions from the primitive equations corresponds to the fact that the limiting process (9,3), with (9,4), *filters out the internal gravity waves*[†].

The boundary condition, that must be suplied on the ground p=1, can be derived (in the *adiabatic nonviscous case*) from (7,11), if we take into account (9,12),

(9,16)
$$\frac{\partial \mathcal{H}_{qg}}{\partial t} + \frac{1}{Bo} \frac{T_0(1)}{K_0(1)} \left[ \frac{\partial}{\partial t} + \frac{\lambda_0}{S} (\frac{\partial \mathcal{H}_{qg}}{\partial x} \frac{\partial}{\partial y} - \frac{\partial \mathcal{H}_{qg}}{\partial y} \frac{\partial}{\partial x}) \right] \frac{\partial \mathcal{H}_{qg}}{\partial p} = 0 \ ,$$

on p=1.

We don't specify the boundary conditions that must be applied at the upper end of the atmosphere, p=0, and at infinity in the horizontal plane. These may be, for example, that

$$\frac{p^2}{K_0(p)} (\frac{\partial \mathcal{H}_{qg}}{\partial p})^2 + \frac{\lambda_0}{S} (\vec{D}\mathcal{H}_{qg})^2$$

*decay sufficiently rapidly at infinity* .

---

[†] In the Chapter III we consider the wave phenomena in the atmosphere and in the Chapter IV the problem of filtering internal acoustic and gravity waves.

*BACKGROUND READING*

For a extensive treatment of concepts of the Mechanics of Fluids the reader is referred to :

GOLDSTEIN ,S. (1960) _ *Lectures on Fluid Mechanics* .
                        Interscience Publishers, LTD,
                        London (Chapters 1 and 2),

and

MEYER,R.E. (1971) _ *Introduction to Mathematical Fluid Dynamics*.
                        Wiley-Interscience,
                        New York (Chapters 3 and 6).

The equations of motions in several useful curvilinear coordinate frames can be in :

BATCHELOR,G.K. (1967) _ *An Introduction to Fluid Dynamics*.
                        Cambridge University Press (Appendix 2).

Concerning the $\beta$-plane approximation the reader is referred to :

LEBLOND,P.H. and MYSAK,L.A. (1978) _ *Waves in the Ocean*.
                        Elsevier Scientific Pulishing Company,
                        Amsterdam (Chapter 1).

For the Boussinesq and quasi-geostrophic model equations see, for instance :

ZEYTOUNIAN,R.Kh. (1985) _ *Recent Advances in Asymptotic Modelling of tangent Atmospheric motions*.
                        Int.J.Engng.Sci.,Vol.23, n$^{\circ}$ 11, pp.1239-1288,

and

PEDLOSKY,J. (1979) _ *Geophysical Fluid Dynamics*.
                        Springer-Verlag,
                        New York (Chapter 6).

Finally, concerning the various model equations for Weather Forecasting see :

MONIN, A.S. (1972° _ *Weather Forecasting as a Problem in Physics*.

      The MIT Press. Cambridge Mass., U.S.A,

this last book is an indispensable one-volume reference and text for Fluid Mechanicians (Students and Researchers) involved in Meteorology,

and

KIBEL, I.A. (1963) _ *An Introduction to the Hydrodynamical Method of Short period Weather Forecasting* (Translation).

      The Mac Millan Company.

# CHAPTER III

## WAVE PHENOMENA IN THE
## ATMOSPHERE

### 10 . WAVE EQUATION FOR INTERNAL WAVES

In order to construct a fluid dynamics theory for atmospheric flows it is
important to clarify first of all what are the possible types of atmospheric
motions in adiabatic, nonviscous processes.

All these atmospheric motions have the character of *waves*, and for their
classification it is sufficient to consider the case of *small-amplitudes*
waves, i.e. small oscillations of the atmosphere relative to the state of
rest. Such a state, called the standard atmosphere, is specified by a zero
velocity and by pressure, density and temperature functions of but one
variable, namely the "standard altitude", $z_\infty$=Boz (in dimensionless form),
where Bo is the Boussinesq number.

This state of rest is assumed to prevail far from the domain, where the
waves are studied. We start, in this section 10, of fo-plane
equations(5,5)-(5,9), with (5,10), where Re≡∞. If we look for solutions of
these equations by expansions of the form:

$$(10,1) \quad \begin{cases} \vec{v} = \vec{v}' + \ldots \ ; \\[2mm] w = w' + \ldots \ ; \\[2mm] p = p_\infty(z_\infty) \ (1 + \pi' + \ldots) \ ; \\[2mm] \rho = \rho_\infty(z_\infty) \ (1 + \omega' + \ldots) \ ; \\[2mm] T = T_\infty(z_\infty) \ (1 + \theta' + \ldots) \ ; \end{cases}$$

and performing the classical linearization process (products of the
perturbation quantities $\vec{v}', w', \pi', \omega'$ and $\theta'$ with themselves or with their
derivatives are neglected) we can easily show that the functions $\vec{v}', w', \pi', \omega'$

and θ' satisfy the following linear system of equations:

(10,2)

$$\begin{cases}
S \dfrac{\partial \vec{v}'}{\partial t} + \dfrac{1}{Ro}(\vec{k} \wedge \vec{v}') + \dfrac{\varepsilon_0}{Ro} \dfrac{1}{\tan \varphi_0} w' \vec{i} + \dfrac{T_\infty(z_\infty)}{\gamma M_0^2} \vec{D}\pi' = 0 \; ; \\[3mm]
\varepsilon_0^2 S \dfrac{\partial w'}{\partial t} - \dfrac{\varepsilon_0}{Ro} \dfrac{1}{\tan \varphi_0} \vec{v}' . \vec{i} + \dfrac{T_\infty(z_\infty)}{\gamma M_0^2} \dfrac{\partial \pi'}{\partial z} = \dfrac{Bo}{\gamma M_0^2} \theta' \; ; \\[3mm]
S \dfrac{\partial}{\partial t}(\pi' - \theta') + \vec{D}.\vec{v}' + \dfrac{\partial w'}{\partial z} = \dfrac{Bo}{T_\infty(z_\infty)} \left[1 + \dfrac{dT_\infty(z_\infty)}{dz_\infty}\right] w' \; ; \\[3mm]
\omega' = \pi' - \theta' \; ; \\[3mm]
S \dfrac{\partial}{\partial t}\left(\theta' - \dfrac{\gamma-1}{\gamma}\pi'\right) + \alpha_\infty^0 N_\infty^2(z_\infty) w' = 0 \; ,
\end{cases}$$

where 

$$\alpha_\infty^0 N_\infty^2(z_\infty) = \dfrac{Bo}{T_\infty(z_\infty)} \left[\dfrac{\gamma-1}{\gamma} + \dfrac{dT_\infty}{dz_\infty}\right] \; , \quad z_\infty = Boz,$$

according to (3,11) and (3,12).

Now, if we take into consideration that we have *two* main atmospheric situations:

      (1) $\varepsilon_0 \ll 1$, but $Ro = O(1)$ or $\varepsilon \ll Ro \ll 1$,

and       (2) $\varepsilon = O(1)$, but $Ro \equiv \infty$,

we can neglect the terms proportional to: $\dfrac{\varepsilon_0}{Ro \tan \varphi_0}$ in the linearized equations (10,2), corresponding to (5,5)-(5,9). Otherwise, we assume that

(10,3a)
$$\dfrac{dT_\infty(z_\infty)}{dz_\infty} \equiv 0 \; ,$$

considering the case of an *isothermal* standard atmosphere. In this case, consistently with the fact that we are dealing with nondimensional quantities, we may take

(10,3b)
$$T_\infty(z_\infty) \equiv 1 \; , \quad N_\infty^2(z_\infty) \equiv 1 \quad \text{and} \quad \alpha_\infty^0 \equiv Bo \dfrac{\gamma-1}{\gamma} \; ,$$

in the equations (10,2).

It is possible to write the equations (10,2), taking into account (10,3), in matrix form in the case where $\varepsilon_0/Ro \equiv 0$ and for a non-trivial solution of this matrix equation the determinant of the coefficient matrix must vanish. Then we may obtain, for the relative pressure perturbation $\pi'$, a wave equation which reads:

$$(10,4) \qquad \left\{ S^2 \frac{\partial^2}{\partial t^2} \left[ \varepsilon_0^2 M_0^2 S^2 \frac{\partial^2}{\partial t^2} - \hat{D} \right] - \hat{B} \right\} S \frac{\partial \pi'}{\partial t} = 0$$

where

$$(10,5a) \qquad \hat{D} = \varepsilon_0^2 \vec{D}^2 + \frac{\partial^2}{\partial z^2} - Bo \frac{\partial}{\partial z} - \left( \frac{\varepsilon_0 M_0}{Ro} \right)^2$$

and

$$(10,5b) \qquad \hat{B} = \frac{Bo}{\gamma M_0^2} \alpha_\infty^0 \vec{D}^2 + \frac{1}{Ro^2} \left( \frac{\partial^2}{\partial z^2} - Bo \frac{\partial}{\partial z} \right).$$

If we set, in (10,4), $Bo \equiv 0$, $Ro \equiv \infty$ and $\varepsilon_0 \equiv 1$, the wave equation (10,4) reduces to the classical one for acoustics.

For the *full* wave equation (10,4) it is necessary to specify *four* initial conditions for $\partial \pi'/\partial t$, when $t=0$. Namely:

$$(10,6) \qquad t=0 : \quad \frac{\partial \pi'}{\partial t}, \ \frac{\partial^2 \pi'}{\partial t^2}, \ \frac{\partial^3 \pi'}{\partial t^3}, \ \text{and} \ \frac{\partial^4 \pi'}{\partial t^4}$$

are given functions of $\vec{\mathcal{P}} = x\vec{i} + y\vec{j}$ and $z$.

Otherwise this wave equation (10,4) is of second order in $z$, and for its solution it is necessary to specify *two* boundary conditions. In particular, on the *flat* ground we obtain the following condition:

$$(10,7) \qquad z=0: \ w'=0 \implies \left[ \frac{\partial}{\partial z} - Bo \frac{\gamma-1}{\gamma} \right] S \frac{\partial \pi'}{\partial t} = 0.$$

A fairly general solution of (10,4) may be expressed as a superposition of elementary waves as follows:

$$(10,8) \qquad \frac{\partial \pi'}{\partial t} = e^{Boz/2} \iint\limits_{-\infty}^{+\infty} \left[ \int_0^\infty \Pi_{mn}(\tau;\lambda) \mathcal{L}(z;\lambda) d\lambda \right] e^{i\vec{k}.\vec{\mathcal{P}}} \, dmdn,$$

where $|\vec{k}|^2 = m^2 + n^2 \equiv k^2$ and $i\vec{k}.\vec{\mathcal{P}} = i(mx+ny)$.

The eigenvalue $\lambda$ and eigenfunction $\mathcal{L}$ are determined from the *Sturm-Liouville* type equation:

$$(10,9) \qquad \frac{d^2\mathcal{L}}{dz^2} + (\lambda - \frac{Bo^2}{4})\mathcal{L} = 0$$

with the boundary condition (see (10,7)):

$$(10,10) \qquad \frac{d\mathcal{L}}{dz} + \frac{2-\gamma}{2\gamma}Bo\mathcal{L} = 0, \text{ on } z=0.$$

But, the differential equation (10,9) is of second order in z, and for its solution it is necessary to specify a second boundary condition on the upper limit of the troposphere when $z \to +\infty$. For large z every solution of (10,9) is bounded if

$$(10,11) \qquad \left(\frac{Bo}{2}\right)^2 < \lambda < \infty$$

and the spectrum is a continuous one with just one discrete eingenvalue embedded within it, namely

$$(10,12) \qquad \lambda_0 = (\gamma-1)\left(\frac{Bo}{\gamma}\right)^2 ,$$

and the corresponding eigenfunction is :

$$(10,13) \qquad \mathcal{L}_0(z,\lambda_0) \equiv e^{-(2-\gamma)Boz/2\gamma}.$$

The functions $\Pi_{mn}(\tau,\lambda)$, where $\tau \equiv t/S$, must satisfy the following differential equation:

$$(10,14) \qquad \left\{\frac{\partial^4}{\partial\tau^4} + 2\mathcal{A}_0\left(\frac{\partial^2}{\partial\tau^2} + \frac{\Delta_0}{2}\mathcal{A}_0\right)\right\}\Pi_{mn} = 0,$$

where

$$(10,15) \qquad \begin{cases} \mathcal{A}_0 = \dfrac{1}{2M_0^2}\left\{\left(\dfrac{M_0}{Ro}\right)^2 + k^2 + \dfrac{\lambda}{\varepsilon_0^2}\right\} \equiv \mathcal{A}_0(k;\lambda), \\[3mm] \Delta_0 = \dfrac{1}{\mathcal{A}_0^2 M_0^2}\left\{\dfrac{\lambda}{(\varepsilon_0 Ro)^2} + k^2\dfrac{\alpha_\infty^0}{\varepsilon_0^2}\dfrac{Bo}{\gamma M_0^2}\right\} \equiv \Delta_0(k;\lambda), \end{cases}$$

with $\alpha_\infty^0 \equiv Bo\,\dfrac{\gamma-1}{\gamma}$ .

If the system of eigenfunctions $\mathcal{L}(z,\lambda)$ is *complete* (corresponding at the eigenvalues $Bo^2/4<\lambda<\infty$ ), then the initial conditions (10,6), for the equation (10,4), can be represented in the form (10,8) in terms of the eigenfunctions $\mathcal{L}(z,\lambda)$. Therefore we obtain, for the equation (10,14), well-posed initial values problem and frequencies of the internal waves (for isothermal standard atmosphere) are determined from the relation

(10,16) $$\sigma^2_{a,g} = \mathcal{A}_0 \left\{ 1 \mp (1 - \Delta_0)^{1/2} \right\} .$$

We note that $\sigma_a$ (sign +) corresponds to internal *acoustic* waves (modified by gravity) and $\sigma_g$ (sign -) to *gravity* waves (modified by compressibility).
If specially we consider the limit:

$$Bo \rightarrow 0 \quad \text{and} \quad \lambda \rightarrow 0,$$

we obtain two-dimensional acoustic waves, in which there are no vertical oscillations of the air particles, with the frequencies

(10,17) $$\sigma^2_{p,a} = \lim_{\substack{Bo \to 0 \\ \lambda \to 0}} \equiv \frac{1}{Ro^2} + \frac{k^2}{M_0^2} ,$$

and in this case : $\lim\limits_{\substack{Bo \to 0 \\ \lambda \to 0}} \sigma^2_g \equiv 0$. These are the only *possible* waves when the atmosphere is a *barotropic* one (in this case, of course, $\lambda_0 \equiv 0$).

## 11 . THE WIND DIVERGENCE EQUATION FOR TWO-DIMENSIONAL INTERNAL WAVES

It is occasionally advantageous, especially in Meteorological applications, to find the differential equation that describes the wind divergence, rather than the one that describes pressure perturbation $\pi'$. We consider in this case a *non*-isothermal standard atmosphere

(11,1) $$- \frac{dT_\infty(z_\infty)}{dz_\infty} \neq 0 ,$$

and a two-dimensional motion ($\vec{D} \equiv \frac{\partial}{\partial x}\vec{i}$) with $Ro \equiv \infty$. In this case, instead of (10,2) we have the following linear system of equations, for $u'$, $w'$, $\pi'$ and $\theta'$,

(11,2a)
$$S \frac{\partial u'}{\partial t} + \frac{T_\infty(Boz)}{\gamma M_0^2} \frac{\partial \pi'}{\partial x} = 0 \; ;$$

(11,2b)
$$\varepsilon_0^2 S \frac{\partial w'}{\partial t} + \frac{T_\infty(B_0 z)}{\gamma M_0^2} \frac{\partial \pi'}{\partial z} - \frac{Bo}{\gamma M_0^2} \theta' = 0 \; ;$$

(11,2c)
$$S \frac{\partial \pi'}{\partial t} + \gamma \chi_2' - \frac{Bo}{T_\infty(B_0 z)} w' = 0 \; ;$$

(11,2d)
$$S \frac{\partial \theta'}{\partial t} + (\gamma-1)\chi_2' + \frac{1}{T_\infty(Boz)} \frac{dT_\infty(Boz)}{dz} w' = 0 \; ,$$

since $Bo \, dT_\infty(z_\infty)/dz_\infty \equiv dT_\infty(Boz)/dz$ and the wind divergence $\chi_2'$ is given by

(11,3)
$$\chi_2' \equiv \frac{\partial u'}{\partial x} + \frac{\partial w'}{\partial z} \; .$$

Instead of (10,4) we are able to obtain, from (11,2a)–(11,2d), the following single wave equation for $\chi_2'$:

$$\left\{ \varepsilon_0^2 M_0^2 S^4 \frac{\partial^4}{\partial t^4} - T_\infty(Boz) \left[ \varepsilon_0^2 \frac{\partial^2}{\partial x^2} + \frac{\partial^2}{\partial z^2} \right] S^2 \frac{\partial^2}{\partial t^2} + \left[ Bo - \frac{dT_\infty(Boz)}{dz} \right] \frac{\partial}{\partial z} \left[ S^2 \frac{\partial^2}{\partial t^2} \right] \right.$$

(11,4)
$$\left. - \frac{Bo}{\gamma M_0^2} \alpha_\infty^0 N_\infty^2(Boz) T_\infty(Boz) \frac{\partial^2}{\partial x^2} \right\} \chi_2' = 0.$$

For the equation (11,4) we may consider as a tentative solution a plane wave, i.e.,

(11,5)
$$\chi_2'(t,x,z) = \mathfrak{X}_2(z) e^{ikx - i\omega\tau},$$

with $\tau = t/S$ , and we set

(11,6)
$$Z(\zeta) = e^{-\zeta/2} \mathfrak{X}_2(\varphi^{-1}(\zeta)) \; ,$$

where[†]

(11,7)
$$\zeta = Bo \int_0^z \frac{dz'}{dT_\infty(Boz')} \equiv \varphi(z) \; .$$

---

[†] In particular case, for the isothermal standard atmosphere we have
$$T_\infty(Boz) \equiv 1, \quad \zeta = Boz \equiv z_\infty, \quad \text{and } z = \varphi^{-1}(\zeta) = \zeta/Bo.$$

Finally, for $Z(\zeta)$ we obtain the following equation:

(11,8)
$$\frac{d^2Z}{d\zeta^2} + \left\{-\frac{1}{4} - \left[\frac{\varepsilon_0 k}{Bo}\right]^2 T_\infty(Boz) + \mathcal{G}(\omega^2;Boz)\right\}Z = 0 \ ,$$

with

(11,9)
$$\mathcal{G}(\omega^2;Boz) = T_\infty(Boz)\left\{\left[\frac{\varepsilon_0 M_0}{Bo}\right]^2 \omega^2 + \frac{k^2}{\gamma M_0^2}\frac{\alpha_\infty^0}{Bo} N_\infty^2(Boz)T_\infty(Boz)\right\}.$$

To the equation (11,8), for $Z(\zeta)$, we must add the following two boundary conditions :

(11,10a)
$$\frac{dZ}{d\zeta} - \left\{\frac{1}{2} - \frac{1}{\gamma M_0^2}\frac{k^2}{\omega^2}\right\} Z = 0, \text{ on } \zeta = 0 \ ;$$

and

(11,10b)
$$\lim_{\zeta\to+\infty} |Z(\zeta)| < \infty \ .$$

The homogeneous equation (11,8), with the homogeneous boundary condition (11,10a), have in general only the zero solution, unless $\omega$ is equal to one of the set of eigenfrequencies appearing in a quite difficult "non-standard" eigenvalue problem for which we refer to Dikij's book (1969). Here, for the present, we merely point out that this problem is substantially simplified by the limiting process: $\varepsilon_0 \to 0$ , assuming that, $M_0$, Bo and $\gamma$ are $O(1)$. In this *hydrostatic* case, instead of (11,8) and (11,10) we obtain the following hydrostatic eigenvalue problem:

(11,11)
$$\begin{cases} \dfrac{d^2Z_0}{d\zeta^2} - \dfrac{1}{4} Z_0 + \mu^2\psi(\zeta)Z_0 = 0 \ ; \\[2em] \dfrac{dZ_0}{d\zeta} - (\dfrac{1}{2} - \mu^2)Z_0 = 0 \ , \text{ on } \zeta = 0, \\[2em] |Z_0(\infty)| < \infty \ , \end{cases}$$

where

$$Z_0(\zeta;\mu) \equiv \lim_{\substack{\varepsilon_0 \to 0 \\ \zeta \text{ fixed}}} Z(\zeta;\varepsilon_0,M_0,Bo,\gamma,\omega,k),$$

with

(11,12)    $$\mu^2 = \frac{k^2}{\gamma M_0^2}\frac{1}{\omega^2}$$

and

(11,13)    $$\psi(\zeta) = \frac{\alpha_\infty^0}{Bo} N_\infty^2(Bo\varphi^{-1}(\zeta))T_\infty^2(Bo\varphi^{-1}(\zeta)).$$

Besides in the case of *very short horizontal waves*, when $k\to\infty$, we obtain the following singular perturbation problem, instead of (11,8), (11,10):

(11,14)

$$\begin{cases} \nu^2\dfrac{d^2Z_\infty}{d\zeta^2} + \left\{\lambda^2\psi(\zeta) - \dfrac{\varepsilon_0^2}{Bo^2}T_\infty(Bo\varphi^{-1}(\zeta))\right. \\[2mm] \left. + \nu^2\left[\dfrac{\varepsilon_0^2}{\gamma Bo^2}\dfrac{1}{\lambda^2}T_\infty(Bo\bar\varphi^{-1}(\zeta)) - \dfrac{1}{4}\right]\right\}Z_\infty = 0 \ , \\[4mm] \nu^2\dfrac{dZ_\infty}{d\zeta} + (\lambda^2 - \dfrac{\nu^2}{2})Z_\infty = 0 \ , \ \text{on } \zeta = 0, \\[4mm] |Z_\infty(\infty)| < \infty \ , \ \lambda^2 \equiv \dfrac{1}{\gamma M_0^2}\dfrac{1}{\omega^2} \end{cases}$$

when    $$\nu^2 \equiv \frac{1}{k^2} \to 0 \ .$$

This problem (11,14) is amenable to a treatment by the double-scale technique,[†] but presently it is not resolved.

At this point we must add that it is possible to consider other limiting process. For instance:

(11,15)    $$\gamma \to \infty \ \text{and} \ M_0 \to 0, \ \text{with} \ \gamma M_0^2 = \hat M = O(1),$$

which lead to the so-called *isochoric* model.

In this case we obtain, instead (11,9),

(11,15)    $$\mathcal{G}(\omega^2; Bo\varphi^{-1}(\zeta)) \equiv \hat\lambda^2\psi(\zeta), \ \hat\lambda^2 \equiv \frac{k^2}{\hat M}\frac{1}{\omega^2} \ .$$

On the other hand we may consider the Boussinesq case (8,5), with (8,6), but for this it is necessary to start of generalized Boussinesq equations (see, equations (12,3) below).

---

† See the Appendix 2 for a brief discussion of this technique .

## 12 . BOUSSINESQ GRAVITY WAVES

When  Bo<<1 one slow scale height

$$(12,1) \qquad \zeta = \frac{z_\infty}{\hat{B}} = M_0 z$$

in built into the full exact equations (8,4) and we can now directly set in (8,4):

$$(12,2) \qquad \pi = M_0^2 \tilde{\pi} \quad , \quad \omega = M_0 \tilde{\omega} \quad , \quad \theta = M_0 \tilde{\theta}$$

Then (8,4) takes the form:

$$(12,3a) \qquad S \frac{D\vec{v}}{Dt} + \frac{T_\infty(\hat{B}\zeta)}{\gamma} \vec{D} \, \tilde{\pi} = -M_0 \tilde{\omega} S \frac{D\vec{v}}{Dt} \; ;$$

$$(12,3B) \qquad S \frac{Dw}{Dt} + \frac{T_\infty(\hat{B}\zeta)}{\gamma} \frac{\partial \tilde{\pi}}{\partial z} - \frac{\hat{B}}{\gamma} \tilde{\theta} = -M_0 \tilde{\omega} \left[ S \frac{Dw}{Dt} - \frac{\hat{B}}{\gamma} \tilde{\theta} \right];$$

$$(12,3c) \qquad \vec{D}.\vec{v} + \frac{\partial w}{\partial z} = -M_0 \left[ S \frac{D\tilde{\omega}}{Dt} + \tilde{\omega}(\vec{D}.\vec{v} + \frac{\partial w}{\partial z}) - (\hat{B} + \frac{dT_\infty(\hat{B}\zeta)}{d\zeta}) \frac{w}{T_\infty(\hat{B}\zeta)} \right]$$

$$+ M_0^2 \tilde{\omega} \left[ \hat{B} + \frac{dT_\infty(\hat{B}\zeta)}{d\zeta} \right] \frac{w}{T_\infty(Bo\zeta)} \; ;$$

$$(12,3d) \qquad \tilde{\omega} + \tilde{\theta} = M_0(\tilde{\pi} - \tilde{\omega}\tilde{\theta}) \; ;$$

$$(12,3e) \qquad S \frac{D\tilde{\theta}}{Dt} + \gamma \Gamma_\infty(\hat{B}\zeta) w = -M_0 \left[ \tilde{\omega} S \frac{D\tilde{\theta}}{Dt} - \frac{\gamma-1}{\gamma} S \frac{D\tilde{\pi}}{Dt} \right] - M_0^2 \gamma \Gamma_\infty(\hat{B}\zeta) w \; ,$$

where

$$(12,4) \qquad \Gamma_\infty(\hat{B}\zeta) \equiv \frac{\hat{B}}{\gamma} \left\{ \frac{\gamma-1}{\gamma} + \frac{1}{\hat{B}} \frac{dT_\infty(\hat{B}\zeta)}{d\zeta} \right\} \frac{1}{T_\infty(\hat{B}\zeta)} \; .$$

The Boussinesq system (8,8) is nothing but the full exact system (12,3) written at *order zero* with respect  $M_0 \to 0$, when  $\hat{B}=O(1)$ (in this case  $\zeta \to 0$, as  $M_0 \to 0$).

The *generalized Boussinesq equations* (12,3) are amenable to a treatment by the method of double height scales[†].

---

[†] The two-variable expansion procedure is briefly discussed in the Appendix 2.

Following Bois (1976) we consider the linearization of (12,3) around the state of rest:

(12,5)

$$
\begin{cases}
S\dfrac{\partial \vec{v}'}{\partial t} + \dfrac{T_\infty(\hat{B}\zeta)}{\gamma}\,\vec{D}\,\pi' = 0 \; ; \\[2ex]
S\dfrac{\partial w'}{\partial t} + \dfrac{T_\infty(\hat{B}\zeta)}{\gamma}\,\dfrac{\partial \pi'}{\partial t} - \dfrac{\hat{B}}{\gamma}\,\theta' = 0 \; ; \\[2ex]
\vec{D}.\vec{v}' + \dfrac{\partial w'}{\partial z} = -M_0 S\,\dfrac{\partial \omega'}{\partial t} + M_0\Big(\hat{B} + \dfrac{d}{d\zeta}(T_\infty(\hat{B}\,\zeta))\Big)\dfrac{w'}{T_\infty(\hat{B}\,\zeta)} \; ; \\[2ex]
S\dfrac{\partial \theta'}{\partial t} + \gamma\Gamma_\infty(\hat{B}\zeta)w' = M_0\dfrac{\gamma-1}{\gamma}\,S\,\dfrac{\partial \pi'}{\partial t} \; ; \\[2ex]
\theta' + \omega' = M_0\pi' .
\end{cases}
$$

We consider a plane progressive wave solution, in an unbounded atmosphere, for the linear system (12,5).
Writing

(12,6)

$$
\begin{pmatrix} \vec{v}' \\ w' \\ \pi' \\ \omega' \\ \theta' \end{pmatrix} = \begin{pmatrix} \vec{v} \\ w \\ \pi \\ \omega \\ \theta \end{pmatrix} (z;M_0)e^{i\psi} \; ,
$$

where $\psi = \sigma_0 t/S - k_0 x$, and substituting this form (12,6) into (12,5) gives a linear system of ordinary differential equations for $\vec{v}, w, \pi, \omega$ and $\theta$ which can be reduced to one equation for $w(z;M_0)$, namely:

(12,7)
$$
\dfrac{d^2 w}{dz^2} + \left(\dfrac{k_0}{\sigma_0}\right)^2\Big[\phi_\infty(\hat{B}\zeta) - \sigma_0^2\Big]w = M_0\dfrac{1}{T_\infty(\hat{B}\zeta)}\Big[\hat{B} + \dfrac{d}{d\zeta}(T_\infty(\hat{B}\,\zeta))\Big]\dfrac{dw}{dz} \; ,
$$

where $\zeta=M_0 z$ and $\phi_\infty(\hat{B}\zeta)\equiv\hat{B}\Gamma_\infty(\hat{B}\zeta)$.
The structure of solutions of equation (12,7) can be examined by considering the case

$$\frac{dT_\infty(\hat{B}\zeta)}{d\zeta} \equiv 0 \implies T_\infty(\hat{B}\zeta) \equiv 1$$

and in this case $\phi_\infty(\hat{B}\zeta) \equiv \hat{B}^2 \frac{\gamma-1}{\gamma^2}$ = constant. At order zero in $M_0$ two cases

appear: if $\sigma_0^2 < \left(\frac{\hat{B}}{\gamma}\right)^2 (\gamma-1)$ the solutions of (12,7) are oscillatory and if

$\sigma_0^2 > \left(\frac{\hat{B}}{\gamma}\right)^2 (\gamma-1)$ they are exponential.

In the case where $\phi_\infty(\hat{B}\zeta)$ is *not constant* a solution of (12,7) can be determined by a *singular perturbation method*. For the problem under study the most adequate method is that of *two scales*. We introduce the two variables:

$$(12,8) \qquad \xi = \frac{\varphi(\zeta)}{M_0} \quad \text{and} \quad \zeta = M_0 z$$

and write that the function

$$(12,9) \qquad w(z, M_0) = w^*(\xi, \zeta; M_0)$$

satisfies, with respect to the two variables $\xi$ and $\zeta$, an equation which must be identified to (12,7) if $\xi$ and $\zeta$ are related to z by (12,8). Thus:

$$\frac{d}{dz} = \frac{d\varphi}{d\zeta} \frac{\partial}{\partial\xi} + M_0 \frac{\partial}{\partial\zeta}$$

$$(12,10)$$

$$\frac{d^2}{dz^2} = \left(\frac{d\varphi}{d\zeta}\right)^2 \frac{\partial^2}{\partial\xi^2} + M_0 \left[\frac{d^2\varphi}{d\zeta^2} \frac{\partial}{\partial\xi} + 2\frac{d\varphi}{d\zeta} \frac{\partial^2}{\partial\xi\partial\zeta}\right],$$

so that (12,7) reads

$$\left(\frac{d\varphi}{d\zeta}\right)^2 \frac{\partial^2 w^*}{\partial\xi^2} + \left(\frac{k_0}{\sigma_0}\right)^2 \left[\phi_\infty(\hat{B}\zeta) - \sigma_0^2\right] w^*$$

$$(12,11)$$

$$= M_0 \frac{d\varphi}{d\zeta} \left\{-2\frac{\partial^2 w^*}{\partial\xi\partial\zeta} - \frac{d^2\varphi/d\zeta^2}{d\varphi/d\zeta} \frac{\partial w^*}{\partial\xi} + \frac{1}{T_\infty(\hat{B}\zeta)} \left[\hat{B} + \frac{d}{d\zeta}(T_\infty(\hat{B}\zeta))\right] \frac{dw^*}{d\xi}\right\}.$$

Let us seek $w^*$ in the form:

(12,12) $\qquad \overset{*}{w} = \overset{*}{w_0}(\xi,\zeta) + M_0 \overset{*}{w_1}(\xi,\zeta) + \dots ,$

and at order 0 in $M_0$, $\overset{*}{W_0}$ satisfies the equation

(12,13) $\qquad \left(\dfrac{d\varphi}{d\zeta}\right)^2 \dfrac{\partial \overset{*}{w_0}}{\partial \xi^2} + \left(\dfrac{k_0}{\sigma_0}\right)^2 \left[\phi_\infty(\hat{B}\zeta) - \sigma_0^2\right] \overset{*}{w_0} = 0 .$

This equation is a differential equation with constant coefficients (concerning the dependance in $\xi$) and without less in generality we can search the elementary solution in the form:

$$\overset{*}{w_0}(\xi,\zeta) = A_0(\zeta) e^{i\xi} ,$$

which gives for $\dfrac{d\varphi}{d\zeta}$ the equation:

(12,9) $\qquad \left(\dfrac{d\varphi}{d\zeta}\right)^2 = \left(\dfrac{k_0}{\sigma_0}\right)^2 \left[\phi_\infty(\hat{B}\zeta) - \sigma_0^2\right] .$

Thus

(12,10) $\qquad \overset{*}{w_0}(\xi,\zeta) = A_0^+ e^{i\xi} + A_0^- e^{-i\xi}$

where $\xi$ has the value:

(12,11) $\qquad \xi = \dfrac{1}{M_0} \displaystyle\int_0^\zeta \dfrac{d\varphi(\zeta')}{d\zeta'} d\zeta'$

and $\dfrac{d\varphi}{d\zeta}$ is one of the two roots of (12,9).
From (12,9), if $\phi_\infty(\hat{B}\zeta) < \sigma_0^2$, the solutions (12,10) are real exponentials. If $\phi_\infty(\hat{B}\zeta) > \sigma_0^2$, these solutions are oscillatory.
At this stage the two functions $A_0^\mp$ are undetermined and in order to obtain these functions we must write down the equations satisfied by $\overset{*}{w_1}$. This equation reads:

(12,12) $\qquad \left(\dfrac{d\varphi}{d\zeta}\right)^2 \dfrac{\partial \overset{*}{w_1}}{\partial \xi^2} + \left(\dfrac{k_0}{\sigma_0}\right)^2 \left[\phi_\infty(\hat{B}\zeta) - \sigma_0^2\right] \overset{*}{w_1}$

$$= -i \dfrac{d\varphi}{d\zeta} \left\{ \left[2\dfrac{dA_0^+}{d\zeta} + \left(\dfrac{d\ell n\rho_\infty(\hat{B}\zeta)}{d\zeta} + \dfrac{d^2\varphi/d\zeta^2}{d\varphi/d\zeta}\right) A^+ \right] e^{i\xi} \right.$$

$$\left. - \left[2\dfrac{dA_0^-}{d\zeta} + \left(\dfrac{d\ell n\rho_\infty(\hat{B}\zeta)}{d\zeta} + \dfrac{d^2\varphi/d\zeta^2}{d\varphi/d\zeta}\right) A^- \right] e^{-i\xi} \right\} .$$

In order that $w_1^*(\xi,\zeta)$ be *no more* singular than $w_0^*(\xi,\zeta)$ it is necessary that[†]:

(12,13)
$$\begin{cases} 2\,\dfrac{dA_0^+}{d\zeta} + \dfrac{d}{d\zeta}\left\{\ell n\left[\rho_\infty(\hat{B}\zeta)\,\dfrac{d\varphi}{d\zeta}\right]\right\} A_0^+ = 0 \ , \\[3em] 2\,\dfrac{dA_0^-}{d\zeta} + \dfrac{d}{d\zeta}\left\{\ell n\left[\rho_\infty(\hat{B}\zeta)\,\dfrac{d\varphi}{d\zeta}\right]\right\} A_0^- = 0 \ . \end{cases}$$

Hence

(12,14)
$$A_0^{\mp}(\zeta) = \frac{C_{00}^{\mp}}{\sqrt{\rho_\infty(\hat{B}\zeta)\dfrac{d\varphi}{d\zeta}}} \quad ,$$

and

(12,15)
$$w_0^*(\xi,\zeta) = \frac{1}{\sqrt{\rho_\infty(\hat{B}\zeta)\dfrac{d\varphi}{d\zeta}}}\left\{C_{00}^+ e^{i\xi} + C_{00}^- e^{-i\xi}\right\} ,$$

where $C_{00}^{\mp}$ are two arbitrary constants.

In the case of a *non oscillatory* solution (12,15) must be replaced by

(12,16)
$$w_0^*(\xi,\zeta) = \frac{1}{\sqrt{\rho_\infty(\hat{B}\zeta)\dfrac{d\varphi}{d\zeta}}}\left\{D_{00}^+ e^{\xi} + D_{00}^- e^{-\xi}\right\} ,$$

where $D_{00}^{\mp}$ are also two arbitrary constants.

In both formulae (12,15) and (12,16), $\xi$ is given by the relation

(12,17)
$$\xi = \frac{k_0}{M_0}\int_0^\zeta \left|\frac{\phi_\infty(\hat{B}\zeta')}{\sigma_0^2} - 1\right|^{1/2} d\zeta' .$$

First the formulae (12,15) and (12,16) allow one to examine the structure of the solution of (12,7).

---

[†] We eliminate the "secular" terms; see the Appendix 2. We note, also, that:
$$\frac{1}{T_\infty(\hat{B}\zeta)}(\hat{B}+\frac{dT_\infty(\hat{B}\zeta)}{d\zeta}) \equiv -\frac{d\ell n\rho_\infty(\hat{B}\zeta)}{d\zeta}.$$

If $\phi_\infty(\hat{B\zeta})$ has a moderate variation (in fact if $\frac{d\phi}{d\zeta}$ does not vanish) the behaviour of oscillatory solutions when $z \to +\infty$, is that of $\left[\rho_\infty(\hat{B\zeta})\right]^{-1/2}$. But

$$(12,18) \qquad \rho_\infty(\hat{B\zeta}) = \exp\left\{-\int_0^\zeta \left[\frac{dT_\infty(\hat{B\zeta}')}{d\zeta'} + \hat{B}\right]\frac{d\zeta'}{T_\infty(\hat{B\zeta}')}\right\}.$$

It can be easily verified that if $T_\infty(\hat{B\zeta}) \sim (\hat{B\zeta})^\alpha$, where $\alpha \geq 0$, when $\zeta \to +\infty$, then $\rho_\infty(\hat{B\zeta}) \to 0$, when $\zeta \to +\infty$. Hence the only oscillations of the form (12,15) which can satisfy a damping condition at infinity, are those such $C^+_{00} = C^-_{00} \equiv 0$, i.e. these is no motion. For motions such that $C^+_{00}$ or $C^-_{00}$ is not zero, the damping proceeds only from dissipation (non adiabatic effects, with $Re \neq \infty$).

## TURNING POINT LOCAL PROBLEM

Now consider the case where $d\varphi/d\zeta$ vanisches at a point $\zeta = \zeta_0$. This point is defined by the equation

$$(12,19) \qquad \sigma_0^2 = \phi_\infty(\hat{B\zeta_0}).$$

In the neighborhood of $\zeta_0$ the solutions of the form (12,15) or (12,16) are no more valid: $\zeta_0$ is called a *turning point* for equation (12,7). In order to study the solution of (12,7) in the neighborhood of $\zeta_0$, it is necessary to make a *local* expansion in the form:

$$(12,20) \qquad \begin{aligned} \phi_\infty(\hat{B\zeta}) &= \sigma_0^2 + (\zeta - \zeta_0)\left.\frac{d\phi_\infty}{d\zeta}\right|_{\zeta=\zeta_0} + \cdots \\ &= \sigma_0^2 + M_0(z - z_0)\left.\frac{d\phi_\infty}{d\zeta}\right|_{\zeta=\zeta_0} + \cdots, \end{aligned}$$

since $\zeta = M_0 z \quad (z_0 \equiv \frac{\zeta_0}{M_0})$.

In the neighborhood of $\zeta_0$ the function $w(z, M_0)$ satisfy the following equation, instead of (12,7),

$$(12,21) \qquad \frac{d^2w}{dz^2} + M_0\frac{k_0^2}{\sigma_0^2}\chi_\infty(\zeta_0)(z - z_0)w + M_0\psi_\infty(\zeta_0)\frac{dw}{dz} = o(M_0),$$

where

$$(12,22) \qquad \chi_\infty(\zeta_0) \equiv \left.\frac{d\phi_\infty(\hat{B\zeta})}{d\zeta}\right|_{\zeta=\zeta_0}, \qquad \psi_\infty(\zeta_0) \equiv \left.\frac{d\ell n\rho_\infty(\hat{B\zeta})}{d\zeta}\right|_{\zeta=\zeta_0}.$$

The equation (12,21), when $M_0 \rightarrow 0$, can be studied by the Method of Matched Asymptotic Expansion. The expansion (12,12) is an outer asymptotic expansion and it is necessary to consider an inner asymptotic expansion in neighborhood of $\zeta_0$. We introduce an inner variable[†]

$$(12,23) \qquad \hat{z} = M_0^\alpha(z-z_0) \equiv \frac{\zeta-\zeta_0}{M_0^{1-\alpha}}, \quad \alpha<1,$$

associated with the inner (local) asymptotic expansion

$$(12,24) \qquad w = \hat{w}_0 + M_0^\beta \hat{w}_1 + \ldots, \quad \beta>0.$$

Substituting (12,23) and (12,24) into (12,21) gives:

$$(12,25) \qquad \frac{d^2\hat{w}_0}{d\hat{z}^2} + M_0^{1-3\alpha}\left(\frac{k_0}{\sigma_0}\right)^2 \chi_\infty(\zeta_0)\hat{z}\hat{w}_0 + M_0^{1-\alpha}\psi_\infty(\zeta_0)\frac{d\hat{w}_0}{d\hat{z}} + M_0^\beta \frac{d^2\hat{w}_1}{d\hat{z}^2} + \ldots = 0.$$

We wish to choose $\alpha$ so that first term, in (12,25), which is the most highly differentiated leading term, is a dominant term in balance with at least one of the other two leading terms (index "o").

Obviously the only good choice is the balance between the first and second terms in (12,25) and that requires

$$(12,26) \qquad 1-3\alpha = 0 \implies \alpha = \frac{1}{3}.$$

But in this case it is necessary that

$$(12,27) \qquad 1-\alpha = \beta \implies \beta = \frac{2}{3}.$$

Finally, for $\hat{w}_0(\hat{z})$ we have the following equation

$$(12,28) \qquad \frac{d^2\hat{w}_0}{d\hat{z}^2} + \left(\frac{k_0}{\sigma_0}\right)^2 \chi_\infty(\zeta_0)\hat{z}\hat{w}_0 = 0,$$

of which the general solution is

$$(12,29) \qquad \hat{w}_0(\hat{z}) = a_{00}^+ \text{Ai}(Z) + a_{00}^- \text{Bi}(Z) \equiv \hat{w}_0^*(Z),$$

---

† See the Appendix 1 for the application of the MMAE.

where

$$(12,30) \qquad Z = \left[\frac{k_0}{\sigma_0}\right]^{2/3} |\chi_\infty(\zeta_0)|^{1/3} \hat{z} ,$$

and Ai and Bi are the Airy functions of first and second kinds. The constants $a_{00}^{\mp}$ can be related to the constants $C_{00}^{\mp}$ and $D_{00}^{\mp}$ of (12,15) and (12,16) by the means of an asymptotic matching. In order to fix ideas, we assume that $\chi_\infty^*(\zeta_0)<0$. The two outer solutions $w_0^*(\xi,\zeta)$ has the form (see, (12,15) and (12,16)) :

$$(12,31a) \qquad w_0^*(\xi,\zeta) = \frac{1}{\sqrt{\rho_\infty(\hat{B}\zeta)\dfrac{d\varphi}{d\zeta}}} \left\{ C_{00}^+ e^{i\xi} + C_{00}^- e^{-i\xi} \right\} ,$$

$$\equiv w_{Osc}^*(\xi,\zeta) , \text{ if } \zeta<\zeta_0,$$

and

$$(12,31b) \qquad w_0^*(\xi,\zeta) = \frac{1}{\sqrt{\rho_\infty(\hat{B}\zeta)\dfrac{d\varphi}{d\zeta}}} D_{00}^+ e^{-\xi}$$

$$\equiv w_{exp}^*(\xi,\zeta) , \text{ if } \zeta>\zeta_0,$$

since we assume that $w_0^* \to 0$ at infinity.

If we note that

$$\hat{z} = \frac{\zeta-\zeta_0}{M_0^{2/3}} \text{ and } Z = \left[\frac{k_0}{\sigma_0}\right]^{2/3} |\chi_\infty(\zeta_0)|^{1/3} \frac{\zeta-\zeta_0}{M_0^{2/3}} ,$$

then[†]

$$(12,32) \qquad \lim_{Z \to -\infty} \hat{w}_0^*(Z) = \lim_{\substack{\zeta \to \zeta_0 \\ \zeta<\zeta_0}} w_{Osc}^*(\xi,\zeta),$$

and

$$(12,33) \qquad \lim_{Z \to +\infty} \hat{w}_0^*(Z) = \lim_{\substack{\zeta \to \zeta_0 \\ \zeta>\zeta_0}} w_{exp}^*(\xi,\zeta),$$

---

[†] The Van Dyke matching criterion is (see, Appendix 1):
Outer expansion [inner expansion]≡Inner expansion [outer expansion]. For the *leading* order terms the relations (12,32) and (12,33) are of good matching conditions

By using the asymptotic expansion of the Airy functions for *large* Z:

$$(12,34) \quad \begin{cases} Ai(Z) \sim \dfrac{1}{\sqrt{\pi}} \, (-Z)^{-1/4} \sin(\psi_0 + \dfrac{\pi}{4}) \; ; \\[2mm] Bi(Z) \sim \dfrac{1}{\sqrt{\pi}} \, (-Z)^{-1/4} \cos(\psi_0 + \dfrac{\pi}{4}) \; ; \\[2mm] \psi_0 = \dfrac{2}{3} \, (-Z)^{3/2}, \end{cases}$$

we obtain for the $\zeta < \zeta_0$ the following outer expansion :

$$(12,35) \quad \hat{W}_0^*(Z) \sim \mathscr{L}_0 \frac{1}{\sqrt{\pi}} \, M_0^{1/6} |\zeta - \zeta_0|^{-1/4} a_{00}^+ \sin(\psi_0 + \frac{\pi}{4}),$$

if we take into accont that the only function which can be matched to (12,31a) then is the function Ai(Z).

In (12,35) we have :

$$(12,36) \quad \begin{cases} \mathscr{L}_0 = \left\{ k_0^{2/3} \sigma_0^{-2/3} |\chi_\infty(\zeta_0)|^{1/3} \right\}^{-1/4}, \\[3mm] \psi_0 = \dfrac{2}{3} \, k_0 \sigma_0^{-1} |\chi_\infty(\zeta_0)|^{1/2} \dfrac{|\zeta - \zeta_0|^{3/2}}{M_0} \, , \quad \zeta < \zeta_0. \end{cases}$$

On the other hand, we have in the neighborhood of $\zeta_0$, for $\zeta < \zeta_0$:

$$(12,37) \quad \xi = M_0^{-1} \left\{ \left[ \int_0^{\zeta_0} + \int_{\zeta_0}^{\zeta} \right] \frac{d\varphi(u)}{du} \, du \right\} = M_0^{-1} \left\{ \varphi_0 - \frac{k_0}{\sigma_0} |\chi_\infty(\zeta_0)|^{1/2} \frac{|\zeta - \zeta_0|^{3/2}}{3/2} \right\},$$

where $\varphi_0 \equiv k_0 \int_0^{\zeta_0} |1 - \phi_\infty(u) \sigma_0^{-2}|^{1/2} du$ , with $\phi_\infty(u) \equiv \sigma_0^2 + (u - \zeta_0) \frac{d\phi_\infty}{du} \Big|_{u = \zeta_0}$.

Therefore:

$$(12,38) \quad \xi = \frac{\varphi_0}{M_0} - \psi_0 \, , \quad (\zeta < \zeta_0).$$

So that we deduce the following inner expansion for $w_{Osc}^*$:

$$(12,39) \quad w_{Osc}^*(\xi, \zeta) \sim \left[ \frac{k_0}{\sigma_0} \rho_\infty(\hat{B}\zeta_0) \right]^{-1/2} |\chi_\infty(\zeta_0)|^{-1/4} |\zeta - \zeta_0|^{-1/4} \left\{ C_{00}^+ e^{i(\varphi_0/M_0) - i\psi_0} \right.$$
$$\left. + C_{00}^- e^{-i(\varphi_0/M_0) + i\psi_0} \right\},$$

since

$$\left[ \rho_\infty(\hat{B}\zeta) \frac{d\varphi}{d\zeta} \right]^{1/2} \sim \left[ \rho_\infty(\hat{B}\zeta_0) \right]^{1/2} \left( \frac{\sigma_0}{k_0} \right)^{-1/2} |\chi_\infty(\zeta_0)|^{1/4} |\zeta - \zeta_0|^{1/4}.$$

Identification of (12,35) (valid for $Z \to \infty$) and (12,39) (valid for $\zeta \to \zeta_0$, with $\zeta < \zeta_0$) yields :

$$(12,40) \qquad \begin{cases} C_{00}^+ = \Lambda_0 e^{-i\varphi_0/M_0 + i\pi/4} & ; \\ C_{00}^- = \Lambda_0 e^{+i\varphi_0/M_0 - i\pi/4} & , \end{cases}$$

with

$$(12,41) \qquad \Lambda_0 = \left(\frac{k_0}{\sigma_0}\right)^{1/3} \frac{1}{2\sqrt{\pi}} \left[\rho_\infty(\hat{B}\zeta_0)\right]^{1/2} |\chi_\infty(\zeta_0)|^{1/6} M_0^{1/6} a_{00}^+.$$

The writing of the expansions for $\zeta > \zeta_0$ would give the relation between $a_{00}^-$ and $D_{00}^-$ in the same manner (here we have $D_{00}^+ \equiv 0$ a priori).

By eliminating $a_{00}^+$ between $C_{00}^+$ and $C_{00}^-$ in (12,40) we obtain

$$(12,42) \qquad C_{00}^- = C_{00}^+ e^{2i\chi_0} \quad , \quad \chi_0 = \frac{\varphi_0}{M_0} - \frac{\pi}{4} \, .$$

It is possible to place the meaning of the turning point $\zeta_0$ and of the formula (12,40) in evidence :

the existence of $\zeta_0$ involves the disappearing of the wave which propagates if $\zeta < \zeta_0$, the amplitude of the wave in the neighborhood of $\zeta_0$ becoming $O(M_0^{-1/6})$, from (12,40) and (12,39), in an interval of length $|\zeta - \zeta_0| = O(M_0^{2/3})$, from (12,23) and (12,26), (12,27) .

For $\zeta > \zeta_0$ the waves decreases as $\exp(-\xi)$, i.e. as $\exp(-\zeta/M_0)$ with respect to $\zeta$. Hence beyond $\zeta_0$ there is no more oscillation, and $\zeta_0$ appears as a reflection level of the oscillation. The relation (12,40) then is a relation giving the reflection coefficient between the incident wave $C_{00}^+ e^{i\xi}$ and the reflected wave $C_{00}^- e^{-i\xi}$.

From (12,42) the ratio of amplitudes in the reflection is $|C_{00}^-/C_{00}^+| = 1$. But the reflection is accompanied of a change of phase.

Note that by taking the dissipative effects into account in the linearized equations (12,5), it is possible to solve the problem of the *upper boundary condition* in a simplified manner.


## 13 . THE ROSSBY WAVES


For the derivation of the equation satisfied by Rossby waves it is necessary to take into account the $\beta$-effect (see (5,12) and (5,13)), according to which variation of the Coriolis parameter is taken into account through the term: $\beta \frac{K1}{S} y \ (\vec{k} \wedge \vec{v})$ in the equation for $\vec{v}$ (see, the system (9,1)).

After linearization of (9,14), relative to the state of rest, we may look a plane wave solutions, namely:

$$(13,1) \qquad \mathcal{H}_{qg} = \mathcal{R}eal\left\{\phi_{qg}(p)\exp\left[i\vec{k}.\vec{r} - \frac{\sigma}{S}t\right]\right\}$$

and this leads to the dispersion relation

$$(13,2) \qquad \sigma = \frac{-m\beta}{Bo\lambda_0(m^2+n^2)+\mu} \equiv \sigma_R ,$$

$$m^2+n^2 = |\vec{k}|^2 \equiv k^2.$$

In (13,2) the scalar $\mu$ is the eigenvalue corresponding to the Sturn-Liouville equation

$$(13,3) \qquad \frac{d}{dp}\left[\frac{p^2}{K_0(p)}\frac{d\phi_{qg}}{dp}\right] + \mu\phi_{qg} = 0 ,$$

with suitable homogeneous boundary conditions.

The Rossby waves frequency $\sigma_R$ is always less than $\beta$, thus existence of Rossby waves requires that both m and $\beta$ be non zero.

The phase speed of a Rossby wave (in the absence of a basic current) in the x-direction is

$$(13,4) \qquad c_x = \frac{\sigma_R}{m} = \frac{-\beta}{Bo\lambda_0 k^2 + \mu} ,$$

and retrogresses toward negative x, i.e., the West. But the phase speed in the y-direction

$$(13,5) \qquad c_y = \frac{\sigma_R}{n} = \frac{-\beta\frac{m}{n}}{Bo\lambda_0 k^2 + \mu} ,$$

can be positive or negative depending on the orientation of the wave vector, i.e., on the sign of $\frac{m}{n}$.

*SLOW VARIATION OF THE WAVE AMPLITUDE*

If $\Delta$ is a small parameter that is a measure of the slowness of the temporal and spatial variations of Rossby waves, $\mathcal{H}_{qg}$ can be written (instead of (13,1)):

$$(13,6) \qquad \mathcal{H}_{qg} = \phi_{qg}(p)\left\{\chi_0 + \Delta\chi_1 + \ldots\right\} ,$$

where

$$\chi_0 = \mathcal{R}eal\left\{A_0(T,X,Y)\exp\left[i\left(mx + ny - \frac{\sigma_R}{S}t\right)\right]\right\}$$

and $T=\Delta t$, $X=\Delta x$, $Y=\Delta y$ are the *slow* variables, while $t,x,y$, are the *fast* variables.

The function $\chi_1(t,x,y;T,X,Y)$ is solution of the order $-\Delta$ equation and the calculation of $\chi_1$ presents us with an apparent dilemma.

The right-hand side of this order $-\Delta$ equation is, as far as its dependence on $t$, $x$ and $y$ is concerned, oscillating with the frequency of homogeneous solution of the equation for $\chi_1$. That is, for a fixed $m$ and $n$ the forcing term on the right-hand side of the order $-\Delta$ equation for $\chi_1$ oscillates at the natural frequency of oscillation of the system.

In a manner precisely analogous to the resonant forcing of an undamped oscillator (see Appendix 2), the solution for $\chi_1$ would then grow linearly with $t$, i.e., would contain a secular growth with $t$, in which case

$$\Delta \frac{\chi_1}{\chi_0} = O(\Delta t) \cong T$$

so that in a time $t=O(\frac{1}{\Delta})$, for which $T=O(1)$, the second term in the expansion (13,6) would become as large as the first. Of course it is precisely for times $t=O(\frac{1}{\Delta})$ that we wish to discribes the evolution of the wave amplitude, so we must insist that the expansion (13,6) remain valid for this length of time.

To accomplish this we must remove the resonant forcing term from the right-hand side of the equation for $\chi_1$ by insisting that (elemination of secular terms):

(13,7)
$$S\frac{\partial A_0}{\partial T} - \frac{2\sigma_R m + \beta/Bo\lambda_0}{m^2+n^2+ \mu/Bo\lambda_0}\frac{\partial A_0}{\partial X} - \frac{2\sigma_R n + \beta/Bo\lambda_0}{m^2+n^2+ \mu/Bo\lambda_0}\frac{\partial A_0}{\partial Y} = 0 \ ,$$

and that equation (13,7) rules the dependency of $A_0$ on the *slow* variables $T,X$ and $Y$.

The vector form of (13,7) is simply

(13,8)
$$S\frac{\partial A_0}{\partial T} + (\vec{V}_g.\vec{D})A_0 = 0,$$

where

$$\begin{cases} \vec{V}_g = u_g \vec{i} + v_g \vec{j} \; ; \\[2mm] \vec{\mathcal{D}} = \dfrac{\partial}{\partial X}\vec{i} + \dfrac{\partial}{\partial Y}\vec{j} \; ; \\[2mm] u_g = \dfrac{\beta}{Bo\lambda_0}\ \dfrac{m^2 - n^2 - \mu/Bo\lambda_0}{(m^2 + n^2 + \mu/Bo\lambda_0)^2}\; ; \\[2mm] v_g = \dfrac{\beta}{Bo\lambda_0}\ \dfrac{2mn}{(m^2 + n^2 + \mu/Bo\lambda_0)^2}\cdot \end{cases}$$

(13,9)

Therefore $A_0$ is constant for an observer moving with velocity $\vec{V}_g$, or :

(13,10)
$$A_0 = A_0(\vec{X} - \frac{T}{S}\vec{V}_g)$$

where $\vec{X} = X\vec{i} + Y\vec{j}$ is the position vector in the $(X,Y)$ plane[†].

To the first approximation the envelope of the Rossby wave packet moves with $\vec{V}_g$, which is called the *group velocity*, the X and Y components of which are given by (13,9).

In contrast to the phase velocity, the group velocity does satisfy the usual vector rules of projection. It can be verified directly from (13,9) and (13,2) that:

$$u_g = \frac{\partial \sigma_R}{\partial m} \quad \text{and} \quad v_g = \frac{\partial \sigma_R}{\partial n}$$

(13,11)
$$\vec{V}_g = \frac{\partial \sigma_R}{\partial m}\vec{i} + \frac{\partial \sigma_R}{\partial n}\vec{j} \equiv \frac{\partial \sigma_R}{\partial \vec{k}}\ ,$$

where $\dfrac{\partial}{\partial \vec{k}}$ denotes the vector gradient with respect to wave number .

Since

(13,11)
$$\sigma_R = k v_R \ ,$$

where $v_R$ is the speed of the advance of the Rossby wave crests in the

---

[†] The chief virtue of the method of multiple scales is that it systematically separates the problem for local dynamics in space and time from the problem of the slow larger-scale variations. In the present problem the method only enriches our understanding of the approximations leading to (13,7). It does, however give a device for the systematic derivation to higher-order corrections if desired. What is much more important is that the method can be efficiently applied to problems for which the heuristic methods are inadequate, for example the problem of nonlinear interaction of Rossby waves.

direction of the wave vector,

$$(13,13) \qquad \vec{V}_g = \frac{\partial}{\partial \vec{k}}(kv_R) = \frac{\vec{k}}{k} v_R + k \frac{\partial v_R}{\partial \vec{k}} = \vec{V}_R + k \frac{\partial v_R}{\partial \vec{k}} .$$

Hence unless the phase speed is independent of $\vec{k}$, the group velocity and the phase speed will be different in both magnitude and direction.

Waves for which $\vec{V}_g \neq \vec{V}_R$ are called "*dispersive*", and the *Rossby wave is particularly striking case of a dispersive wave*.

*BAROTPOPIC CASE*

The linearized form of (9,14) is

$$(13,14) \qquad \frac{\partial}{\partial t}\left\{ B_0 \lambda_0 \vec{D}^2 \mathcal{H}'_{qg} + \frac{\partial}{\partial p}\left[ \frac{p^2}{K_0(p)} \frac{\partial \mathcal{H}'_{qg}}{\partial p} \right] \right\} + \frac{\beta}{S} \frac{\partial \mathcal{H}'_{qg}}{\partial x} = 0 .$$

But on the flat ground we have, from (9,16), the following linearized bondary condition :

$$(13,15) \qquad \frac{\partial \mathcal{H}'_{qg}}{\partial t} + \frac{1}{Bo} \frac{T_0(1)}{K_0(1)} \frac{\partial}{\partial t}\left( \frac{\partial \mathcal{H}'_{qg}}{\partial p} \right) = 0, \text{ on } p=1.$$

Integrating (13,14) from the level p=0 (infinity) at the ground p=1 yields

$$(13,16) \qquad B_0 \lambda_0 \int_0^1 \frac{\partial}{\partial t}(\vec{D}^2 \mathcal{H}'_{qg})dp + \frac{\beta}{S}\int_0^1 \frac{\partial \mathcal{H}'_{qg}}{\partial x}dp + \frac{\partial}{\partial t}\left[ \frac{p^2}{K_0(p)} \frac{\partial \mathcal{H}'_{qg}}{\partial p} \right]\Big|_0^1 = 0.$$

If we suppose that on the upper limite (p→0) of the atmosphere, when p→0, we have :

$$(13,17) \qquad \frac{\partial}{\partial t}\left[ \frac{p^2}{K_0(p)} \frac{\partial \mathcal{H}'_{qg}}{\partial p} \right] \to 0 ,$$

then we obtain, in place of (13,16) and with (13,15), the following equation :

$$(13,18) \qquad Bo\lambda \int_0^1 \frac{\partial}{\partial t}(\vec{D}^2 \mathcal{H}'_{qg})dp + \frac{\beta}{S}\int_0^1 \frac{\partial \mathcal{H}'_{qg}}{\partial x}dp = \frac{Bo}{T_0(1)} \frac{\partial \mathcal{H}'_{qg}}{\partial t} .$$

In the *barotropic case, we suppose that* :

$$(13,19) \qquad \mathcal{H}'_{qg} \equiv \mathcal{H}_b(t,x,y)$$

*only*, and we obtain from (13,18) the following *barotropic linearized equation* for $\mathcal{H}_b(t,x,y)$

(13,20)
$$\frac{\partial}{\partial t}\left\{\vec{D}^2\mathcal{H}_b - \frac{1}{\lambda_0 T_0(1)}\mathcal{H}_b\right\} + \frac{\beta}{Bo\lambda_0 S}\frac{\partial \mathcal{H}_b}{\partial x} = 0.$$

If the linear wave equation (13,20) is multiplied by $\mathcal{H}_b$, a little manipulation yields

(13,21)
$$\frac{\partial}{\partial t}\left[\frac{1}{2}\left\{\left(\vec{D}\mathcal{H}_b\right)^2 + \frac{1}{\lambda_0 T_0(1)}\mathcal{H}_b^2\right\}\right] + \vec{D}\cdot\left\{-\mathcal{H}_b\vec{D}\left(\frac{\partial \mathcal{H}_b}{\partial t}\right) - \frac{\beta}{2Bo\lambda_0 S}\mathcal{H}_b^2\vec{i}\right\} = 0 \;,$$

which has a direct interpretation in terms of energy.

It is apparent that the term in the square bracket [] in (13,21) is the nondimensional form of sum of the kinetic plus potential energy, whose rate of increase with time balances the divergence of the flux vector

(13,22)
$$\vec{\mathcal{F}}_b = -\mathcal{H}_b\vec{D}\left(\frac{\partial \mathcal{H}_b}{\partial t}\right) - \frac{\beta}{2Bo\lambda_0 S}\mathcal{H}_b^2\vec{i} \;,$$

so that the energy $\mathcal{E}_b(t,x,y)$ for the barotropic quasigeostrophic flow (linearized) satisfies the *conservation law* :

(13,23)
$$\frac{\partial \mathcal{E}_b}{\partial t} + \vec{D}\cdot\vec{\mathcal{F}}_b = 0 \;.$$

For a Rossby wave packet

(13,24)
$$\mathcal{H}_b = \phi_b\cos\theta \quad,\quad \theta = mx + ny - \sigma t$$

where $\phi_b$ can be assumed as a constant during the derivation relative to $t, x$ and $y$ .

In this case the energy to the lowest order is:

(13,25)
$$\mathcal{E}_b = \frac{\phi_b^2 k^2}{2}\sin^2\theta + \frac{1}{\lambda_0 T_0(1)}\phi_b^2\cos^2\theta \;,$$

and as written, at any fixed point, $\mathcal{E}_b$ varies rapidly with half the period of the wave about the *average value*

(13,26)
$$<\mathcal{E}_b> = \left[k^2 + \frac{1}{\lambda_0 T_0(1)}\right]\frac{\phi_b^2}{4} \;,$$

so that

(13,27)
$$\mathcal{E}_b = \langle \mathcal{E}_b \rangle + \frac{\phi_b^2}{4}\left[k^2 - \frac{1}{\lambda_0 T_0(1)}\right]\cos 2\theta \ .$$

It is the average over the period, $\langle \mathcal{E}_b \rangle$, which gives a stable definition of the local wave energy and is the appropriate definition of the *wave energy* $\langle \mathcal{E}_b \rangle$. Note that $\langle \mathcal{E}_b \rangle$ varies *slowly* over the packet with $\phi_b^2$.
The energy flux vector $\vec{\mathcal{F}}_B$, from (13,22) and (13,24) is :

(13,28)
$$\vec{\mathcal{F}}_b = -\phi_b^2 \sigma \ \vec{k} \ \cos^2\theta - \frac{\beta}{2Bo\lambda_0 S} \ \phi_b^2 \cos^2\theta \ \vec{i} \ ,$$

whose average over a period is

(13,29)
$$\langle \vec{\mathcal{F}}_b \rangle = \frac{1}{2} \ \phi_b^2 \left[-\sigma\vec{k} - \frac{\beta}{2Bo\lambda_0 S} \ \vec{i}\right].$$

But for the barotropic Rossby waves, we have,

$$\sigma = \frac{-m\beta}{Bo\lambda_0 S\left[m^2 + n^2 + \frac{1}{\lambda_0 T_0(1)}\right]} \equiv \sigma_{Rb} \ ,$$

and

$$u_{gb} = \frac{\beta}{Bo\lambda_0 S} \ \frac{m^2 - n^2 - \frac{1}{\lambda_0 T_0(1)}}{\left[k^2 + \frac{1}{\lambda_0 T_0(1)}\right]^2} \ ;$$

$$v_{gb} = \frac{2\beta}{Bo\lambda_0 S} \ \frac{mn}{\left[k^2 + \frac{1}{\lambda_0 T_0(1)}\right]^2} \ ,$$

and therefore (13,29) can be written as :

(13,30)
$$\langle \vec{\mathcal{F}}_b \rangle = \langle \mathcal{E}_b \rangle \ \vec{V}_{gb} \ .$$

Hence the average energy flux vector is equal to the wave energy multiplied by the group velocity. Or since $\vec{V}_{gb}$ is independent of space (13,23) becomes, when averaged over a wave period,

(13,31)
$$\frac{\partial}{\partial t}\langle \mathcal{E}_b \rangle + \vec{V}_{gb} . \vec{D}\langle \mathcal{E}_b \rangle = 0.$$

*The wave energy*, defined by (13,26), *is conserved and propagates with the packet at the group velocity.*

## 14 . THE ISOCHORIC[†] NONLINEAR WAVE EQUATION
##         (LONG'S EQUATION)

We consider the equations (5,5)-(5,9), with :

$$\beta\equiv0, \quad \varepsilon_0\equiv1, \quad Ro\equiv\infty, \quad \text{and} \quad Re\equiv\infty ;$$

the corresponding equations are the usual Euler's equations and we can write
the following system, with the dimensions :

$$(14,1) \quad \begin{cases} \rho\dfrac{D\vec{u}}{Dt} + \vec{\nabla}p + \rho g\vec{k} = 0 ; \\[2mm] \dfrac{D\rho}{Dt} + \rho\vec{\nabla}.\vec{u} = 0 ; \\[2mm] p = R\rho T ; \\[2mm] \dfrac{Ds}{Dt} = 0 , \end{cases}$$

where $s=c_v \text{Log } p/\rho^\gamma$ , $\gamma= c_p/c_v$ , for a perfect gas .

The equations of *isochoric* motion may be obtained, formally, from (14,1), in
the limit : $\gamma\to\infty$ ($c_p=O(1)$ *but* $c_v\to0$). The formal $\gamma\to\infty$ limit provides an
answer:

$$(14,2) \quad \begin{cases} s\to s_{is} \equiv -c_p \text{Log}\rho , \quad T\to T_{is} \equiv \dfrac{1}{c_p} p/\rho ; \\[2mm] \dfrac{Ds}{Dt} = 0 \to \dfrac{D\rho}{Dt} = 0 \implies \vec{\nabla}.\vec{u} = 0 . \end{cases}$$

The resulting equations are the equations for the isochoric motions :

$$(14,3) \quad \begin{cases} \rho\dfrac{D\vec{u}}{Dt} + \vec{\nabla}p + \rho g\vec{k} = 0 ; \\[2mm] \vec{\nabla}.\vec{u} = 0 ; \\[2mm] \dfrac{D\rho}{Dt} = 0 . \end{cases}$$

For the *steady* flows the equations (14,3) are written in the following form :

$$(14,4) \quad \begin{cases} \vec{\nabla} \dfrac{q^2}{2} - \vec{u} \wedge \vec{\nabla} \wedge \vec{u} + \dfrac{1}{\rho} \vec{\nabla}p + g\vec{k} = 0 , \\[2mm] \vec{\nabla}.\vec{u} = 0 , \qquad \vec{u}.\vec{\nabla}\rho = 0 , \end{cases}$$

where

$$q = |\vec{u}| .$$

---

[†] For the full discussion of isochoric motions, see the book of Yih (1980) .

*FUNCTIONS OF SPACE CURRENT $\psi$ AND $\chi$ .*

*REPRESENTATION OF THE MOTION BY TWO EQUATIONS IN $\psi$ AND $\chi$ .*

In the general case of motion which is steady, three-dimensional and rotational[†] , we are led to introduce two stream functions. It is evident that the equations of continuity in the system (14,4) is satisfied by putting:

(14,5)              $\vec{u} = \vec{\nabla}\psi \wedge \vec{\nabla}\chi$ ;

The two stream functions $\psi$ and $\chi$ constituting the three-dimensional generalisation of the notion of function of plane stream. The third equation of the system (14,4) gives us then:

(14,6)              $\rho = \rho(\psi,\chi)$ ,

that is to say that the specific mass is *conserved* throughout the lenght of each stream line.

It is well known also that one of the integrals of the system (14,4) is the *equation of Bernoulli*:

(14,7)              $\dfrac{q^2}{2} + \dfrac{p}{\rho} + gz = \mathcal{F}(\psi,\chi)$  ,

the function $\mathcal{F}(\psi,\chi)$ being itself also conserved throughout the lenght of each stream line.

Using relations (14,6) and (14,7) the first equation of (14,4) is written in the form

(14,8)              $\vec{u} \wedge \vec{\nabla} \wedge \vec{u} = \vec{\nabla}\mathcal{F} + \dfrac{p}{\rho^2}\vec{\nabla}\rho$ ,

or better still

(14,9)              $\vec{u} \wedge \vec{\nabla} \wedge \vec{u} = (\dfrac{\partial \mathcal{F}}{\partial \psi} + \dfrac{p}{\rho^2}\dfrac{\partial \rho}{\partial \psi})\vec{\nabla}\psi$

$= (\dfrac{\partial \mathcal{F}}{\partial \chi} + \dfrac{p}{\rho^2}\dfrac{\partial \rho}{\partial \chi})\vec{\nabla}\chi.$

---

[†] See the book of Zeytounian (1974) : *Notes sur les écoulements rotationnels de fluides parfaits* . Lecture Notes in Physics, vol.27. Springer-Verlag, Heidelberg.

If relation (14,5) is taken into account we obtain on the left hand side of the equation (14,9) a double vectorial product: $(\vec{\nabla}\psi \wedge \vec{\nabla}\chi) \wedge (\vec{\nabla} \wedge \vec{u})$ which can be put in the form

$$(14,10) \qquad \left[ (\vec{\nabla} \wedge \vec{u}).\vec{\nabla}\psi \right] \vec{\nabla}\chi - \left[ (\vec{\nabla} \wedge \vec{u}).\vec{\nabla}\chi \right] \vec{\nabla}\psi.$$

The last two relations, (14,9) and (14,10), allow the two following scalar equations to be written:

$$(14,11) \qquad \begin{cases} (\vec{\nabla} \wedge \vec{u}).\vec{\nabla}\psi = \dfrac{\partial \mathcal{F}}{\partial \chi} + \dfrac{p}{\rho^2} \dfrac{\partial \rho}{\partial \chi} \; ; \\[4mm] (\vec{\nabla} \wedge \vec{u}).\vec{\nabla}\chi = -\left[ \dfrac{\partial \mathcal{F}}{\partial \psi} + \dfrac{p}{\rho^2} \dfrac{\partial \rho}{\partial \psi} \right]. \end{cases}$$

These two expressions (14,11) are the two other first integrals of the system of equations of the isochoric steady motion (14,4).

We state, at once, that the arbitrary functions $\mathcal{F}$ and $\rho$ must be determined from boundary conditions at infinity upstream far flow non-disturbed by the obstacle and supposed known.

Here we suppose that at infinity upstream the non-disturbed flow is *two-dimensional* (in the planes $(x,z)$) and uniquely a function of altitude, denoted by $z_\infty$ at infinity upstream. Letting

$$(14,12) \qquad \vec{u}_\infty = U_\infty(z_\infty)\vec{i}$$

be the speed far upstream of the obstacle, we will write that:

$$(14,13) \qquad \text{for } x \longrightarrow -\infty, \quad u = U_\infty(z_\infty), \quad v = w = o \; .$$

## DETERMINATION OF THE FUNCTIONS $\mathcal{F}$ AND $\rho$

In the framework of hypothesis (14,13), the relation (14,5) for $\vec{u}$ enables it to be written that at infinity upstream:

$$(14,14) \qquad \psi = - \int_0^{z_\infty} U_\infty(\xi)d\xi \equiv \psi_\infty(z_\infty),$$

$z_\infty$ being, therefore, the altitude of a stream line in the basic non-disturbed flow. In this particular case the second stream function $\chi$ at infinity upstream is a plane

(14,15) $\qquad \chi_\infty = y = $ constant.

We will suppose implicitly that the solution of the problem considered ought to be uniformly bounded at all points of the infinite plane $(x,y)$. We mention that $\psi$ is uniform function of $z_\infty$, but that $z_\infty(\psi)$ is only a uniform function of $\psi$ while $U_\infty(z_\infty)$ is strictly positive.

If, now $\psi=0$ determine the wall of the obstacle, we will have, as a direct consequence of the nature of the flow at infinity upstream, that in all the region occupied by the moving fluid

(14,16) $\qquad \mathcal{I} = \mathcal{I}_\infty(\psi)$ and $\rho = \rho_\infty(\psi)$ .

In this case the first equation of the system (14,11) becomes homogeneous: the stream surface $\psi$=constant being then also a vortex surface, and the second equation of the system (14,11) has a second member, a unique function of $\psi$ :

(14,17) $\qquad -(\dfrac{d\mathcal{I}_\infty}{d\psi} + \dfrac{p}{\rho_\infty^2}\dfrac{d\rho_\infty}{d\psi})$ .

If we note that from relation (14,7) for $\mathcal{I}_\infty(\psi)$ we can write :

$\dfrac{p}{\rho_\infty(\psi)} \equiv \mathcal{I}_\infty(\psi) - (\dfrac{q^2}{2} + gz)$ , we will obtain, in place of expression (14,17),

(14,18) $\qquad \dfrac{1}{\rho_\infty(\psi)}\dfrac{d\rho_\infty}{d\psi}(\dfrac{q^2}{2} + gz) - \dfrac{1}{\rho_\infty}\dfrac{d}{d\psi}(\rho_\infty\mathcal{I}_\infty)$ .

But the second equation of the system (14,11) in taking into account (14,18) as also (14,14) and (14,15) gives at infinity upstream

$$\dfrac{1}{\rho_\infty(\psi)}\dfrac{d}{d\psi}(\rho_\infty\mathcal{I}_\infty) = \dfrac{1}{\rho_\infty}\dfrac{d\rho_\infty}{d\psi}(\dfrac{U_\infty^2}{2} + gz_\infty) - \dfrac{dU_\infty}{dz_\infty} .$$

Finally, we will obtain in place of (14,11) the following system of equations for $\psi$ and $\chi$:

64

$$(14,19) \begin{cases} \vec{\nabla} \wedge (\vec{\nabla}\psi \wedge \vec{\nabla}\chi) . \vec{\nabla}\psi = 0; \\[4mm] \vec{\nabla} \wedge (\vec{\nabla}\psi \wedge \vec{\nabla}\chi) . \vec{\nabla}\chi = \dfrac{1}{\rho_\infty} \dfrac{d\rho_\infty}{d\psi} \dfrac{(\vec{\nabla}\psi \wedge \vec{\nabla}\chi)^2}{2} \\[4mm] \qquad\qquad - U_\infty \dfrac{dU_\infty}{d\psi} - \dfrac{1}{\rho_\infty} \dfrac{d\rho_\infty}{d\psi} \left\{ \dfrac{U_\infty^2}{2} + g(z - z_\infty(\psi)) \right\} . \end{cases}$$

The two equations (14,19) are the generalisation of the three-dimensional case of the equation of Long (1955) obtained for a plane stream function in a two-dimensional, stratified, incompressible flow.

*PLANE FLOW: LONG'S EQUATION*

If we suppose in system(14,19) that

$$\psi = \psi_p(x,z) \quad \text{and} \quad \chi \equiv y$$

we will obtain for the plane stream function $\psi_p(x,z)$ the classical equation of Long[†]:

$$(14,20) \begin{aligned} & \frac{\partial^2 \psi_p}{\partial x^2} + \frac{\partial^2 \psi_p}{\partial z^2} + \frac{1}{2} \frac{1}{\rho_\infty(\psi_p)} \left[ (\frac{\partial \psi_p}{\partial x})^2 + (\frac{\partial \psi_p}{\partial z})^2 \right] \\[3mm] & - U_\infty \frac{dU_\infty}{d\psi_p} = \frac{1}{\rho_\infty(\psi_p)} \frac{d\rho_\infty}{d\psi_p} \left\{ \frac{U_\infty^2}{2} + g(z - z_\infty(\psi_p)) \right\}, \end{aligned}$$

where $U_\infty(z_\infty(\psi_p)) \equiv U_\infty(\psi_p)$ .

Let $z_\infty + \delta(x,z)$ represent the altitude of a stream line in the disturbed flow in such a way that the expression

$$z - \delta(x,z) = z_\infty(\psi_p)$$

remains constant along the length of the stream line (see the Fig.1 below).

---

[†] This equation has been derived at first by Dubreil-Jacotin in 1935.

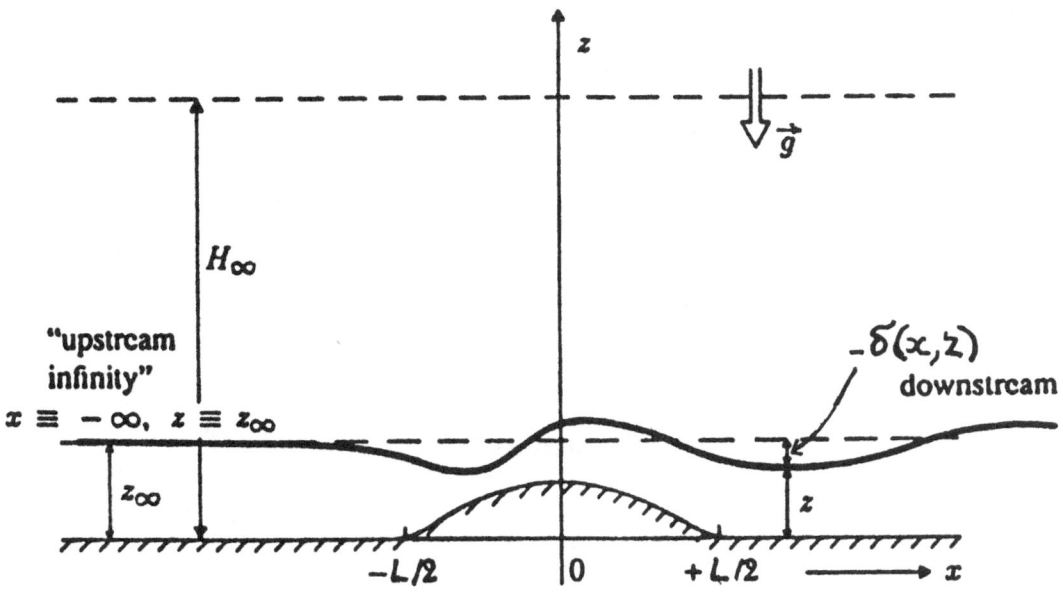

*Fig. 1: Lee waves problem.*

We obtain, in place of the equation (14,20), for the function δ(x,z):

$$\frac{\partial^2 \delta}{\partial x^2} + \frac{\partial^2 \delta}{\partial z^2} + \frac{1}{2}\left\{2\frac{\partial \delta}{\partial z} - \left[(\frac{\partial \delta}{\partial x})^2 + (\frac{\partial \delta}{\partial z})^2\right]\right\}\frac{d}{dz_\infty}(\text{Log}(U_\infty^2 \rho_\infty))$$

(14,21)

$$= \frac{g}{U_\infty^2 \rho_\infty}\frac{d\rho_\infty}{dz_\infty}\delta \;,$$

where $z_\infty \equiv z_\infty(\psi_p) = z - \delta(x,z)$ .

Long(1955) has remarked that the nonlinear terms in equation (14,21) disappear if:

(14,22)                 $U_\infty^2 \rho_\infty$ =constant,  $\dfrac{d\rho_\infty}{dz_\infty}$ = constant .

The equation which results from this reduces to the equation of Helmoltz:

(14,24)                 $\dfrac{\partial^2 \delta}{\partial x^2} + \dfrac{\partial^2 \delta}{\partial z^2} + \sigma_0^2 \delta = 0$ ,

with

(14,24)                 $\sigma_0^2 = \dfrac{g}{U_\infty^2 \rho_\infty} \left| \dfrac{d\rho_\infty}{dz_\infty} \right| =$ constant .

The dominant feature from the mathematical point of view is that the linearity of equation (14,23) is not related to any one hypothesis of small perturbations. But an important difficulty remains: it is that the boundary condition on the wall of the obstacle[†],

(14,25)                 $z = h_p(x) \implies \delta(x, h_p(x)) = h_p(x)$ ,

is *non* linear and cannot be linearized without invoking the hypothesis of small disturbances.

## 15 . BOUSSINESQ'S THREE-DIMENSIONAL LINEARIZED WAVE EQUATION AND RESULTS OF THE CALCULATIONS

The full system of equations (14,4) is nonlinear; the purpose of the theory of small perturbations is to linearise this system by supposing that the disturbed flow does not contain strong perturbations in relation to the basic non disturbed flow. We introduce the perturbations p', $\rho$', u', v' and w' of the corresponding hydrodynamical element and we note[††]:

---

[†] $z=h_p(x)$ is the trace of the obstacle, $z=h(x,y)$, three-dimensional in the plane of $(x,z)$ for y=constant.

[††] We assume, a priori, that all the perturbations are of the same order. We recall that at infinity, upstream, $z \equiv z_\infty$ and

$\dfrac{dp_\infty}{dz_\infty} + \rho_\infty g = 0.$

$$(15,1) \quad \begin{cases} p = p_\infty(z_\infty) + p'(x,y,z) \; ; \\[2mm] \rho = \rho_\infty(z_\infty) + \rho'(x,y,z) \; ; \\[2mm] u = U_\infty(z_\infty) + u'(x,y,z) \; ; \\[2mm] v = V_\infty(z_\infty) + v'(x,y,z) \; ; \\[2mm] w = 0 + w'(x,y,z) \; , \end{cases}$$

and the basic motion at infinity, upstream, having one speed
$$\vec{V}_\infty(z_\infty) = U_\infty(z_\infty)\vec{i} + V_\infty(z_\infty)\vec{j}.$$

We obtain, from equation (14,4), taking into account (15,1) and neglecting the higher order terms, the following linear system:

$$(15,2) \quad \begin{cases} \rho_\infty(U_\infty \dfrac{\partial u'}{\partial x} + V_\infty \dfrac{\partial u'}{\partial y} + \dfrac{dU_\infty}{dz} w') + \dfrac{\partial p'}{\partial x} = 0 \; ; \\[4mm] \rho_\infty(U_\infty \dfrac{\partial v'}{\partial x} + V_\infty \dfrac{\partial v'}{\partial y} + \dfrac{dV_\infty}{dz} w') + \dfrac{\partial p'}{\partial y} = 0 \; ; \\[4mm] \rho_\infty(U_\infty \dfrac{\partial w'}{\partial x} + V_\infty \dfrac{\partial w'}{\partial y}) + \dfrac{\partial p'}{\partial z} + g\rho' = 0 \; ; \\[4mm] U_\infty \dfrac{\partial \rho'}{\partial x} + V_\infty \dfrac{\partial \rho'}{\partial y} + \dfrac{d \, \text{Log} \, \rho_\infty}{dz} \rho_\infty w' = 0 \; ; \\[4mm] \dfrac{\partial u'}{\partial x} + \dfrac{\partial v'}{\partial y} + \dfrac{\partial w'}{\partial z} = 0 \; , \end{cases}$$

since $\dfrac{dz}{dz_\infty} \equiv 1$.

We note that: $\rho_\infty u' = \bar{u}$, $\rho_\infty v' = \bar{v}$, and $\rho_\infty w' = \bar{w}$; from the first two equations and the last equation of system (15,2) we obtain an equation in $\bar{w}$ and p':

$$(15,3) \quad \begin{aligned} (U_\infty \frac{\partial}{\partial x} + V_\infty \frac{\partial}{\partial y})(\frac{d \, \text{Log} \, \rho_\infty}{dz} \bar{w} - \frac{\partial \bar{w}}{\partial z}) + \frac{dU_\infty}{dz} \frac{\partial \bar{w}}{\partial x} + \frac{dV_\infty}{dz} \frac{\partial \bar{w}}{\partial y} \\[2mm] = -(\frac{\partial^2 p'}{\partial x^2} + \frac{\partial^2 p'}{\partial y^2}) \; . \end{aligned}$$

We obtain a second equation relating $\bar{w}$ and p' from the third and fourth equations of system (15,2), which gives us:

$$(15,4) \qquad g \, \frac{d \, \mathrm{Log} \, \rho_\infty}{dz} \, \bar{w} - (U_\infty \frac{\partial}{\partial x} + V_\infty \frac{\partial}{\partial y})^2 \, \bar{w} = (U_\infty \frac{\partial}{\partial x} + V_\infty \frac{\partial}{\partial y}) \frac{\partial p'}{\partial z} \, .$$

By eliminating p' from (15,3) and (15,4) we obtain the following equation in $\bar{w}$:

$$(U_\infty \frac{\partial}{\partial x} + V_\infty \frac{\partial}{\partial y})^2 \left[ \frac{\partial^2 \bar{w}}{\partial x^2} + \frac{\partial^2 \bar{w}}{\partial y^2} + \frac{\partial^2 \bar{w}}{\partial z^2} \right] - g \, \frac{d \, \mathrm{Log} \, \rho_\infty}{dz} \left[ \frac{\partial^2 \bar{w}}{\partial x^2} + \frac{\partial^2 \bar{w}}{\partial y^2} \right]$$

$$(15,5) \qquad - (U_\infty \frac{\partial}{\partial x} + V_\infty \frac{\partial}{\partial y}) \left[ \frac{d^2 U_\infty}{dz^2} \frac{\partial \bar{w}}{\partial x} + \frac{d^2 V_\infty}{dz^2} \frac{\partial \bar{w}}{\partial y} \right]$$

$$= - (U_\infty \frac{\partial}{\partial x} + V_\infty \frac{\partial}{\partial y}) \frac{d}{dz} \left\{ (U_\infty \frac{\partial}{\partial x} + V_\infty \frac{\partial}{\partial y}) \frac{d \, \mathrm{Log} \, \rho_\infty}{dz} \, \bar{w} \right\} \, .$$

*BOUSSINESQ'S APPROXIMATION*

We see the case where

$$(15,6) \qquad \rho_\infty(z) = \rho_0 \exp(-\beta z) \, ,$$

and we pass to the non-dimensional variables:

$$\xi = \frac{x}{H_0}, \quad \eta = \frac{y}{H_0}, \quad \zeta = \frac{z}{H_0}$$

and to the non-dimensional speeds

$$\bar{U}_\infty = \frac{U_\infty}{U_\infty^0}, \quad \bar{V}_\infty = \frac{V_\infty}{V_\infty^0}, \quad \omega = \frac{\bar{w}}{\rho_0 U_\infty^0} \, ,$$

where $U_\infty^0$ is a constant speed characteristic of the flow. We obtain from equation (15,5)

$$(\bar{U}_\infty \frac{\partial}{\partial \xi} + \bar{V}_\infty \frac{\partial}{\partial \eta})^2 \left[ \frac{\partial^2 \omega}{\partial \xi^2} + \frac{\partial^2 \omega}{\partial \eta^2} + \frac{\partial^2 \omega}{\partial \zeta^2} \right] + \mathcal{D}_0 \left[ \frac{\partial^2 \omega}{\partial \xi^2} + \frac{\partial^2 \omega}{\partial \eta^2} \right]$$

(15,7)

$$- (\bar{U}_\infty \frac{\partial}{\partial \xi} + \bar{V}_\infty \frac{\partial}{\partial \eta}) \left[ \frac{d^2 \bar{U}_\infty}{d\zeta^2} \frac{\partial \omega}{\partial \xi} + \frac{d^2 \bar{V}_\infty}{d\zeta^2} \frac{\partial \omega}{\partial \eta} \right]$$

$$= 2\lambda_0 (\bar{U}_\infty \frac{\partial}{\partial \xi} + \bar{V}_\infty \frac{\partial}{\partial \eta}) \frac{d}{d\zeta} (\bar{U}_\infty \frac{\partial \omega}{\partial \xi} + \bar{V}_\infty \frac{\partial \omega}{\partial \eta}) \ ,$$

where

(15,8)

$$\mathcal{D}_0 = \beta H_0^2 \frac{g}{(U_\infty^0)^2} \quad ; \quad \lambda_0 = \frac{\beta H_0}{2} \ .$$

We observe that relation

(15,9)

$$\frac{2\lambda_0}{\mathcal{D}_0} = Fr^2 \ , \quad \text{where } Fr = \frac{U_\infty^0}{\sqrt{gH_0}}$$

is the Froude number (vertical) for our flow. If $H_0$ is the thickness of the troposphere $Fr^2$ will be of the order of $10^{-2} - 10^{-3}$ , which shows that the terms with $2\lambda_0$, in (15,7), can be neglected as a first approximation in relation to the other terms of the equation (15,7). This approximation is, in fact, the approximation of Boussinesq (for isochoric motion) and in this case the derivatives of $\rho_\infty(z_\infty)$ are neglected *except* when they intervene in the calculation of the force of Archimedes.

Hence, if we again seek a solution of equation (15,5) in the form of an asymptotic development of the type

$$\omega = \omega_0 + \lambda_0 \omega_1 + \ldots$$

we will obtain for $\omega_0$, as a first approximation, the following equation:

$$(\bar{U}_\infty \frac{\partial}{\partial \xi} + \bar{V}_\infty \frac{\partial}{\partial \eta})^2 \left[ \frac{\partial^2 \omega_0}{\partial \xi^2} + \frac{\partial^2 \omega_0}{\partial \eta^2} + \frac{\partial^2 \omega_0}{\partial \zeta^2} \right] + \mathcal{D}_0 \left[ \frac{\partial^2 \omega_0}{\partial \xi^2} + \frac{\partial^2 \omega_0}{\partial \eta^2} \right]$$

(15,10)

$$- (\bar{U}_\infty \frac{\partial}{\partial \xi} + \bar{V}_\infty \frac{\partial}{\partial \eta}) \left[ \frac{d^2 \bar{U}_\infty}{d\zeta^2} \frac{\partial \omega_0}{\partial \xi} + \frac{d^2 \bar{V}_\infty}{d\zeta^2} \frac{\partial \omega_0}{\partial \eta} \right] = 0 \ .$$

*PARTICULAR CASE*

When $U_\infty \equiv U_\infty^0$ = constant and $V_\infty^0 \equiv 0$ we obtain, in place of equation (15,10), the equation

$$(15,11) \qquad \left[ \frac{\partial^2}{\partial\xi^2} + \frac{\partial^2}{\partial\eta^2} + \frac{\partial^2}{\partial\zeta^2} \right] \frac{\partial^2\omega_0}{\partial\xi^2} + \mathcal{D}_0 \left[ \frac{\partial^2\omega_0}{\partial\xi^2} + \frac{\partial^2\omega_0}{\partial\eta^2} \right] = 0$$

which has been investigated by Kibel (1955), Wurtele (1957) and Crapper (1959).

We can again seek $\omega_0$ in the form:

$$\omega_0 = \omega_{00}(\zeta)\exp\{i(k\xi+\ell\eta)\}$$

which gives for $\omega_{00}(\zeta)$ the following equation:

$$(15,12) \qquad \frac{d^2\omega_{00}}{d\zeta^2} + A_0(\zeta)\omega_{00} = 0 \ ,$$

where

$$(15,13) \qquad A_0(\zeta) = \mathcal{D}_0 \frac{k^2+\ell^2}{(\bar{U}_\infty k+\bar{V}_\infty \ell)^2} - (k^2+\ell^2)$$
$$- \frac{(d^2\bar{U}_\infty/d\zeta^2)k + (d^2\bar{V}_\infty/d\zeta^2)\ell}{\bar{U}_\infty k + \bar{V}_\infty \ell} \ ;$$

this equation (15,12), with (15,13), which originates from the general equation (15,10) has been obtained and analysed by Sawyer (1962) and also by a different method by Veltichev (1965).

*CONSIDERATION OF THE GENERAL EQUATION* (15,10)

We introduce, in (15,10),

$$(15,14) \qquad \begin{cases} G_\infty(\zeta) = U_\infty^0 \sqrt{\bar{U}_\infty^2 + \bar{V}_\infty^2} \ ; \\ \\ \tan\alpha_\infty(\zeta) = \dfrac{\bar{V}_\infty}{\bar{U}_\infty} \ , \end{cases}$$

and let

(15,15)
$$
\begin{cases}
\hat{A}(\zeta) = \left( \dfrac{d\alpha_\infty}{d\zeta} \right)^2 - \dfrac{1}{G_\infty} \dfrac{d^2 G_\infty}{d\zeta^2} \; ; \\[4mm]
\hat{B}(\zeta) = \dfrac{d^2\alpha_\infty}{d\zeta^2} + 2\dfrac{1}{G_\infty} \dfrac{dG_\infty}{d\zeta} \dfrac{d\alpha_\infty}{d\zeta} \; .
\end{cases}
$$

The equation (15,10) may be written in the following form:

(15,16)
$$
\begin{cases}
\left( \cos \alpha_\infty \dfrac{\partial}{\partial\xi} + \sin \alpha_\infty \dfrac{\partial}{\partial\eta} \right)^2 \left\{ \dfrac{\partial^2\omega_0}{\partial\xi^2} + \dfrac{\partial^2\omega_0}{\partial\eta^2} + \dfrac{\partial^2\omega_0}{\partial\zeta^2} + \hat{A}(\zeta)\omega_0 \right\} \\[4mm]
+ \hat{B}(\zeta)\left( \cos \alpha_\infty \dfrac{\partial}{\partial\xi} + \sin \alpha_\infty \dfrac{\partial}{\partial\eta} \right)\left( \sin \alpha_\infty \dfrac{\partial\omega_0}{\partial\xi} - \cos \alpha_\infty \dfrac{\partial\omega_0}{\partial\eta} \right) \\[4mm]
+ \hat{D}\left( \dfrac{\partial^2\omega_0}{\partial\xi^2} + \dfrac{\partial^2\omega_0}{\partial\eta^2} \right) = 0 \; .
\end{cases}
$$

The solution of equation (15,16) will be a function, therefore, of three non-dimensional parameters $\hat{A}(\zeta)$, $\hat{B}(\zeta)$ and

(15,17)
$$
\hat{D} = \beta H_0^2 \dfrac{g}{G_\infty^2(\zeta)} \equiv \hat{D}(\zeta) \; .
$$

## SOLUTION OF THE EQUATION (15,11). ZEYTOUNIAN'S WORK

We will write the boundary conditions which must be associated with equation (15,11); considering the altitude of the relief h' as being also a small quantity of the same order as the hydrodynamic perturbations, we obtain the conditions:

(15,18)
$$
\begin{cases}
\text{for } \zeta = 0 : \quad \omega_0 = \dfrac{\partial \bar{h}'}{\partial\xi} \; ; \\[4mm]
\text{for } \zeta = 1 : \quad \omega_0 = 0 \; ,
\end{cases}
$$

with $\bar{h}'(\xi,\eta) \equiv \dfrac{1}{H_0} h'(H_0\xi, H_0\eta)$ , and if suppose that the tropopause can be represented by a rigid horizontal plane found at an altitude $H_0$.

Now the new non-dimensional variables

(15,19) . $\qquad X = a_0 \xi , \quad Y = a_0 \eta ,$

with $a_0 = H_0 / L_0$ and $L_0$ an horizontal lenght, are introduce and the new unknown function

(15,20) $\qquad \Omega = \omega_0 - (1-\zeta) a_0 \dfrac{\partial \delta}{\partial X}$

where $\delta(X,Y) \equiv \dfrac{1}{H_0} h'(L_0 X, L_0 Y) .$

We obtain for $\Omega(X,Y,\zeta)$ the following non-homogenous equation:

(15,21) $\qquad \left\{ a_0^2 \left( \dfrac{\partial^2}{\partial X^2} + \dfrac{\partial^2}{\partial Y^2} \right) + \dfrac{\partial^2}{\partial \zeta^2} \right\} \dfrac{\partial^2 \Omega}{\partial X^2} + \mathcal{D}_0 \left( \dfrac{\partial^2 \Omega}{\partial X^2} + \dfrac{\partial^2 \Omega}{\partial Y^2} \right) = -(1-\zeta) \mathcal{F}(X,Y) ,$

with $\mathcal{F}(X,Y) \equiv a_0 \left( \dfrac{\partial^2}{\partial X^2} + \dfrac{\partial^2}{\partial Y^2} \right) \left[ a_0^2 \dfrac{\partial^3 \delta}{\partial X^3} + \mathcal{D}_0 \dfrac{\partial \delta}{\partial X} \right] .$

The solution $\Omega(X,Y,\zeta)$ having to satisfy also the homogeneous bondary conditions:

(15,22) $\qquad \Omega = 0 , \quad$ for $\zeta = 0$ and $\zeta = 1 .$

First we develop $(1-\zeta)$ in the interval $(0,1]$ in series of $\sin(n\pi\zeta)$:

$$1-\zeta = \dfrac{2}{\pi} \sum_{n=1}^{\infty} \dfrac{1}{n} \sin(n\pi\zeta) , \quad \zeta \in (0,1] ,$$

and seek the solution $\Omega$ in the form :

(15,23) $\qquad \Omega = \sum_{n=1}^{\infty} \Omega_n(X,Y) \sin(n\pi\zeta) ,$

which satisfies conditions(15,22). We obtain for $\Omega_n$ the equation:

(15,24) $\quad a_0^2(\dfrac{\partial^2}{\partial X^2} + \dfrac{\partial^2}{\partial Y^2})\dfrac{\partial^2 \Omega_n}{\partial X^2} + (\mathcal{D}_0 - n^2\pi^2)\dfrac{\partial^2 \Omega_n}{\partial X^2} + \mathcal{D}_0\dfrac{\partial^2 \Omega_n}{\partial y^2} = -\dfrac{2}{n\pi}\mathcal{F}(X,Y)$ .

Assuming that $\delta(X,Y)$ is a symmetric function in Y we seek the solution of equation (15,24) in the semi-plane $\left\{ -\infty<X<+\infty \ , \ 0<Y<1 \right\}$, by imposing a condition of the type :

(15,25) $\quad \dfrac{\partial \Omega_n}{\partial Y} \propto 0$ , for Y=1 .

We construct solution of equation (15,24) satisfying also the conditions in X

(15,26) $\quad \Omega_n = \dfrac{\partial \Omega_n}{\partial X} = \dfrac{\partial^2 \Omega_n}{\partial X^2} = \dfrac{\partial^3 \Omega_n}{\partial X^3} \rightarrow 0$ , for $X \rightarrow -\infty$.

For this reason, in the (X,Y) plane we construct, between the straight lines Y=0 and Y=1, q-1 equidistant straight lines :

$$Y_j = \dfrac{j}{q} \ , \ j=1,2,\ldots,q-1,$$

and we develop the function $\mathcal{F}(X,Y)$ as a limited Fourier series in relation to $Y_j$, by supposing that $\dfrac{\partial \mathcal{F}}{\partial Y} \rightarrow 0$, for $Y \rightarrow Y_q \equiv 1$ :

(15,27) $\quad \mathcal{F}(X,Y_j) = \mathcal{F}_j(X) = \displaystyle\sum_{m=0}^{j} \hat{\mathcal{F}}_m(X)\cos(\dfrac{m\pi}{q}j)$

with

$$\hat{\mathcal{F}}_m(X) = \dfrac{2}{q}\sum_{r=1}^{q-1}\mathcal{F}(X,\dfrac{r}{q})\cos(\dfrac{m\pi r}{q}) + \dfrac{1}{q}\mathcal{F}_0 + \dfrac{(-1)^m}{q}\mathcal{F}_q \ .$$

We note that : $\Omega_n(X,Y_j) = \Omega_n(X,\dfrac{j}{q}) \equiv \Omega_{nj}(X)$ and we seek the solution $\Omega_{nj}(X)$ in the form :

(15,28) $\quad \Omega_{nj}(X) = \displaystyle\sum_{m=0}^{q}\Omega_{nm}(X)\cos(\dfrac{m\pi}{q}j)$ ;

then, after expressing $\dfrac{\partial^2 \Omega_n}{\partial Y^2}\Big|_j$ by a sum of the type (15,28) :

$$\frac{\partial^2 \Omega_n}{\partial Y^2}\Big|_j \cong q^2 \sum_{m=0}^{q} \sin^2(\frac{m\pi}{q})\Omega_{nm}(X)\,\cos(\frac{m\pi}{q}j),$$

we obtain for $\Omega_{nm}(X)$ the following differential equation :

(15,29)
$$a_0^2 \frac{d^4\Omega_{nm}}{dX^4} + \left[\mathcal{D}_0 - n^2\pi^2 - a_0^2 q^2 \sin^2(\frac{m\pi}{q})\right]\frac{d^2\Omega_{nm}}{dX^2}$$

$$- q^2 \sin^2(\frac{m\pi}{q})\mathcal{D}_0\Omega_{nm} = -\frac{2}{\pi n}\,\hat{\mathcal{F}}_m.$$

We will consider the following two cases :

(I) $\mathcal{D}_0 > 0$ : this is the case of a *stable* stratification of the atmosphere; we then obtain for $\Omega_{nm}(X)$ the solution :

(15,30)
$$\Omega_{nm}(X) = \frac{1}{\pi n a_0^2 Q_{nm}\lambda_{nm}}\left\{\int_{-\infty}^{X}\exp[-\lambda_{nm}(X-X')]\,\hat{\mathcal{F}}_m(X')dX'\right.$$

$$+ \int_{X}^{+\infty}\exp[\lambda_{nm}(X-X')]\,\hat{\mathcal{F}}_m(X')dX'$$

$$\left.+ 2\frac{\lambda_{nm}}{\mu_{nm}}\int_{-\infty}^{X}\sin\left[\mu_{nm}(XX')\right]\hat{\mathcal{F}}_m(X')dX'\right\},$$

with

(15,31)
$$\begin{cases}
\lambda_{nm} = \sqrt{\dfrac{1}{2}\left[Q_{nm} - \left\{\dfrac{\mathcal{D}_0 - n^2\pi^2 - a_0^2 q^2\sin^2(\frac{m\pi}{q})}{a_0^2}\right\}\right]} \quad ; \\[4ex]
\mu_{nm} = \sqrt{\dfrac{1}{2}\left[Q_{nm} + \left\{\dfrac{\mathcal{D}_0 - n^2\pi^2 - a_0^2 q^2\sin^2(\frac{m\pi}{q})}{a_0^2}\right\}\right]} \quad ; \\[4ex]
Q_{nm} = \sqrt{\left[\dfrac{\mathcal{D}_0 - n^2\pi^2 - a_0^2 q^2\sin^2(\frac{m\pi}{q})}{a_0^2}\right]^2 + \dfrac{4}{a_0^2}q^2\sin^2(\frac{m\pi}{q})\mathcal{D}_0} \quad .
\end{cases}$$

(II) $\mathcal{D}_0 < 0$ : this is the case of an atmosphere having an *unstable* stratification; we obtain for $\Omega_{nm}(X)$ the solution :

(15,32)

$$
\Omega_{nm}(X) = \frac{1}{\pi n a_0^2 \mathcal{R}_{nm} \rho_{nm}} \left\{ \int_{-\infty}^{X} \exp[-\rho_{nm}(X-X')] \, \hat{\mathcal{F}}_m(X')dX' \right.
$$

$$
+ \int_{X}^{+\infty} \exp[\rho_{nm}(X-X')] \, \hat{\mathcal{F}}_m(X')dX'
$$

$$
- \frac{\rho_{nm}}{\nu_{nm}} \int_{X}^{+\infty} \exp[\nu_{nm}(X-X')] \, \hat{\mathcal{F}}_m(X')dX'
$$

$$
\left. - \frac{\rho_{nm}}{\nu_{nm}} \int_{X}^{+\infty} \exp[-\nu_{nm}(X-X')] \, \hat{\mathcal{F}}_m(X')dX' \right\},
$$

with

(15,33)

$$
\left\{
\begin{array}{l}
\rho_{nm} = \sqrt{\dfrac{1}{2}\left[\dfrac{-\mathcal{D}_0 + n^2\pi^2 + a_0^2 q^2 \sin^2(\frac{m\pi}{q})}{a_0^2} + \mathcal{R}_{nm}\right]} \quad ; \\[2em]
\nu_{nm} = \sqrt{\dfrac{1}{2}\left[\dfrac{-\mathcal{D}_0 + n^2\pi^2 + a_0^2 q^2 \sin^2(\frac{m\pi}{q})}{a_0^2} - \mathcal{R}_{nm}\right]} \quad ; \\[2em]
\mathcal{R}_{nm} = \sqrt{\left[\dfrac{-\mathcal{D}_0 + n^2\pi^2 + a_0^2 q^2 \sin^2(\frac{m\pi}{q})}{a_0^2}\right]^2 + \dfrac{4}{a_0^2} q^2 \sin^2(\frac{m\pi}{q})\mathcal{D}_0} \quad .
\end{array}
\right.
$$

The two solutions obtained : (15,30) for $\mathcal{D}_0 > 0$ and (15,32) for $\mathcal{D}_0 < 0$ evidently satisfy the condition at infinity upstream :

$$
\Omega_{nm} = \frac{d\Omega_{nm}}{dX} = \frac{d^2\Omega_{nm}}{dX^2} = \frac{d^3\Omega_{nm}}{dX^3} \to 0,
$$

for $X \to -\infty$ , which follows from (15,26) .

*ANALYSIS OF THE SOLUTIONS* (15,30) *AND* (15,32)

In the first case, for $\mathcal{D}_0 > 0$ we have systems of stationary waves of two kinds which are well defined:

(1) waves whose amplitudes decrease exponentially as they move away from the obstacle;

(2) sinusoidal periodic waves which are caused by the obstacle and uniquely downstream of an above the obstacle; these periodic waves, which do not decay give rise to zones, downstream of the obstacle, with vertical speeds alternatively positive or negative.

These zones being, in general, perpendicular to the basic non disturbed flow, come from infinity upstream (in the planes $\zeta$=constant parallel to the ground). In the second case, when $\mathcal{D}_0 < 0$, the solution (15,32) is uniquely comprised of waves of the first type.

The solution for the perturbations, w', of the vertical speed will be obtained by the formula

$$(15,34) \qquad w'_j = \frac{\rho_0 U^0_\infty}{\rho_\infty} \left\{ (1-\zeta)a_0 \frac{\partial \delta}{\partial X} + \sum_{n=1}^{\infty} \sum_{m=0}^{q} \Omega_{nm}(X)\sin(n\pi\zeta)\cos(m\pi\frac{j}{q}) \right\} .$$

In a pratical way the obtained solutions may be realised in different ways, but it is often preferable to apply these solutions for a simple obstacle, defined in such a way that we can easily calculate(analytically) the integrals which comprise the solutions obtained in (15,30) and (15,32). As we are treating the linear problem here, the sum of these simple solutions will give the more general solution for a more complex relief.

We will therefore write that

$$\delta(X, Y) = \sum_{i} \delta_i(X, Y)$$

and we will take $\delta_i(X, Y)$ in the form of a paraboloid of revolution having its summit at the point $\zeta = \zeta_0$ and a base radius(on the plane $\zeta=0$) equal to $1/q$, that is to say that

$$(15,35) \qquad \begin{cases} \delta_i(X, Y) = \zeta_0 \left[ 1-q^2(X^2+Y^2) \right], \text{ for } X^2+Y^2 < \dfrac{1}{q^2} ; \\[2ex] \delta_i(X, Y) = 0, \text{ for all the other values } X, Y . \end{cases}$$

We give here the formulae for the caculations in the case of a typical model with only a single intermediary level in altitude ($\zeta=1/2$).

In this case we can write, in place of solutions (15,30) (for $\mathcal{D}_0 > 0$), the formula:

$$\chi_m = \frac{-2}{a_0^2 Q_m} \left\{ \int_{-\infty}^{X} \exp[-\lambda_m (X-X')] \frac{d\hat{\Gamma}_m}{dX'} dX' - \int_{X}^{+\infty} \exp[+\lambda_m (X-X')] \frac{d\hat{\Gamma}_m}{dX'} dX' \right.$$

(15,36)

$$\left. -2 \int_{-\infty}^{X} \cos[\mu_m (X-X')] \frac{d\hat{\Gamma}_m}{dX'} dX' \right\},$$

with

$$\hat{\Omega} = \Omega|_{\zeta=\frac{1}{2}} \quad , \quad \hat{\Omega}_j = \sum_{m=0}^{q} \chi_m (X) \cos(\frac{m\pi}{q} j);$$

$$\Gamma(X,Y) \equiv a_0 \frac{\partial \delta_1}{\partial X} \quad , \quad \Gamma_j = \sum_{m=0}^{q} \hat{\Gamma}_m (X) \cos(\frac{m\pi}{q} j),$$

where $\Gamma_j \equiv \Gamma(X,\frac{j}{q})$ .

If for $j=1,2,\dots q$ , $\Gamma_j \equiv 0$, then there will remain only the term $\Gamma_0/q$ for $\hat{\Gamma}_m (X)$. It is also evident that $\frac{d\hat{\Gamma}_m}{dX} \to 0$, for $X \to -\infty$.
Using formulae (15,35) for $\delta_1 (X,Y)$ we obtain:

$$\Gamma(X,Y) = -2a_0 q^2 \zeta_0 X,$$

and

$$\hat{\Gamma}_m \equiv \frac{\Gamma_0}{q} = -2a_0 q \zeta_0 X.$$

Therfore $\frac{d\hat{\Gamma}_m}{dX} = -2a_0 q \zeta_0 \equiv K_0$ and we denote by M the quantity

$$-\frac{2}{a_0^2 Q_m} K_0 \equiv \frac{4q}{a_0} \frac{\zeta_0}{Q_m} = \text{constant}.$$

We have that

$$\lambda_m = \frac{1}{a_0} \sqrt{\frac{1}{2} \left[ a_0^2 Q_m - (\mathcal{D}_0 - 8 - a_0^2 q^2 \sin^2 \frac{m\pi}{q}) \right]};$$

$$\mu_m = \frac{1}{a_0} \sqrt{\frac{1}{2} \left[ a_0^2 Q_m + \mathcal{D}_0 - 8 - a_0^2 q^2 \sin^2 \frac{m\pi}{q} \right]},$$

$$Q_m = \lambda_m^2 + \mu_m^2 .$$

We now perform the integration in relation to X (see the sketch below), and we obtain respectively for $\chi_m(X)$ the following formulae :

(15,37a)
$$\chi_m = -\frac{8q}{a_0}\frac{\zeta_0}{Q_m\lambda_m} e^{\lambda_m X} sh(\frac{\lambda_m}{q}) \quad , \quad X \leq -\frac{1}{q} ;$$

(15,37b)
$$\chi_m = \frac{8q}{a_0}\frac{\zeta_0}{Q_m\lambda_m}\left\{ e^{-\lambda_m/q} sh(\lambda_m X) - \frac{\lambda_m}{\mu_m}sin\left[\mu_m(X+\frac{1}{q})\right]\right\}, \quad -\frac{1}{q} \leq X \leq +\frac{1}{q} ;$$

(15,37c)
$$\chi_m = \frac{8q}{a_0}\frac{\zeta_0}{Q_m\lambda_m}\left\{ e^{-\lambda_m X} sh\left(\frac{\lambda_m}{q}\right) - 2\frac{\lambda_m}{\mu_m}sin\left(\frac{\mu_m}{q}\right)cos(\mu_m X)\right\}, \quad X \geq +\frac{1}{q}.$$

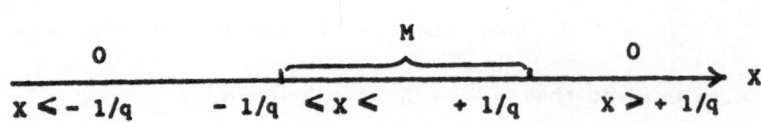

These formulae (15,37) are uniquely valid when $m \neq 0$ and $m \neq q$. Let us look at the case for $m=0$ and $m=q$; we obtain for $\chi_m(X)$ the following formula (for $\mathcal{D}_0 > 0$):

(15,38)
$$\chi_m = -\frac{4}{a_0^2 \hat{\mu}_m}\int_{-\infty}^{X} sin[\hat{\mu}(X-X')]\,\hat{\Gamma}_m(X')dX',$$

with
$$\hat{\mu} \equiv \frac{1}{a_0}\sqrt{\mathcal{D}_0 - 8} \quad ,$$

which gives us the formulae (for $m=0$ and $m=q$):

(15,39)
$$\begin{cases} \chi_m \equiv 0 \;, \quad \text{when } X < -\frac{1}{q} ; \\[2ex] \chi_m = -\frac{16}{a_0}q\frac{\zeta_0}{\hat{\mu}_m^3}sin\left(\frac{\hat{\mu}_m}{q}\right)cos\left(\frac{\hat{\mu}_m}{q}X\right) \;, \quad \text{when } X \geq +\frac{1}{q} ; \\[2ex] \chi_m = \frac{8q}{a_0}\frac{\zeta_0}{\hat{\mu}_m^2}X - \frac{8q}{a_0}\frac{\zeta_0}{\hat{\mu}_m^3}sin\left[\hat{\mu}_m(X+\frac{1}{q})\right] \;, \quad \text{when } -\frac{1}{q} \leq X \leq +\frac{1}{q} . \end{cases}$$

*RESULTS OF THE CALCULATIONS*

In Fig.2 we have represented the range of vertical speeds $\rho_\infty w'/\rho_0 U_\infty^0 \zeta_0$ on the plane $\zeta=1/2$ The values of the parameters are the following:

$$H_0 = 8.10^3 \text{ m}, \quad L_0 = 96.10^3 \text{ m}, \quad a_0 = \frac{1}{12};$$

$$\zeta_0 = 0,1 \quad \text{and} \quad q = 24,$$

which gives us a parabolic of 800m in height and $8.10^3$m in diameter. Finally $U_\infty^0 = 17$m/s, $\alpha_\infty \equiv 0$, the basic flow being in the direction x>0 has a stable stratification, with $g\beta \cong 10^{-3}$ $1/s^2$, which gives $\mathcal{D}_0 = 25$.

By superimposing typical solutions corresponding to an obstacle in the form of a paraboloid of revolution we can obtain ranges of vertical speeds in the planes z=constant above varying sites; in particular, calculations have been made above the region of Cantal in the Massif Central of France and for the region near the Basin of Arcachon, which, according to Trochu (1967), predict the distributions of rain in these regions very well. The calculated rain distributions given by Trochu are very fascinating indeed (see, Fig.3 and Fig.4).

We note that, although linear equations for steady three-dimensional stratified flow over barriers have been solved before, Zeytounian's work (1969) is more systematic and, in any case, his calculations for the barrier form of a paraboloid of revolution give fascinating results.

For the calculation of rain zones above the Arcachon Basin (Fig.3) we have represented the coast in the form of an algebraic sum of typical simple reliefs as paraboloids of revolution; we note that the coast, which separates the sea from the land, plays the role of an obstacle due to the difference in roughness of the land in relation to that of the sea. With the coast represented by such a sum, we have calculated the influence of each paraboloid at each point of the region investigated and have obtained the sum of these influence.in the case of the Cantal region (Fig.4) the relief of the region has been represented by means of a simple relief of the type (15,35) with values of $\zeta_0$ and q variables.

For the calculations on the Cantal (Fig.4) a detailed analysis has enabled us to conclude that they reflect quite well the local structure of rain in this region.

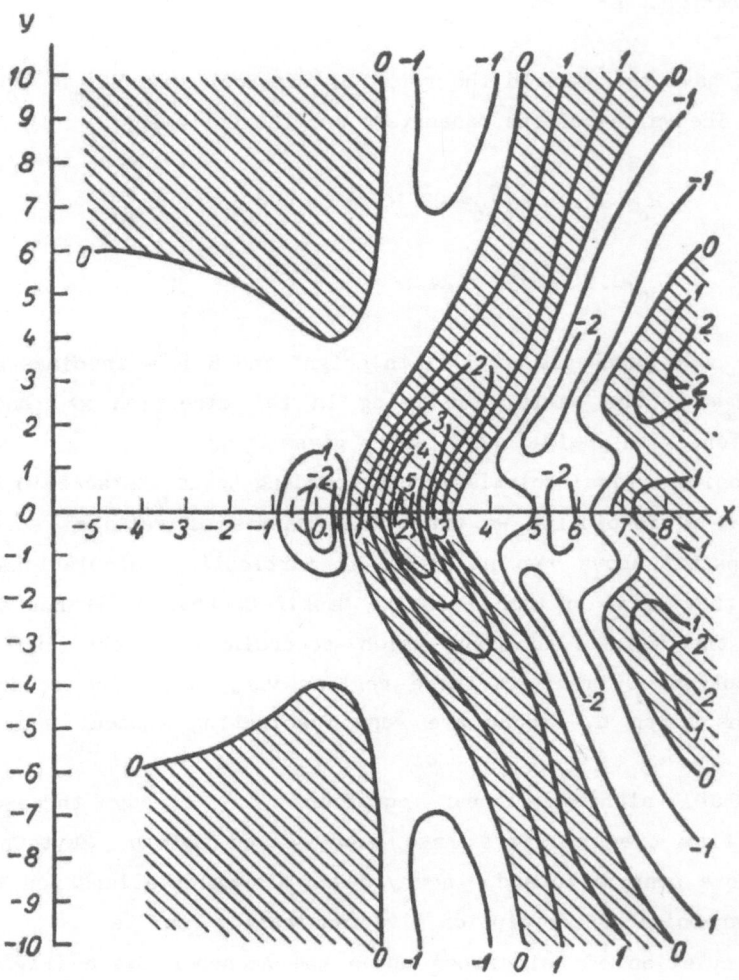

Fig. 2 : Standard flow above a typical obstacle
of form (15, 35). Range of vertical speeds in
the plane z=4 km.

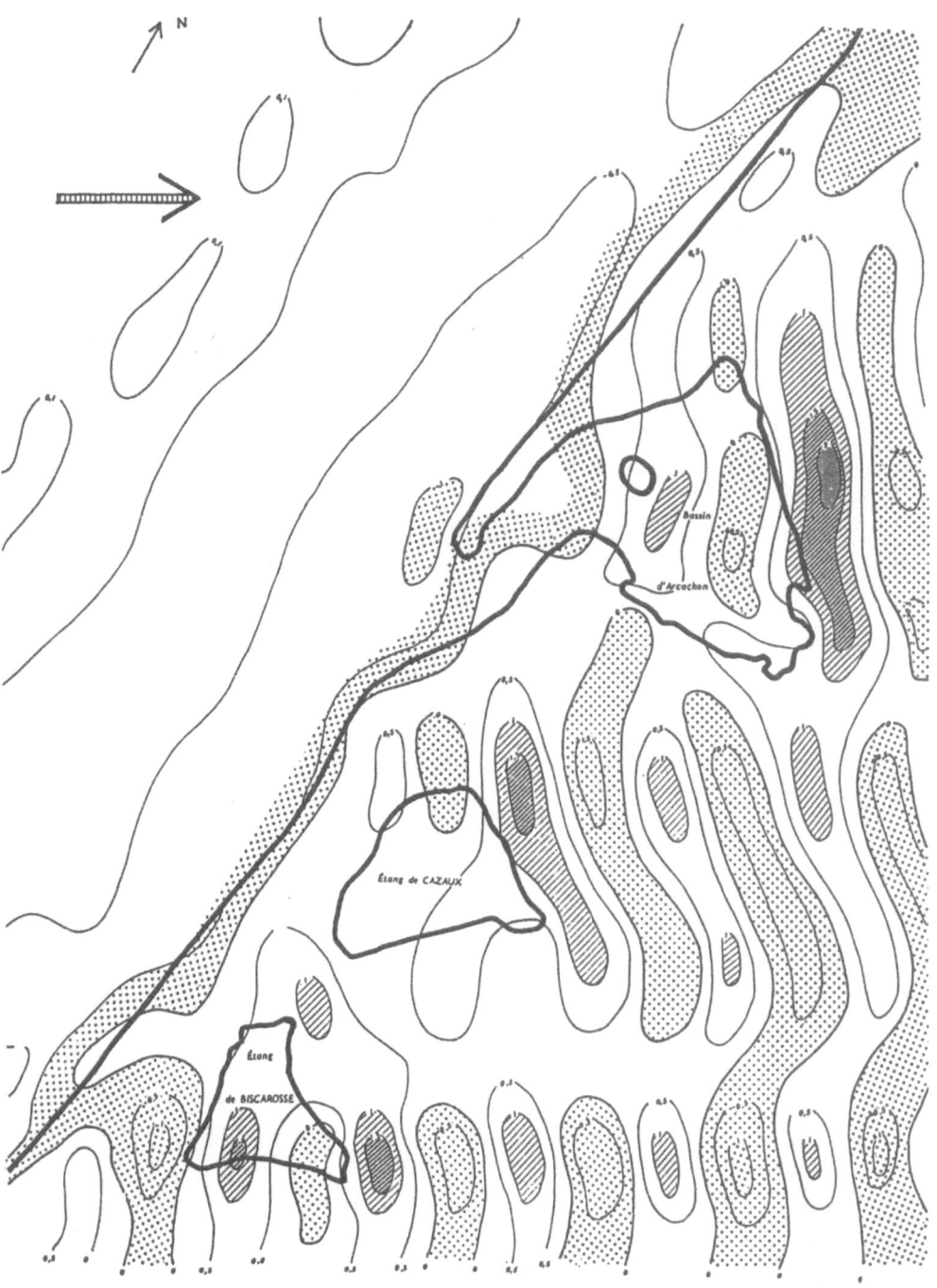

Fig. 3 : Wet zones above Arcachon Basin for a
stable stratification; $\alpha_\infty = 240°$ and $D_0 = 50$.
Negative: brokenline.
Positive: hatched.

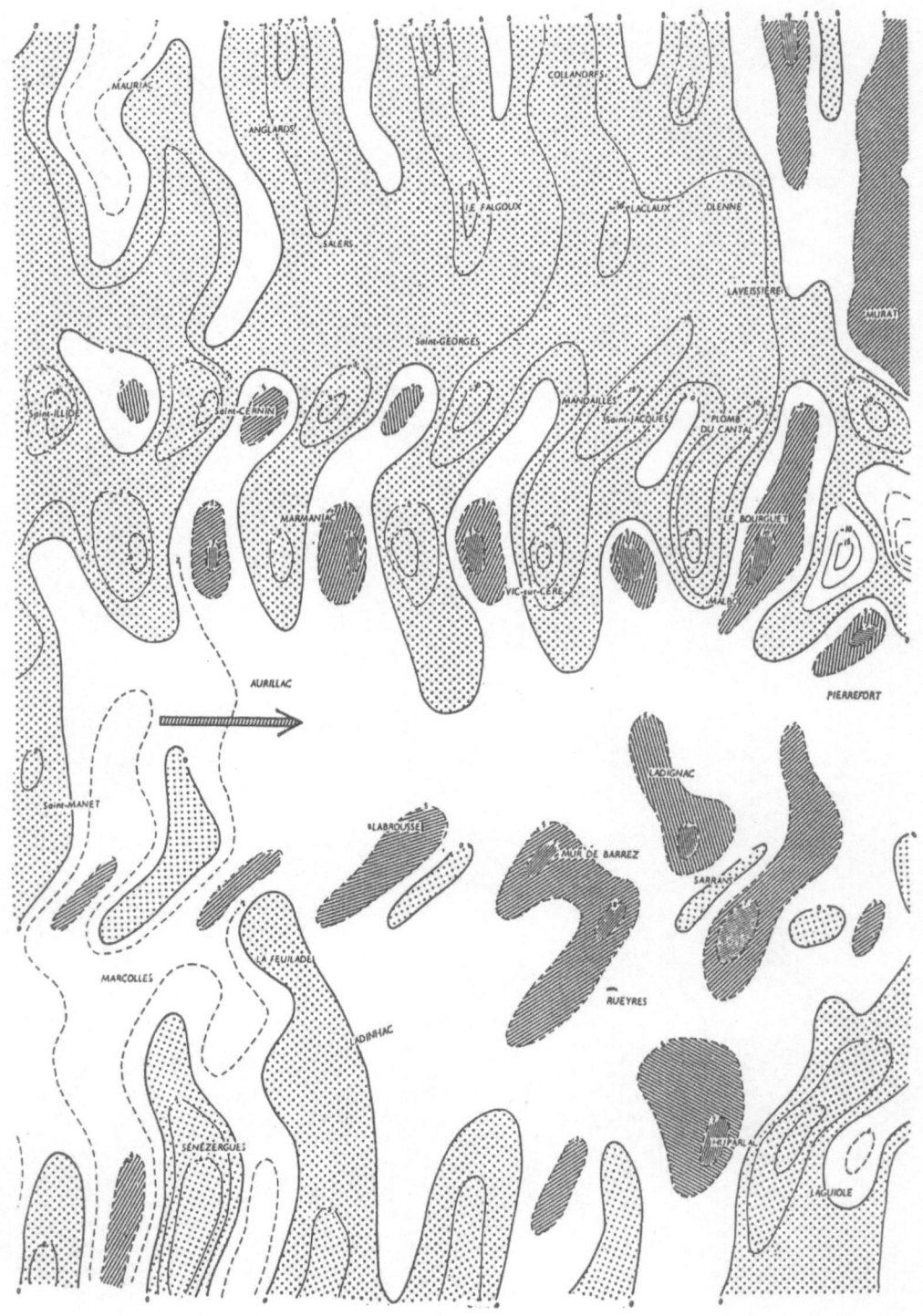

Fig. 4 : Wet zones over a part of the Massif
Central (Cantal); $\alpha_\infty = 270°$ and $D_0 = 50$.

*BACKGROUND READING*

For a extensive treatement of the concept of waves in fluids the reader is referred to:

LIGHTHILL, J. (1978) _ *Waves in Fluids.*
                    Cambridge University Press.

Concerning the waves in the atmosphere, see:

BEER, T. (1974) *Atmospheric waves.*
                Adam Hilger, London.

and

DIKIJ, L. A. (1969) _ *The theory of oscillations of the Earth's atmosphere*
                    (in Russian). Guidrometeo-Izdat, Moscow.

For the nonlinear aspects of waves, see:

WHITHAM, G. B. (1974) _ *Linear and nonlinear waves.*
                      J. Wiley et sons.

*REFERENCES TO WORKS CITED IN THE TEXT*

BOIS, P. A. (1976) - Journal de Mécanique, 15, 781.
CRAPPER, G. D. (1959) - J. Fluid Mech., vol. 6, part 1, 51.
DUBREIL-JACOTIN, M. L. (1935) - Atti Accad. Lincei Rend. Cl. Sci. Fis. Mat. Nat (6)
                              21, 344-346.
KIBEL, I. A. (1955) - Doklady Akad. Nauk, 100, n°2, 247-250.
LONG, R. R. (1955) - Tellus, 7, n°3, 342-357.
SAWYER, J. S. (1962) - Quart. J. Roy. Met. Soc. vol. 88 , n°378, 412.
TROCHU, M. (1967) - Calcul d'un champ de vitesse verticale en mésométérologie :
                  Application. "Etude de Stage", Ecole de la métérologie,
                  Paris.
VELTICHEV, I. (1965) - Travaux du Centre Mondial Métérologique de Moscou,
                     n°8, p. 45 (in Russian).

WURTELE,M.G. (1957) - Aero-revue. 32, n°12; see also: Beitr.Phys.Atmos.
                 29,242-252.

YIH,C.S (1980) - *Stratified Flows*. Academic Presss, London.

ZEYTOUNIAN,R.Kh. (1969) - *Study of wave Phenomena in the steady Flow of an inviscid stratified fluid.*
            Royal Aircraft Establishment. Library translation
            n°1404, December 1969.

ZEYTOUNIAN,R.Kh. (1974) - *Notes sur les Ecoulements Rotationnels de Fluides Parfaits.*
            Lecture Notes in Physics, vol.27. Springer-Verlag,
            Heidelberg.

# CHAPTER IV

# FILTERING
# OF
# INTERNAL WAVES

We have already noted that the basic model equations (see, the Chapter II) are formulated a view to *filtering acoustic waves* out of the solutions of "exact" equations for atmospheric motions, because such waves are of *no importance concerning weather prediction*.

## 16 . HYDROSTATIC FILTERING

In the hydrostatic approximation, whan $\varepsilon_0 \rightarrow 0$, we see that the general wave equation (10,4) is highly degenerate, and the consequences of this will be:

$$\lim_{\varepsilon_0 \rightarrow 0} (\sigma_a^2) \equiv \infty \quad \text{and} \quad \lim_{\varepsilon_0 \rightarrow 0} (\sigma_g^2) \equiv \sigma_{ghs}^2$$

with

$$(16,1) \qquad \sigma_{ghs}^2 = \frac{1}{Ro^2} + \frac{k^2}{\lambda} \frac{\gamma-1}{\gamma} \left(\frac{Bo}{M_0}\right)^2 ,$$

i.e., all internal acoustic waves are filtred, and the internal gravity waves are severely distorted. The frequencies of the gravity waves are in this case slightly over-estimated, but the smaller is k (i.e. the longer are the waves), the closer will be the estimate; namely, those internal gravity waves, which correspond to $k^2 \ll 1$ or $\lambda \gg 1$, remain.

The frequencies of two-dimensional acoustic waves (see, (10,17)) remain unchanged and this justifies the use of the primitive equations (7,9) for the description of synoptic processes.

The reduction of the order of (10,4) from four to two with respect to time t also indicates the non-uniqueness of the double limiting process $\varepsilon_0 \rightarrow 0$, $t \rightarrow 0$. Near t=0 we must formulate the problem of adjustment to hydrostatic balance (see, the section 19 in the Chapter V). This problem of adjustment to hydrostatic balance makes it possible to solve the fundamental problem of the relationship between the initial conditions for primitive equations and the true initial conditions for full adiabatic, nonviscous atmospheric equations. We note that the hydrostatic filtering in these full equations altered the hyperbolic character of these equations; see for instance the work of Oliger and Sundstrom (1978).

Let us note that even in the steady-state case (Strouhal number ≡ 0)the limiting process $\varepsilon_0 \to 0$ leads to the singular perturbation problem in connection with the fact that the horizontal short internal gravity waves (for exemple, lee waves downstream of the barrier) are filtered out. In order to investigate how these horizontal short internal gravity waves are filtered out as $\varepsilon_0 \to 0$ we must obviously use the method of multiple scales. But for now, unfortunately, the question of the conversion of the process with short gravity waves to the process with long gravity waves, when $\varepsilon_0 \to 0$, remains unanswered in spite of the fact that this quetion is very important for a complete understanding of the theory of long waves in the atmospheric meso-scale (regional) motions.

Finally, we note that the adjustment to the hydrostatic balance is brought about by the generation and scattering of internal acoustic waves. The duration of this process is approximately the same as the time required for a front of internal acoustic waves to traverse (with the sound speed $a_\infty^0 \approx 20$km/min) the main thickness of the atmosphere (troposphere)-a process requiring only a few minutes in all. After this, the atmosphere continues to adapt to the state of geostrophic equilibrium (9,7). This statement will be made more precise in the sections 19 and 21.


## 17 . BOUSSINESQ FILTERING

It is clear that in the Boussinesq approximation, when

$$Bo = \hat{B}M_0 \quad , \quad \hat{B} = O(1) \quad \text{but} \quad M_0 \to 0,$$

the wave equation (10,4) is again highly degenerate and as a consequence of this:

$$\sigma_a^2 \to \infty,$$

$$(17,1) \qquad \sigma_g^2 \to \sigma_{gB}^2 = \frac{\lambda/R\hat{o}^2 + (\gamma-1)\hat{B}^2/\gamma^2}{\varepsilon_0^2 k^2 + \lambda} .$$

Thus, all the internal acoustic waves are filtered again, and the internal gravity waves are severely distorted. The order of the wave equation (10,4) in terms of t is again reduced from four to two, indicating the nonuniqueness of the double limiting process:

$$(17,2) \qquad Bo = \hat{B}M_0 \ , \quad \hat{B} = O(1) \ , \quad M_0 \to 0 \quad \text{and} \quad t \to 0 \ .$$

The problem to adjustement to a Boussinesq state must be formulated near t=0; this problem will make it possible to formulate the initial conditions for the Boussinesq equations correctly (see, the section 20 in the Chapter V).

On the other hand, we see that the Boussinesq approximation is correct only if the characteristic height $H_0$ of the atmospheric motions being considered is such that:

$$(17,3) \qquad \frac{a_\infty^2(0)}{\gamma g} \gg H_0 \simeq \hat{B} \frac{U_0}{g\gamma} a_\infty(0) \equiv H_B \; ,$$

where $a_\infty(0) = \left[ \gamma R T_\infty(0) \right]^{1/2}$ is the speed of sound in the standard atmosphere at the ground level.

When $\varepsilon_0 \equiv 1$ i.e., $L_0 \equiv H_0 \equiv H_B$ then the Boussinesq model equations (8,8) are capable of describing atmospheric processes only locally; therefore the behavior of the solutions of these equations at infinity must be determined.

But, in the general case, a question still remains unresolved: what outside equations supplement to the Boussinesq equations and are joined to them via the radiation conditions in the steady-state case? This was done in the work of Guiraud and Zeytounian (1979) in the steady-state plane adiabatic nonviscous case (see, the Chapter VI of the present Course).

Hence it is necessary to elucidate the behavior of the solution of the Boussinesq model equations (8,8) (obtained initially by Zeytounian (1974)) at infinity; for the three-dimensional model of steady lee waves, this problem was resolved by Guiraud (1979) who showed that the classical radiation condition is satisfied at infinity. In section 20 we find a comprehensive analysis of the problem of adjustement to a Bousinesq state, for the time being we not only that the Boussinesq equations are not of the hyperbolic type.

Finally, when the Boussinesq approximation is used "correctly", we lose the ability to take account of the effect of a change in stratification with altitude on the tropospheric process being considering; in particular, this follows from the fact that for this approximation the full Eulerian energy equation (the third eq. of (8,4)) is replaced by the conservation equation:

$$(17,4) \qquad S \frac{D}{Dt} \left\{ \theta_B + \hat{B} \left[ \frac{\gamma-1}{\gamma} + \left[ \frac{dT_\infty}{dz} \right]_{z_\infty \equiv 0} \right] z \right\} = 0 \; .$$

## 18 . GEOSTROPHIC FILTERING

Superposed to the hydrostatic limiting process, $\varepsilon_0 \rightarrow 0$ (see, the section 16), we can consider the following quasi-geostrophic limiting process (see, (9,3) and (9,4)):

(18,1)      $Ki \rightarrow 0$ and $M_0 \rightarrow 0$, with $\lambda_0 = \frac{1}{\gamma S} \left( \frac{Ki}{M_0} \right)^2 = O(1)$,

the so-called geostrophic filtering. In this case we have

(18,2)                $\sigma^2_{ghs} \rightarrow \infty$ ,

and amongst the fast waves, two of them, the gravity and acoustic ones, are filtered out after $\varepsilon_0 \rightarrow 0$ and (18,1). Thus for forecasting adiabatic nonviscous atmospheric processes at synoptic scales according to the geostrophic approximation (18,1) it is sufficient to prescribe initial value for the $\mathcal{H}_{qg}$ , only, which satisfies to equation (9,14).
The initial values of the other meteorological fields are lost during the limiting process (18,1) and a new initial layer result from the double limit process:
           $Ki \rightarrow 0$  and  $t \rightarrow 0$,
and this initial layer describes a process of adjustement to geostrophy (see, the section 21), i.e., the adaptation of the hydrostatic fields to the geostrophic equilibrium state

$$\vec{v}_{qg} = \lambda_0 Bo(\vec{k} \wedge \vec{D} \mathcal{H}_{qg}) \ .$$

We must add that in the filtered out quasi-geostrophic model we have only one *elliptic* equation (see, (9,14)) for the evolution of $\mathcal{H}_{qg}$, instead of the system of full hyperbolic Euler equations for adiabatic, nonviscous atmospheric motions. In connection with the loss of meteorological field adaptation arises and for each of the model (outer, in time) equations the (inner) initial problem of adjustement must be examined near t=0 in order to obtain the "correct" initial values, which are needed for its unique solution. Finally, we not that adjustement of the meteorological fields as a result of the generation, dispersion, and damping of the *fast* internal waves.

*BACKGROUND READING*

Concerning the problem of filtering of internal waves, see:

MONIN,  S. A  (1972) _ *Weather Forecasting as a Problem in Physics.* The
                    MIT Press, Cambridge, Masschusetts (see, the sections 4-9),
ZEYTOUNIAN,  R. Kh. (1982)  _ *Modelling of Atmospheric Motions as a Problem of*
                    *Filtering out short internal waves.*  Izvestiya,
                    Atmospheric and Oceanics Physics.
                    Vol. 18, n°6, 583-601 (Russian Edition),
and
ZEYTOUNIAN, R. Kh. (1990) _ *Asymptotic Modeling of Atmospheric flows.*
                    Springer-Verlag, Heidelberg (see, the Chapter III).

*REFERENCES TO WORKS CITED IN THE TEXT*

GUIRAUD, J.P. (1979) _ Comptes Rendus Acad. Sci. Paris (A), 288, 435.
GUIRAUD, J.P. and ZEYTOUNIAN, R. Kh. (1979) _ Geophys. Astrophys. Fluid
                    Dynamics , 12, 61.
OLIGER, J. and SUNDSTRÖM, S. (1978) _ SIAM. J. Appl. Math., 35, 419.
ZEYTOUNIAN, R. Kh. (1974) _ Arch. Mech. Stosowanej, 26, 499.

# CHAPTER V

## UNSTEADY
## ADJUSTMENT PROBLEMS

The basic approximations are formulated with a view to filtering acoustic waves out of the solutions of equations for atmospheric motions, because such waves are of no importance concerning weather prediction. On the other hand, when considering the approximate, simplified set of equations (primitive equations, Boussinesq equations or quasi-geostrophic model equation), one is allowed to specify a set of initial conditions *less* in number than for the "exact" equations. This is due to the fact that the limiting process which leads to the approximate model, filters out some time derivatives. Due to this one encounters the problem of deciding *what initial conditions one may prescribe and in what way these are related to the initial conditions associated with the exact, full, equations?* The latter are not in general consistent with the estimates of basic orders of magnitude implied by the asymptotic model. A physical process of time evolution is necessary to bring the initial set to a consistent level as far as the orders of magnitude is concerned. Such a process is called one of ADJUSTMENT of the initial set of data to the asymptotic structure of the model under consideration. The process of adjustment, which occurs in many fields of Fluid Mechanics besides Meteorology, is short on the time scale of the asymptotic model considered, and *at the end of it*, in an asymptotic sence, *we obtain values for the set of initial conditions suitable to the model.*

If we consider our basic model equations (see, Chapter II) in such a case it is necessary to elucidate the problems of the *adjustement to hydrostatic balance, to a Boussinesq state* and *to geostrophy.*

A number of adjustement problems occur in Fluid Mechanics being related to loss of initial conditions as a consequence of loss of time derivatives during some limiting processes leading to a simplified set of equations. We refer to just one of them which is most celebrated and has intrigued many Fluid Dynamicists. It is the loss of initial conditions for the full distribution function when one goes from the Boltzmann equation to the Navier-Stokes equations by letting the ratio of the mean free path to macroscopic length scale (Knudsen number) go to zero; see, for example, Cercignani (1975; Chap. V, §.5).

To the best of our knowledge these problems are solved by rescaling the time and possibly some dependent variables leading to a so-called *initial layer* problem.

Depending on the kind of problems, we may have mainly two kinds of behavior when the rescaled time goes to infinity. Either one may have a tendency towards a limiting steady state or an undamped set of oscillations (think, for example, of the inertial waves in the inviscid problem of spin-up for a rotating fluid; see for that Greenspan (1968; §24)).

For the terminology of the initial layer as adapted to this kind of singular perturbation problems we refer to Nayfeh (1973, p.23).

## 19 . ADJUSTEMENT TO HYDROSTATIC BALANCE

It is obvious that the classical, Kibel, primitive equations (7,1) are obtained through the following limiting process:

(19,1)
$$\begin{cases} \varepsilon_0 \to 0, \text{ keeping } t, z \text{ and the horizontal} \\ \text{position fixed during the process,} \end{cases}$$

applied to the full equations for the tangent non-hydrostatic, adiabatic and nonviscous atmospheric motions:

(19,2)
$$\left[ \begin{aligned} & S\frac{D\rho}{Dt} + \rho\left[\vec{D}.\vec{v} + \frac{\partial w}{\partial z}\right] = 0 \; ; \\[2mm] & \rho\left\{ S\frac{D\vec{v}}{Dt} + \left[\frac{1}{Ro} + \beta y\right](\vec{k} \wedge \vec{v}) + \frac{\varepsilon_0}{\tan\varphi_0}\frac{1}{Ro}w\vec{i} \right\} + \frac{1}{\gamma M_0^2}\vec{D}\,p = 0 \; ; \\[2mm] & \rho\left\{ S\,\varepsilon_0^2\frac{Dw}{Dt} - \frac{\varepsilon_0}{\tan\varphi_0}\frac{1}{Ro}\vec{v}.\vec{i} \right\} + \frac{1}{\gamma M_0^2}\left[\frac{\partial p}{\partial z} + Bo\rho\right] = 0 \; ; \\[2mm] & p = \rho T; \\[2mm] & S\frac{D}{Dt}\left[\frac{p}{\rho\gamma}\right] = 0, \end{aligned} \right.$$

where $\qquad S\frac{D}{Dt} \equiv S\frac{\partial}{\partial t} + \vec{v}.\vec{D} + w\frac{\partial}{\partial z}$ and $\vec{v} = u\vec{i} + v\vec{j}$.

We observe that the initial data, for equations (19,2), need not fit the hydrostatic balance and, in particular, the vertical velocity need not be $O(\varepsilon)$ with respect to the horizontal one, so that in order to consider the most general case, we assume that at the initial time $\varepsilon_0 w$ is of order $O(1)$. Accordingly we get as initial conditions, for the full equations (19,2),

$$(19,3) \qquad t=0 \; : \; \vec{v}=\vec{V}^0, \; \varepsilon w=W^0 \; , \; p=P^0 \text{ and } \rho=R^0,$$

where $\vec{V}^0, W^0, P^0$ and $R^0$ are given functions of z and of the horizontal position. On the other hand when considering the primitive equations (7,1) we must give only the initial values of $\vec{v}$ and $\rho$, since the initial value of $\rho$ yield, from the hydrostatic balance, the initial value of p and the relation $T=p/\rho$ yield the initial value of T. The initial values of $\vec{v}$ and $\rho$ have nothing to do with the corresponding initial conditions for the full equations (19,2) (see, (19,3)). Consequently, we get as initial conditions for the primitive equations (7,1):

$$(19,4) \qquad t=0 \; : \; \vec{v}=\vec{v}^0 \text{ and } \rho=\rho^0,$$

where $\vec{v}^0$ is different from $\vec{V}^0$ and $\rho^0$ is different from $R^0$.

Two of the initial conditions (19,3) have been lost during the process and two questions arises:

      1) How have these initial conditions been lost?

      2) How are $\vec{v}^0$ and $\rho^0$ related to $\vec{V}^0, W^0, P^0$ and $R^0$?

Regarding the first question, the answer is simple. According to the primitive equations model (7,1), p is related to $\rho$ by the equation of hydrostatic balance, while w, as noticed for the first time by Richardson (1922; see the Chapter V of his book), is computed by the process of solution of the primitive equations (7,1). All this hold true at the initial time as well.

As a matter of fact if we consider the primitive equations (7,9), use of pressure coordinates, we observe that the two main variables of the primitive model (7,9) are $\vec{v}$ and T. The situation is slightly reminiscent of the one which occurs in classical boundary layer theory if one considers that $\omega$ is deduced from $\vec{v}$ by a kind of divergence-free condition, namely the second equation of (7,9). We should keep in mind that like $\omega$, $\mathcal{H}$ is known, apart from an integration constant, when T is known. Hence, w, the vertical component of velocity is not a primary variable in the sense that it may be computed afterwards, through use of (7,7), when $\mathcal{H}$ and $\omega$ have been computed themselves from the knowledge of $\vec{v}$, T and p.

Now we intend here to adress ourselves to the second question.

In the present case it is fairly obvious that the proper rescaling, the time and dependent variables, is[†]:

(19,5)          $t = \varepsilon \hat{t}$ , $z \equiv \hat{z}$ , $\vec{v} \equiv \hat{\vec{v}}$ , $\varepsilon w = \hat{w}$ , $p \equiv \hat{p}$ and $\rho \equiv \hat{\rho}$ .

For the adjustement to hydrostatic balance problem we use the following limiting process:

(19,6)          $\varepsilon_0 \rightarrow 0$ , with $\hat{t}$ and $\hat{z}$ fixed,

and the horizontal position is also fixed during the limiting process (19,6). We note that through the process $\hat{w} = O(1)$.

Let us set $\hat{\vec{v}}_0$, $\hat{w}_0$, $\hat{p}_0$ and $\hat{\rho}_0$ for the limiting values (functions of $\hat{t}$, $\hat{z}$ and of the horizontal position), it is straightforward to derive the set of limiting *initial layer hydrostatic, equations*, namely:

(19,7)
$$
\begin{cases}
\gamma M_0^2 \hat{\rho}_0 \left[ \dfrac{\partial \hat{w}_0}{\partial \hat{t}} + \hat{w}_0 \dfrac{\partial \hat{w}_0}{\partial \hat{z}} \right] + \dfrac{\partial \hat{p}_0}{\partial \hat{z}} + \hat{\rho}_0 = 0 \; ; \\[4mm]
\dfrac{\partial \hat{\rho}_0}{\partial \hat{t}} + \dfrac{\partial}{\partial \hat{z}} \left[ \hat{\rho}_0 \hat{w}_0 \right] = 0 \; ; \\[4mm]
\left[ \dfrac{\partial}{\partial \hat{t}} + \hat{w}_0 \dfrac{\partial}{\partial \hat{z}} \right] \left( \dfrac{\hat{p}_0}{\hat{\rho}_0^{\gamma}} \right) = 0 \; ,
\end{cases}
$$

(19,8)          $\dfrac{\partial \hat{\vec{v}}_0}{\partial \hat{t}} + \hat{w}_0 \dfrac{\partial \hat{\vec{v}}_0}{\partial \hat{z}} = 0$ ,

if $S \equiv 1$ ($t_0 \equiv L_0 / U_0$) and $Bo \equiv 1$ ($H_0 \equiv RT_\infty(0)/g$) .

The system (19,7), for $\hat{\rho}_0$, $\hat{p}_0$ and $\hat{w}_0$, is identical to the equations for *one-dimensional unsteady* vertical motion in the atmosphere. Onece $\hat{w}_0$ has been obtained through the solution of (19,7) with the initial conditions (see, (19,3)):

(19,9)   $\hat{t} \equiv 0$ :   $\hat{w} = W^0$, $\hat{p}_0 = P^0$, $\hat{\rho}_0 = R^0$ ,

---

[†] We note that the horizontal positions not play of rôle in the problem of adjustment to hydrostatic balance.

and proper boundary conditions on the ground and at infinity, we may use the transport equation (19,8) in order to compute $\hat{\vec{v}}_0$ using the initial condition:

(19,10)  $\hat{t}=0 : \hat{\vec{v}}_0 = \vec{v}^0$,

The equation (19,8) merely says that $\hat{\vec{v}}_0$ is convected without change, vertically, with the velocity $\hat{w}_0$.

*NUMERICAL SOLUTION OF EQS.(19,7)*

The set of equations (19,7) have been solved numerically by Outrebon (1981) using the slip condition:

(19,11)  $\hat{w}_0 = 0$  at  $\hat{z}_0 = 0$

and enforcing

(19,12)  $\dfrac{\partial \hat{w}_0}{\partial \hat{z}} = 0, \quad \dfrac{\partial \hat{p}_0}{\partial \hat{z}} + \hat{\rho}_0 = 0$

at the maximum altitude z=20km, used in the computational grid[†].

The initial conditions were chosen to be the standard equilibrium atmosphere, concerning the thermodynamic state, while for $\hat{w}_0$ a triangular profile was used, with $\hat{w}_0$ being zero at the ground and above z=10km with a maximum at z=3km.

Two numerical computations were run; the first one, with results shown on the Fig.5, corresponds to a maximum in the initial value of $\hat{w}_0$ equal to 1m/s, while for the second (with results not reproduced here) the maximum was significantly higher, namely 5m/s. The unit of time used for the numerical computations was 26km/$a_\infty(0)$, where $a_\infty(0)$ is the speed of the sound at the ground level for the standard atmosphere.

The process of the adjustement, to hydrostatic balance is composed of three main phases. During the first phase of adjustment typical profiles of which being shown at $\hat{t}=0.058$, are characterised by several inversion of the direction of the vertical velocity and rather strong perturbations in

---

[†] The numerical code used by Outrebon is inspired of one of Lerat and Peyret (1973) which is the generalisation of the work of Fromm (1969) concerning a method for reducing dispersion in convective difference schemes and the pratical investigation of constructive difference approximation of reduced dispersion.

temperatures and pressures. For the second run, with results not reproduced here, a shock wave is formed. There is a second phase during which the vertical velocity decays to zero, while the temperature and pressure profiles approach to the equilibrium ones. Typical such profiles are shown for $\hat{t}=0.180$. The third phase is the ultimate phase of adjustement during which convergence to *steady* state is achieved; we have shown typical profiles at $\hat{t}=0.998$. We observe that, thanks to the unit of time used for numerical computations, the dimensionless time $\hat{t}$ used for the presenting the numerical results may be identifyed with the one used in (19,7).

We note now that the basic enquiry about adjustement to hydrostatic balance is one of assessment whether or not the model of hydrostatic balance is asymptotically stable. As an indication for this let us consider what may be called the *vertical shift* $\hat{\Delta}_0(\hat{t})$ which is a solution to:

$$(19,13) \qquad \frac{d\hat{\Delta}_0}{d\hat{t}} = \hat{w}_0(\hat{t}, \hat{z}^0 + \hat{\Delta}_0(\hat{t})) \;,\; \hat{\Delta}_0(0) \equiv 0.$$

The meaning of $\hat{\Delta}_0(\hat{t})$ is that it allows to integrate at once the last equation in (19,7) and the equation (19,8), namely:

$$(19,14) \qquad \frac{\hat{P}_0}{\hat{\rho}_0^\gamma} \equiv \hat{\Sigma}_0(\hat{t}, \hat{z}) = \Sigma^0(\hat{z} - \hat{\Delta}_0(\hat{t})),$$

where $\qquad \Sigma^0 \equiv \dfrac{P^0}{(R^0)^\gamma} \;,\;$ and

$$(19,15) \qquad \hat{\vec{v}}_0(\hat{t}, \hat{z}) = \vec{v}^0(\hat{z} - \hat{\Delta}_0(\hat{t})).$$

The Fig.6, shows according to Outrebon (1981) the aspect of $\hat{\Delta}_0(t)$ as a function of $\hat{t}$ for various values of $\hat{z}^0$ and for the same initial data as in Fig.5. We definitely see that $\hat{\Delta}_0$ tends to a limit when $\hat{t} \rightarrow \infty$ .

More precisely than that we may even say that most of the adjustement isaccomplished when $\hat{t}>0.5$ and that the final approach to equilibrium is rather slow. If we let $\hat{t} \rightarrow \infty$ we find, through matching and adjustment to hydrostatic balance model, that *the initial conditions concerning horizontal velocity and entropy for the primitive model equations are simply the ones pertinent to the full equations* (19,2), *shifted vertically by amount equal to* $\hat{\Delta}_0^\infty$.

We give on Fig.7, according to Outrebon(1981) a graph of $\hat{\Delta}_0^\infty$ as a function of $\hat{z}_0$.

This shows that the vertical shift is a quite significant phenomenon. On the other hand it is difficult to maintain that the phenomenon is of pratical importance for weather prediction according to the primitive equations (7,1). One reason, amongst a number of good ones, being that it is doubtful that the initial conditions might be sufficiently precise for rendering worthwhile a correction based on the vertical shift!

The best argument for considering adjustment to hydrostatic balance rests on the investigation of stability of the hydrostatic model. What the computations by Outrebon tell us is that there is built into the equations a quite active mechanism which drives back the atmosphere to a state of hydrostatic balance and that the transient time tied to this mechanism is less than the time necessary for a sound wave, starting from the ground, to cross back and forth the whole of the troposphere.

Finally, we are able to write the following relations for the initial values of primitive equations (7,1), namely:

(19,16)
$$
\begin{cases}
p^0 = \left\{ 1 - \frac{\gamma-1}{\gamma} \int_0^z \exp\left[ -\frac{1}{\gamma}\Sigma^0 (\mathfrak{z} - \hat{\Delta}_0^\infty) \right] d\mathfrak{z} \right\}^{\frac{\gamma}{\gamma-1}} , \\[4mm]
\rho^0 = (p^0)^{1/\gamma} \exp\left[ -\frac{1}{\gamma}\Sigma^0 (z - \hat{\Delta}_0^\infty) \right] , \\[4mm]
\vec{v}^0 = \lim_{\hat{t} \to +\infty} \hat{\vec{v}}_0 = \vec{V}^0 (z - \hat{\Delta}_0^\infty) .
\end{cases}
$$

The final result may be stated very simply by saying that:

*the horizontal components of velocity and the specific entropy in the two set of initial conditions are merely shifted vertically by the amount of vertical displacement during the whole process of the vertical, one-dimensional, unsteady motion.*

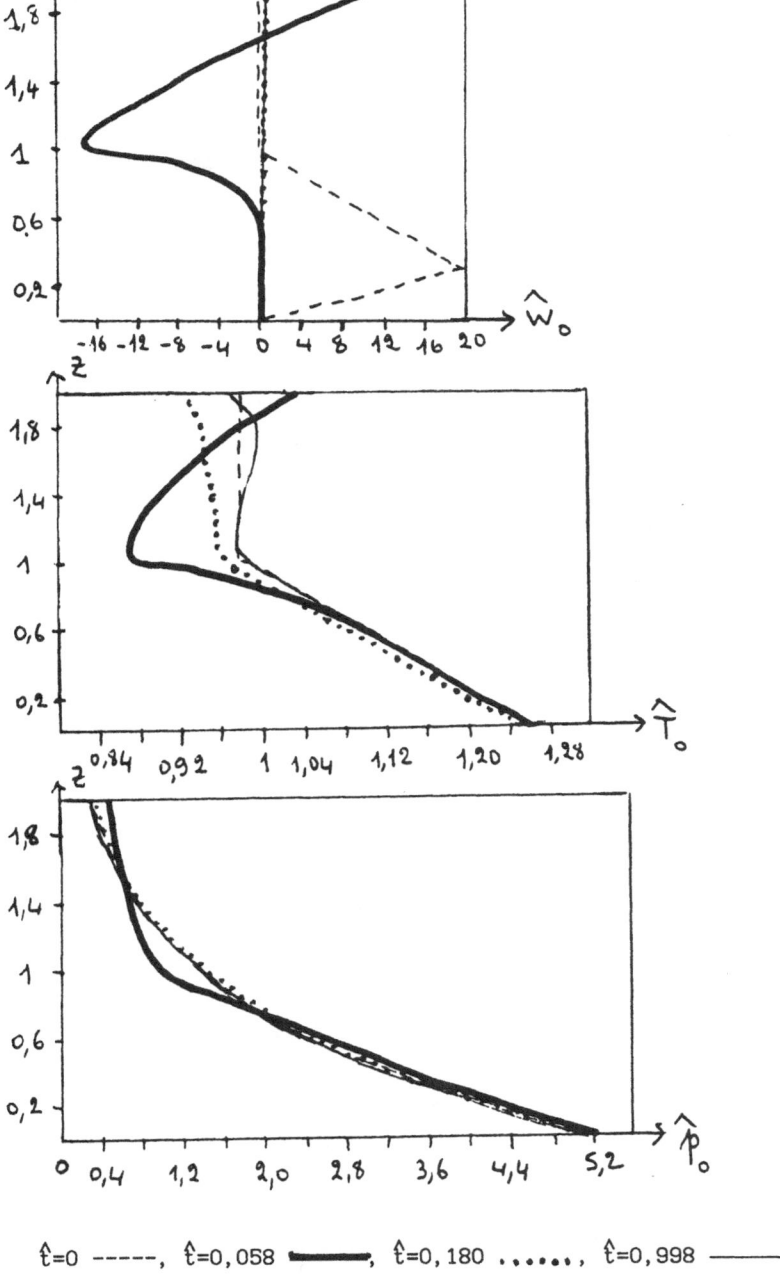

Fig. 5 : Adjustment to hydrostatic balance.

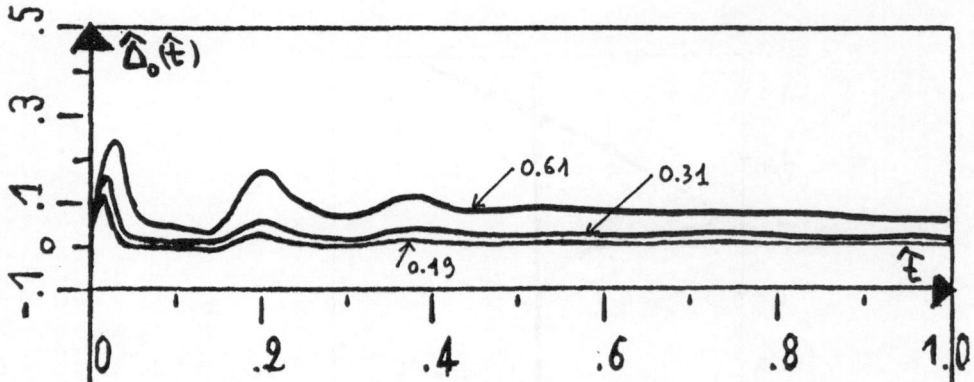

Fig. 6: *Vertical shift as a function of time for three initial altitudes.*

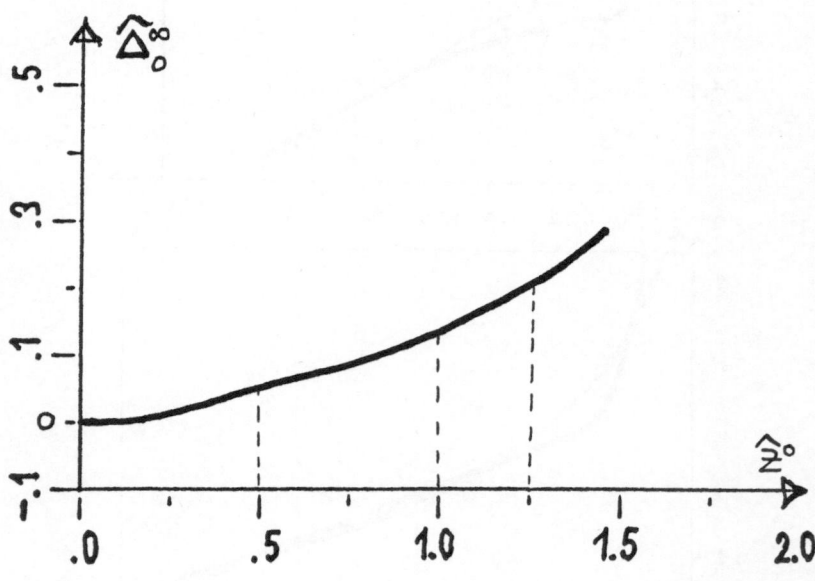

Fig. 7: *Final altitude at the end of adjustment to hydrostatic balance process as a function of initial altitudes.*

*THE ULTIMATE PHASE OF ADJUSTMENT TO HYDROSTATIC BALANCE.*

Let us examine now analytically the ultime decay towards the hydrostatic balance. We assume decay and study the way in the remaining perturbation with respect to the assumed limiting state behaves when $\hat{t}$ goes to infinity according to a *linearized* theory obtained by perturbing (19,7) around the limiting state and retaining only the linear terms.

Setting[†]

$$(19,17) \qquad \hat{p}_0 = \hat{p}_0^\infty(\hat{z})\{1+\pi\}, \quad \hat{\rho}_0 = \hat{\rho}_0^\infty(\hat{z})\{1+\omega\}, \quad \hat{T}_0 = \hat{T}_0^\infty(\hat{z})\{1+\theta\}, \quad \hat{w}_0 \equiv w,$$

we obtain, for $\pi$, $\omega$, $\theta$ and $w$ the following linear version of (19,7):

$$(19,18) \qquad \begin{cases} \gamma M_0^2 \dfrac{\partial w}{\partial \hat{t}} + \hat{T}_0^\infty(\hat{z}) \dfrac{\partial \pi}{\partial \hat{z}} - \theta = 0 \; ; \\[3mm] \dfrac{\partial \omega}{\partial \hat{t}} + \dfrac{\partial w}{\partial \hat{z}} - \dfrac{1}{\hat{T}_0^\infty(\hat{z})}\left[1 + \dfrac{d\hat{T}_0^\infty}{d\hat{z}}\right] w = 0 \; ; \\[3mm] \dfrac{\partial}{\partial \hat{t}}\left[\dfrac{\gamma-1}{\gamma}\pi - \theta\right] - \dfrac{1}{\hat{T}_0^\infty(\hat{z})}\left[\dfrac{\gamma-1}{\gamma} + \dfrac{d\hat{T}_0^\infty}{d\hat{z}}\right] w = 0 \; ; \\[3mm] \pi = \omega + \theta. \end{cases}$$

From (19,18) we get the following equation for $w$:[††]

$$(19,19) \qquad M_0^2 \dfrac{\partial^2 w}{\partial \hat{t}^2} + \dfrac{d\hat{T}_0^\infty}{d\hat{z}} \dfrac{\partial^2 w}{\partial \hat{z}^2} + \dfrac{\partial w}{\partial \hat{z}} = 0 \; .$$

Let us change from $w$ to $W$ defined by

$$(19,20) \qquad W = w \exp\left\{-\dfrac{1}{2} \int_0^{\hat{z}} \dfrac{d\zeta}{\hat{T}_0^\infty(\zeta)}\right\} ,$$

we get the equation:

$$(19,21) \qquad \dfrac{\partial^2 W}{\partial \hat{z}^2} - \dfrac{M_0^2}{\hat{T}_0^\infty(\hat{z})} \dfrac{\partial^2 W}{\partial \hat{t}^2} - \dfrac{1}{4\hat{T}_0^\infty(\hat{z})}\left[1 + 2\dfrac{d\hat{T}_0^\infty}{d\hat{z}}\right] W = 0 ,$$

---

[†] We consider the limit of $\hat{T}_0, \hat{p}_0$ and $\hat{\rho}_0$ when $\hat{t} \to \infty$, with $\hat{z}$ fixed. Through matching with the hydrostatic, primitive equations, model, these $\hat{T}_0^\infty$, $\hat{p}_0^\infty$ and $\hat{\rho}_0^\infty$ are equal to the initial values of $T$, $p$ and $\rho$ according to the primitive equations (7,1).

[††] According to the work of Guiraud and Zeytounian (1982).

which is nothing else but the well-known equation of telegraphy provided that $\hat{\uparrow}_0^\infty$ does not depend on $\hat{z}$ .

Hence, $\hat{\uparrow}_0^\infty \equiv 1$ (with the non-dimensional variables), and we set

$$\hat{z} = \frac{\hat{z}}{2} \ , \quad \tau = \frac{1}{2M_0}\hat{t} \ ,$$

and $\quad W(2M_0\tau, 2\hat{z}) \equiv \chi(\tau, \hat{z})$.

The solution of the equation

(19,22)
$$\frac{\partial^2 \chi}{\partial \hat{z}^2} - \frac{\partial^2 \chi}{\partial \tau^2} - \chi = 0,$$

corresponding to conditions:

(19,23)
$$\begin{cases} \chi(\tau,0) = \chi(\tau,\infty) = 0 \ ; \\ \chi(0,\hat{z}) = \chi^0(\hat{z}) \ ; \\ \left.\frac{\partial \chi}{\partial \hat{z}}\right|_{\tau=0} = \chi^1(\hat{z}) \ , \end{cases}$$

is readily obtained through application of the sine-Fourier transform (refer to Titchmarsch (1948)).

The solution of (19,22), with (19,23) is the following:

(19,24)
$$\chi(\tau,\hat{z}) = \int_0^\infty \Gamma(\tau,\hat{z},\zeta)\chi^1(\zeta)d\zeta + \frac{\partial}{\partial \tau}\int_0^\infty \Gamma(\tau,\hat{z},\zeta)\chi^0(\zeta)d\zeta \ ,$$

with
$$\Gamma(\tau,\hat{z},\zeta) = \frac{2}{\pi}\int_0^\infty \frac{\sin[\sqrt{\nu^2+1}\ \tau]}{\sqrt{\nu^2+1}} \sin(\nu\hat{z})\sin(\nu\zeta)d\nu \ .$$

What we need now is an estimation of $\Gamma(\tau,\hat{z},\zeta)$ when $\tau \to \infty$. This is most easily got by using complex integration in order to transform the integral along the positive real axis $0<\nu<+\infty$ by setting $\nu=1+i\mu$ and integrating along $\mu$ real from $\mu=0$ to $\mu=\infty$. One gets, in this way, for $\tau \to +\infty$, the following asymptotic result

(19,25)
$$\Gamma(\tau,\hat{z},\zeta) \cong \sqrt{\frac{2}{\pi}}\frac{\hat{z}\zeta}{\tau^{3/2}}\cos(\tau+\frac{\pi}{4})$$

when $\quad \tau \to +\infty$ .

This calculation shows that, when $\hat{t} \to +\infty$, w behaves in the following way (the reader must keep in mind that $\hat{T}_0^\infty \equiv 1$):

(19,26)
$$w \cong M_0^{3/2} \cos\left[\frac{\hat{t}}{2M_0} + \frac{\pi}{4}\right] \frac{\hat{z}}{\hat{t}^{3/2}} \, e^{\hat{z}/2} .$$

We see from this last formula that w decays to zero like $(\hat{t})^{-3/2}$ *provided* $\hat{z}$ *remain finite* but there is a *non uniformity at infinity*. We note that the non uniformity reveled by (19,26) is not an artefact of the asymptotic estimation; it is actually related to the double limiting process:

$$\lim_{\hat{t} \to \infty} \text{applied first and then } \lim_{\hat{z} \to +\infty} .$$

The combination of Outrebon's numerical calculations and of the above asymptotic analysis gives a strong evidence that the hydrostatic approximation is asymptotically stable but that this approximation should *fail at high altitude* . As a matter of fact , the hydrostatic model should be considered as an *inner* one to be matched with an outer one. For a discussion in a related context the reader is referred to the recent book of Zeytounian (1990; see the Chapter VII).

## 20 . ADJUSTMENT TO A BOUSSINESQ STATE

As explained in the section 8 (see the Chapter II) to a Boussinesq, adiabatic non viscous, state (8,8) results from the limiting process

(20,1)
$$Bo = \hat{B} M_0 , \quad M_0 \to 0 \quad \text{and} \quad \hat{B}, t, x, y, z \text{ fixed.}$$

The so-called Boussinesq equations (8,8) imply a divergence free flow

(20,2)
$$\vec{D}.\vec{v}_B + \frac{\partial w_B}{\partial z} = 0$$

and the constraint

(20,3)
$$\omega_B + \theta_B = 0 ,$$

where the Boussinesq state is characterised by

(20,4)
$$\vec{v} = \vec{v}_B + .. \quad , \quad w = w_B + .. \quad , \quad \pi = M_0^2 \pi_B + .. \quad , \quad \omega = M_0 \omega_B + .. \quad , \quad \theta = M_0 \theta_B + .. \quad .$$

This constraint (20,3) is tied to the loss of the $\partial\omega/\partial t$ derivative (in (8,4)) by the Boussinesq limiting process (20,1) and we should expect an adjustment process to the Boussinesq state. We discuss this adjustment, unsteady process, below by relying on a work to Zeytounian (1984).

Let us set

$$(20,5) \qquad \hat{t}=t/M_0 \; ,$$

and apply, in place of (20,1), the new, initial, limiting process:

$$(20,6) \qquad Bo = \hat{B}M_0 \; , \; M_0 \rightarrow 0 \; , \; \text{and} \; \hat{B}, \hat{t}, x, y, z \; \text{fixed.}$$

We rewrite the exact full equations (8,4) setting $\hat{f}$ for any quantity $f$ considered as a function of $\hat{t}$ instead of $t$, and we get the following system of equations[†]:

$$(20,7a) \qquad (1+\hat{\omega}) \; S \; \frac{\partial \hat{\vec{u}}}{\partial \hat{t}} + \frac{T_\infty(\hat{B}M_0 z)}{\gamma M_0} \; \vec{\nabla} \; \hat{\pi} - \frac{\hat{B}}{\gamma} \; (1+\hat{\omega}) \; \hat{\theta}\vec{k} + \vec{O}(M_0) = 0 \; ;$$

$$(20,7b) \qquad S \; \frac{\partial \hat{\omega}}{\partial \hat{t}} + M_0(1+\hat{\omega})\vec{\nabla}.\hat{\vec{u}} + O(M_0) = 0 \; ;$$

$$(20,7c) \qquad (1+\hat{\omega}) \; S \; \frac{\partial \hat{\theta}}{\partial \hat{t}} - \frac{\gamma-1}{\gamma} \; S \; \frac{\partial \hat{\pi}}{\partial \hat{t}} + O(M_0) = 0 \; ,$$

$$(20,7d) \qquad \hat{\pi} = \hat{\omega} + \hat{\theta} + \hat{\omega} \; \hat{\theta} \; .$$

For the equations (20,7) we assume as initial conditions:

$$(20,8) \qquad \hat{t}=0 : \hat{\vec{u}}=\vec{u}^0, \; \hat{\pi}=\pi^0, \; \hat{\omega}=\omega^0 \; , \; \hat{\theta}=\theta^0,$$

where $\vec{u}^0$, $\pi^0$, $\omega^0$ and $\theta^0$ are given functions of $x$, $y$ and $z$, and we set up an asymptotic (initial) expansion

$$(20,9) \qquad \begin{pmatrix} \hat{\vec{u}} \\ \hat{\pi} \\ \hat{\omega} \\ \hat{\theta} \end{pmatrix} = \begin{pmatrix} \hat{\vec{u}}_0 \\ \hat{\pi}_0 \\ \hat{\omega}_0 \\ \hat{\theta}_0 \end{pmatrix} + M_0 \begin{pmatrix} \hat{\vec{u}}_1 \\ \hat{\pi}_1 \\ \hat{\omega}_1 \\ \hat{\theta}_1 \end{pmatrix} + \ldots \; .$$

---

[†] We observe that all the dissipative terms are comrised within the terms $O(M)$ so that the result below is the same whether we start from the equations (8,4) or the full viscous, non adiabatic equations ones $(Re \neq \infty)$.

We get at first:

$$(20,10) \quad \begin{cases} \vec{\nabla}\hat{\pi}_0 = 0 \ , \quad \dfrac{\partial\hat{\omega}_0}{\partial\hat{t}} = 0 \ , \quad \dfrac{\partial\hat{\theta}_0}{\partial\hat{t}} - \dfrac{\gamma-1}{\gamma}\dfrac{\partial\hat{\pi}_0}{\partial\hat{t}} = 0 \ , \\[4mm] \hat{\pi}_0 = \hat{\omega}_0 + \hat{\theta}_0 + \hat{\omega}_0\hat{\theta}_0 . \end{cases}$$

Using the initial conditions (20,8) we can prove that:

$$(20,11) \qquad \hat{\pi}_0 \equiv \pi_0^0, \ \hat{\omega}_0 \equiv \omega^0, \ \hat{\theta}_0 \equiv \theta^0,$$

if $\omega^0 \neq 1/(\gamma-1)$ and where $\pi_0^0 \equiv$ constant .

But, according to method of matched asymptotic expansions (see, for example, Van Dyke (1964)) it is necessary that the expansion (20,9), when $\hat{t}\to\infty$, to match with the outer expansion (20,4), when $t\to 0$. This matching is possible only if: $\hat{\pi}_0 = \hat{\omega}_0 = \hat{\theta}_0 \equiv 0$ and consequently it is necessary that we suppose, in (20,8), that

$$(20,12) \qquad \pi^0 = M_0\pi_1^0, \ \omega^0 = M_0\omega_1^0 \text{ and } \theta^0 = M_0\theta_1^0$$

for $\hat{t}=0$, where $\pi_1^0$ , $\omega_1^0$ and $\theta_1^0$ are given functions of x,y, and z.

In this case, going to the next order, we find the following system, for $\hat{\vec{u}}_0, \hat{\pi}_1, \hat{\omega}_1$ and $\hat{\theta}_1$:

$$(20,13) \quad \begin{cases} S\,\dfrac{\partial\hat{\vec{u}}_0}{\partial\hat{t}} + \vec{\nabla}(\hat{\omega}_1 + \dfrac{\pi_1^0}{\gamma} - \omega_1^0) = 0 \ ; \\[4mm] S\,\dfrac{\partial\hat{\omega}_1}{\partial\hat{t}} + \vec{\nabla}.\hat{\vec{u}}_0 = 0 \ ; \\[4mm] \hat{\pi}_1 = \gamma(\hat{\omega}_1 + \dfrac{\pi_1^0}{\gamma} - \omega_1^0) \ ; \\[4mm] \hat{\theta}_1 = \hat{\pi}_1 - \hat{\omega}_1 , \end{cases}$$

and, from (20,12) and (20,8) we have the following initial conditions, for (20,13),

$$(20,14) \qquad \hat{t}=0 \ : \ \hat{\vec{u}}_0 = \vec{u}^0 \ , \ \hat{\omega}_1 = \omega_1^0 \ , \ \hat{\theta}_1 = \theta_1^0 \ .$$

If the ground is flat, then we have the following boundary condition:

$$(20,15) \qquad z=0 : \quad \hat{\vec{u}}_0 \cdot \vec{k} = 0 .$$

*THE UNSTEADY ADJUSTMENT PROBLEM*

Let us set:

$$(20,16a) \qquad \vec{u}^0 = \vec{\nabla}\varphi^0 + \vec{\nabla} \wedge \vec{\psi}^0 ,$$

and

$$(20,16b) \qquad \hat{\vec{u}}_0 = \vec{\nabla} \hat{\varphi}_0 + \vec{\nabla} \wedge \hat{\vec{\psi}}_0 .$$

In this case for $\hat{\varphi}_0$ we obtain the following relation [†]:

$$(20,17) \qquad S \frac{\partial \hat{\varphi}_0}{\partial \hat{t}} + \hat{\omega}_1 + \frac{\pi_1^0}{\gamma} - \omega_1^0 = 0,$$

and for $\hat{\vec{\psi}}_0$ we have: $\hat{\vec{\psi}}_0 \equiv \vec{\psi}^0$.
Therefore we get, for $\hat{\varphi}_0$, the following initial value problem:

$$(20,18) \qquad \begin{cases} S^2 \dfrac{\partial^2 \hat{\varphi}_0}{\partial \hat{t}^2} - \Delta \hat{\varphi}_0 = 0, \\[2mm] \hat{t} = 0: \ \hat{\varphi}_0 = \varphi^0, \ S \dfrac{\partial \hat{\varphi}_0}{\partial \hat{t}} = -\dfrac{\pi_1^0}{\gamma} . \end{cases}$$

The solution of this problem is straightforward and , from it we get:

$$(20,19) \qquad \hat{t} \to \infty : \quad |\vec{\nabla}\hat{\varphi}_0| \to 0 , \quad |\partial \hat{\varphi}_0 / \partial \hat{t}| \to 0 .$$

Coming back to (20,16b) and (20,17) we obtain:

$$\hat{\vec{u}}_0 = \vec{\nabla} \wedge \vec{\psi}^0 , \quad \hat{\omega}_1 = -S \frac{\partial \hat{\varphi}_0}{\partial \hat{t}} + \omega_1^0 - \frac{\pi_1^0}{\gamma} ,$$

---

[†] $S \dfrac{\partial \hat{\varphi}_0}{\partial \hat{t}} + \hat{\omega}_1 + \dfrac{\pi_1^0}{\gamma} - \omega_1^0$ is an harmonic function in the whole of space and we
assume that this harmonic function is actually zero, an assumption that
could be justified only through matching with a solution valid on a large
horizontal scale.

and, from this we get

(20,20)          $\hat{t} \longrightarrow +\infty : \quad \hat{\vec{u}}_0 \longrightarrow \vec{\nabla} \wedge \vec{\psi}^0 , \quad \hat{\omega}_1 \longrightarrow \omega_1^0 - \dfrac{\pi_1^0}{\gamma} .$

This solves the main issue by showing that adjustment to a Boussinesq state, at least for the initial conditions (20,14), with (20,12), is of the decaying type. Furthermore, for the Boussinesq equations (8,8) we get, as initial conditions:

(20,21)          $t=0 : \quad \vec{u}_B = \vec{\nabla} \wedge \vec{\psi}^0 , \quad \theta_B = \dfrac{\pi_1^0}{\gamma} - \omega_1^0 .$

The reader should keep in mind that no initial conditions are required on $\pi_B$ and $\omega_B$. More precisely, from the Boussinesq equations (8,8) the initial value of $\pi_B$ may be computed once the initial value of $\vec{u}_B$ is known and $\omega_B = \omega_1^0 - \pi_1^0/\gamma$, for t=0.

What happens if the initial values $\pi^0$, $\omega^0$ and $\theta^0$ not satisfying the conditions (20,12) is not known?

## 21 . ADJUSTMENT TO GEOSTROPHY

It is not difficult to find, by trial, that the adjustment is good by setting:

(21,1)          $\hat{t} = \dfrac{t}{Ki}$

and applying the limiting process

(21,2)          $Ki \longrightarrow 0 , \quad M_0 \longrightarrow 0 , \quad \lambda_0, \hat{t}, x, y, p$ fixed,

where          $\lambda_0 \equiv \dfrac{1}{\gamma S} \left( \dfrac{Ki}{M_0} \right)^2 .$

Let us set $\hat{f}$ for any quantity f, considered as a function of $\hat{t}$ instead of t. We expand the various quantities according to:

$$(21,3) \qquad \begin{Bmatrix} \hat{\vec{v}} \\ \hat{\mathcal{R}} \\ \hat{\omega} \\ \hat{T} \end{Bmatrix} = \begin{Bmatrix} \hat{\vec{v}}_0 \\ \hat{\mathcal{R}}_0 \\ \hat{\omega}_0 \\ \hat{T}_0 \end{Bmatrix} + Ki \cdot \begin{Bmatrix} \hat{\vec{v}}_1 \\ \hat{\mathcal{R}}_1 \\ \hat{\omega}_1 \\ \hat{T}_1 \end{Bmatrix} + \ldots$$

and we substitue into (see, (9,1)):

$$(21,4) \quad \begin{cases} \lambda_0 Bo \; \vec{D} \; \hat{\mathcal{R}} + Ki\left\{\dfrac{\partial \hat{\vec{v}}}{\partial \hat{t}} + (1+\dfrac{\beta}{S}Ki\; y)(\vec{k} \wedge \hat{\vec{v}})\right\} + \dfrac{Ki^2}{S}\left[\hat{\vec{v}}.\vec{D}\hat{\vec{v}} + \hat{\omega}\; \dfrac{\partial \hat{\vec{v}}}{\partial p}\right] = 0; \\[2mm] \vec{D}.\hat{\vec{v}} + \dfrac{\partial \hat{\omega}}{\partial p} = 0 \; ; \\[2mm] \hat{T} + Bo \; p\dfrac{\partial \hat{\mathcal{R}}}{\partial p} = 0 \; ; \\[2mm] S \; \dfrac{\partial \hat{T}}{\partial \hat{t}} + Ki\left\{\hat{\vec{v}}.\vec{D}\hat{T} + \hat{\omega}\left[\dfrac{\partial \hat{T}}{\partial p} - \dfrac{\gamma-1}{\gamma}\; \dfrac{\hat{T}}{p}\right]\right\} = 0 \; . \end{cases}$$

One finds, at first:

$$(21,5) \qquad \vec{D}\hat{\mathcal{R}}_0 = 0 \; , \quad \dfrac{\partial \hat{T}_0}{\partial \hat{t}} = 0, \quad Bo \; p \; \dfrac{\partial \hat{\mathcal{R}}_0}{\partial p} + \hat{T}_0 = 0$$

and we get, at once, that $\hat{\mathcal{R}}_0$, $\hat{T}_0$ and $\hat{\rho}_0 \equiv p/\hat{T}_0$ are functions of p alone, namely (see, the section 9):

$$(21,6) \qquad \hat{\mathcal{R}}_0 = \mathcal{H}_0(p), \quad \hat{T}_0 = T_0(p) = -Bo \; p\dfrac{\partial \mathcal{H}_0}{\partial p}, \quad \hat{\rho}_0 = \dfrac{p}{T_0(p)} \; .$$

In order to find equations for $\hat{\vec{v}}_0$, $\hat{\omega}_0$, $\hat{\mathcal{R}}_1$ and $\hat{T}_1$, we have to go to higher order:

$$(21,7a) \qquad \dfrac{\partial \hat{\vec{v}}_0}{\partial \hat{t}} + \vec{k} \wedge \hat{\vec{v}}_0 + \lambda_0 Bo \; \vec{D} \; \hat{\mathcal{R}}_1 = 0 \; ;$$

$$(21,7b) \qquad \vec{D}.\hat{\vec{v}}_0 + \dfrac{\partial \hat{\omega}_0}{\partial p} = 0 \; ;$$

(21,7c)
$$S \frac{\partial \hat{T}_1}{\partial \hat{t}} - \frac{K_0(p)}{p} \hat{\omega}_0 = 0 \; ;$$

(21,7d)
$$\hat{T}_1 = -Bo \; p \; \frac{\partial \hat{R}_1}{\partial p} \; ,$$

where $K_0(p) = \frac{\gamma-1}{\gamma} T_0(p) - p \frac{dT_0}{dp}$ (see, (9,13)), must be considered as a given function of pressure alone.

The system of equations (21,7) is the system governing the *unsteady process of adjustment to geostrophy*.

We may, without restricting the analysis, set:

(21,8)
$$\vec{\hat{v}}_0 = \vec{D} \hat{\varphi}_0 + \vec{k} \wedge \vec{D} \hat{\psi}_0 ,$$

and we derive, at once, from the equation (21,7a)

(21,9)
$$\vec{D} \left[ \frac{\partial \hat{\varphi}_0}{\partial \hat{t}} + \lambda_0 Bo \hat{R}_1 - \hat{\psi}_0 \right] + \vec{k} \wedge \vec{D} \left[ \hat{\varphi}_0 + \frac{\partial \hat{\psi}_0}{\partial \hat{t}} \right] = 0 \; .$$

This last relation (21,9) implies that the two expressions

(21,10)
$$\mathcal{F}_0 \equiv \frac{\partial \hat{\varphi}_0}{\partial \hat{t}} + \lambda_0 Bo \hat{R}_1 - \hat{\psi}_0 \text{ and } \mathcal{G}_0 \equiv \hat{\varphi}_0 + \frac{\partial \hat{\psi}_0}{\partial \hat{t}} \; ,$$

are, as functions of the two horizontal coordinates, x, y, harmonic ones which related by the Cauchy-Riemann relations. Liouville theorem tells us that they should be polynomials and physical evidence suggests that they are, indeed, zero[†]:

(21,11)
$$\frac{\partial \hat{\varphi}_0}{\partial \hat{t}} + \lambda_0 Bo \hat{R}_1 = \hat{\psi}_0 \; , \quad \frac{\partial \hat{\psi}_0}{\partial \hat{t}} + \hat{\varphi}_0 = 0 \; .$$

From the last two equations of (21,7) we derive

(21,12)
$$\hat{\omega}_0 = -S \; Bo \; \frac{p^2}{K_0(p)} \frac{\partial^2 \hat{R}_1}{\partial \hat{t} \partial p} \; ,$$

---

[†] As a matter of fact, physical evidence should be replaced by matching with a solution valid on the whole sphere. If we start from primitive equations on the sphere (for, example, see the equations (6,2)-(6,6) , where $Re_\perp \equiv \infty$) and carry over the analysis which follows from (21,1) and (21,2) we shall find a result analogous to the one concerning (21,10) above and then it will not be necessary to call for physical evidence in order to justify (21,11).

and going back to the first two equations of (21,7) we find a couple of equations for $\overset{\wedge}{\vec{v}}_0$ and $\overset{\wedge}{\mathcal{R}}_1$, namely:

$$(21,13) \quad \begin{cases} \dfrac{\partial \overset{\wedge}{\vec{v}}_0}{\partial \hat{t}} + \vec{k} \wedge \overset{\wedge}{\vec{v}}_0 + \lambda_0 Bo \vec{D} \overset{\wedge}{\mathcal{R}}_1 = 0 \; ; \\[4mm] \vec{D}.\overset{\wedge}{\vec{v}}_0 - SBo \dfrac{\partial}{\partial p}\left\{ \dfrac{p^2}{K_0(p)} \dfrac{\partial^2 \overset{\wedge}{\mathcal{R}}_1}{\partial \hat{t} \partial p} \right\} = 0 \; . \end{cases}$$

It is obvious that we have to give initial values for $\overset{\wedge}{\vec{v}}_0$ and for $\overset{\wedge}{\mathcal{R}}_1$. Concerning the initial value of $\overset{\wedge}{\vec{v}}_0$ we may use the initial value of the horizontal velocity according to the primitive equations model (see, the section 19). Concerning the initial value of $\overset{\wedge}{\mathcal{R}}_1$ we may tentatively use the initial value of $\mathcal{H}$ for the primitive equations but this works only if this initial value may be set under the form: $\mathcal{H}_0(p)+Ki\mathcal{H}_1^0$, then we get

$$(21,14) \quad \hat{t}=0 : \overset{\wedge}{\vec{v}}_0 = \vec{v}^0 \; , \; \overset{\wedge}{\mathcal{R}}_1 = \mathcal{H}_1^0 \; .$$

Whenever the initial value appropriate to the primitive equations cannot be put under the form $\mathcal{H}_0(p)+Ki\mathcal{H}_1^0$ we must expect that another adjustment process holds. The second of equations (21,13), according to (21,8), can be written:

$$(21,15) \quad \vec{D}^2 \overset{\wedge}{\varphi}_0 - SBo \dfrac{\partial}{\partial p}\left\{ \dfrac{p^2}{K_0(p)} \dfrac{\partial^2 \overset{\wedge}{\mathcal{R}}_1}{\partial \hat{t} \partial p} \right\} = 0 \; ,$$

but from (21,11) we obtain

$$(21,16) \quad \lambda_0 Bo \dfrac{\partial \overset{\wedge}{\mathcal{R}}_1}{\partial \hat{t}} = -\left( \dfrac{\partial^2 \overset{\wedge}{\varphi}_0}{\partial \hat{t}^2} + \overset{\wedge}{\varphi}_0 \right) \; .$$

Finally, from (21,15) and (21,16), we obtain a single equation for $\overset{\wedge}{\varphi}_0$:

$$(21,17) \quad \dfrac{\partial^2}{\partial \hat{t}^2}\left\{ \dfrac{\partial}{\partial p}\left[ \dfrac{p^2}{K_0(p)} \dfrac{\partial \overset{\wedge}{\varphi}_0}{\partial p} \right]\right\} + \dfrac{\lambda_0}{S} \vec{D}^2 \overset{\wedge}{\varphi}_0 + \dfrac{\partial}{\partial p}\left[ \dfrac{p^2}{K_0(p)} \dfrac{\partial \overset{\wedge}{\varphi}_0}{\partial p} \right] = 0 \; ,$$

where

$$\vec{D}^2 = \dfrac{\partial^2}{\partial x^2} + \dfrac{\partial^2}{\partial y^2} \; .$$

When $K_0(p) \equiv 1$, the equation (21,17) is the one obtained by Kibel (1957). He was able to settle the main issue of the adjustment problem which is to know whether or not $\hat{\vec{v}}_0$ and $\hat{\mathcal{R}}_1$ evolve towards the geostrophic balance (9,7) when $\hat{t} \to \infty$. As a matter of fact one has

(21,18)    $\hat{t} \to +\infty : \hat{\vec{v}}_0 \to \vec{v}_{qg} , \quad \hat{\mathcal{R}}_1 \to \mathcal{H}_{qg} ,$

with the geostrophic relation:

(21,19)    $\vec{k} \wedge \vec{v}_{qg} + \lambda_0 B_0 \vec{D} \mathcal{H}_{qg} = 0.$

There is an important observation, which was known to Kibel and which concerns the way in which

$$\lim_{\hat{t} \to +\infty} \hat{\mathcal{R}}_1 \equiv \hat{\mathcal{R}}_1^\infty$$

is related to the initial values (21,14).

One starts from

(21,20)    $\vec{k} . \left\{ \vec{D} \wedge \left[ \dfrac{\partial \hat{\vec{v}}_0}{\partial \hat{t}} + \vec{k} \wedge \hat{\vec{v}}_0 \right] \right\} = 0,$

which follows the first of (21,13) and we transform it, thanks to the second of (21,7), to

$$\frac{\partial}{\partial \hat{t}} \left[ \vec{k} . \left[ \vec{D} \wedge \hat{\vec{v}}_0 \right] \right] - \frac{\partial \hat{\omega}_0}{\partial p} = 0 ,$$

then, using the last two equations in (21,7), we get

(21,21)    $\dfrac{\partial}{\partial \hat{t}} \left\{ \vec{k} . \left[ \vec{D} \wedge \hat{\vec{v}}_0 \right] + SBo \dfrac{\partial}{\partial p} \left[ \dfrac{p^2}{K_0(p)} \dfrac{\partial \hat{\mathcal{R}}_1}{\partial p} \right] \right\} = 0 .$

Now if we integrate this last equation between $\hat{t}=0$ and $\hat{t}=\infty$, and if we use the geostrophic balance for limiting values of $\hat{\vec{v}}_0$ and $\hat{\mathcal{R}}_1$, when $\hat{t} \to +\infty$, we get:

(21,22)    $Bo\lambda_0 \vec{D} \hat{\mathcal{R}}_1^\infty + SBo \dfrac{\partial}{\partial p} \left[ \dfrac{p^2}{K_0(p)} \dfrac{\partial \hat{\mathcal{R}}_1^\infty}{\partial p} \right]$

$$= \vec{k} . \left[ \vec{D} \wedge \vec{v}^0 \right] + SBo \dfrac{\partial}{\partial p} \left[ \dfrac{p^2}{K_0(p)} \dfrac{\partial \mathcal{H}_1^0}{\partial p} \right] .$$

This is an equation from which, with suitable boundary conditions on p and x, y, we may deduce the value of $\hat{\mathcal{H}}_1^\infty$.

From (21,22) we obtain the initial condition that must be supplied for equation (9,14), namely[†]:

$$(21,23) \qquad \text{Bo } \hat{\Lambda} \, \mathcal{H}_{qg}\Big|_{t=0} = \vec{k}.\left[\vec{D} \wedge \vec{v}^0\right] + \text{SBo } \frac{\partial}{\partial p}\left(\frac{p^2}{K_0(p)} \frac{\partial \mathcal{H}_1^0}{\partial p}\right).$$

where $\hat{\Lambda}$ is the operator (9,15).

We observe that from the solution derived by Kibel, which is restricted to $K_0(p)$=constant, it appears that the differences between $(\hat{\vec{v}}_0, \hat{\mathcal{H}}_1)$ and their limiting values tends to zero like $\frac{1}{\hat{t}^{1/2}}$ osc($\hat{t}$), where osc($\hat{t}$) stands for some bounded functions which oscillate like a cosine function.

We mention for further study that the geostrophic balance occurs in a number of other situations, with various processes of adjustment discussed in Blumen (1972).

The Figure 8 below gives an example of the adjustment of meteorological fieds (after Monin (1958)) in the baroclinic atmosphere.

In this case, there were no pressure perturbations at the initial instant of time, and the velocity field corresponded to a plane-parallel flow of the type of tangential discontinuity along the ordinate axis (the initial distribution of the horizontal velocity $\vec{v}^0$ is given by the dotted line in Figure 8). The velocity field changed only slightly as a result of adjustment; see the limiting distribution of the horizontal velocity $\vec{v}(x)$ (the kinetic energy decreased by 3% from losses due to the generation of fast gravity waves and the formation of inhomogeneities in the pressure field). The pressure field *actively* "*adapted*" to the velocity field: a distinct dip was produced in it (see the limiting distribution of the altitudes of the isobaric surface $\mathcal{H}(x)$ at ground level; it dropped by 4 dkm along the ordinate axis).

The problem of the adjustment to geostrophy in the case of a hydrostatic barotropic atmosphere was first formulated by Rossby (1938) and Cahn (1945) and solved by Obukhov (1949). For the baroclinic atmosphere, this problem was treated by Bolin (1953), Kibel (1955; without taking the two-dimensional waves into-account), Veronis (1956), Fjelstad (1958) and Monin (1958). See also the excelent review by Phillips (1963) dealing with geostrophic motions.

---

[†] From matching with the main outer quasi-geostrophic region this limit value $\hat{\mathcal{H}}_1^\infty$ must coincide with the initial value $\mathcal{H}_{qg}\big|_{t=0}$ for the equation (9,14). This result has been at first obtained by Guiraud and Zeytounian (1980).

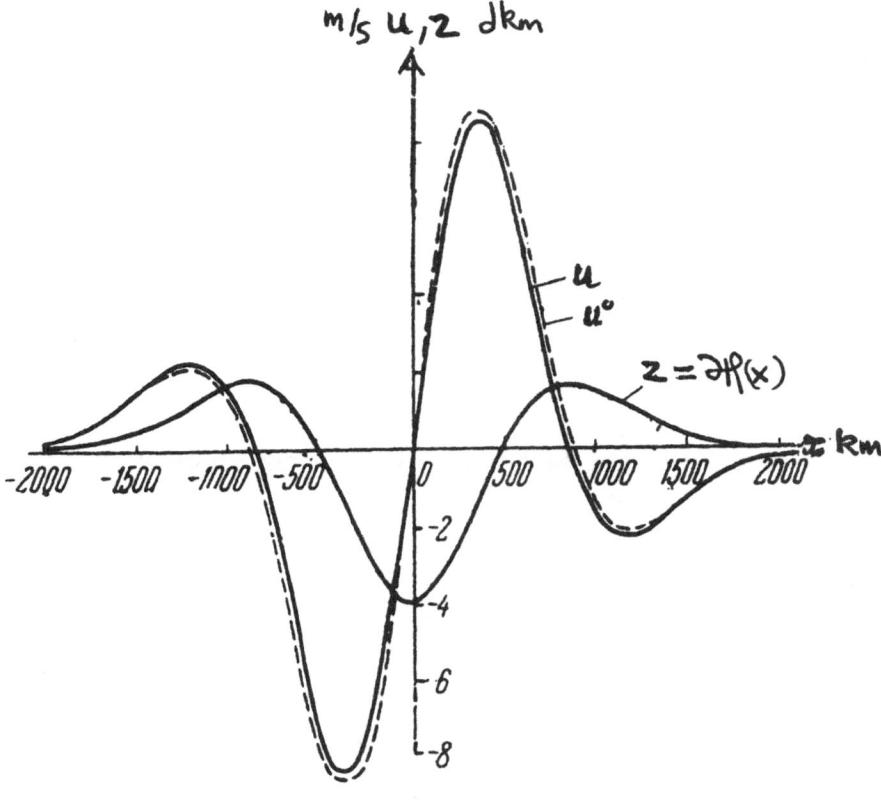

Fig. 8: An example of the adaptation of meteorological fields in the baroclinic atmosphere (After Monin (1958)).

BACKGROUND READING

Concerning the problem of adjustement of meteorological fields, see the lectures I and II of

GUIRAUD, J.P. (1983) _ *Some examples of application of asymptotic techniques to the derivation of models for atmospheric flows.*

Mecanique Theorique, Université de Paris 6.

( Unpublished manuscript; C.I.S.M, Udine, Italy).

REFERENCES TO WORKS CITED IN THE TEXT

CAHN, A.(1945) _ J.Meteorol. 2, 113-119.

CERCIGNANI, C.(1975) _ *Theory and Application of the Boltzmann equation.*

Scottish Academic Press.

BLUMEN, W. (1953) _ Reviews of Geophysics and Space Physics, 10, 485-528.

BOLIN, B.(1953) _ Tellus, 5, 373-385.

FJELSTAD, J.E. (1958) _ Geofys.Publikasjoner Norske Videnskaps-Akad.Oslo,20,1.

FROMM, J.E. (1969) _ Phys. of Fluids, vol.12, suppl.II, pp.II, 3-II, 12.

GREENSPAN, H.P. (1968) _ *The Theory of rotating fluids.* Cambridge Univ.Press

GUIRAUD, J.P. and ZEYTOUNIAN, R.Kh. (1982) _ Tellus, 34, 50-54.

GUIRAUD, J.P. and ZEYTOUNIAN, R.Kh. (1980) _ Geophys.Astrophys.fluid Dynamics 15, 283.

KIBEL, I.A. (1955) _ DAN SSSR, 104, 60-63 (in Russian).

KIBEL, I.A. (1957) _ *An Introduction to the hydrodynamical Method of short period Weather Forecasting,* Moscow (in Russian).

LERAT, A. and PEYRET, R. (1973) _ C.R.Acad. Sci. Paris, t.276 A, 759-762 and t.277 A, 363-366.

MONIN, A.S. (1958) _ Izv.Akad.Nauk SSSR, ser. Geofiz. 497.

MONIN, A.S. (1969) _ *Prognos pogody kak zadacha fiziki.* Izd.Nauka, Moscow (in Russian).

NAYFEH, A.H. (1973) _ *Perturbation methods.* John Wiley and Sons.

OBUKHOV, A.M. (1949) _ Izv.Akad.Nauk SSSR, ser. Geogr.i Geofiz., 13, 281.

OUTREBON, P. (1981) _ Correction de Fromm pour les schémas $\mathscr{S}_{\beta}^{\alpha}$ et applications au phénoméne d'adaptation au quasi-statisme en Météorologie. These de 3$^e$ cycle.Université de Paris 6, Mécanique Théorique.

PHILLIPS, N.A. (1963) _ Rev.Geophys., 1, 2, 123-176.

RICHARDSON, L.F.(1922) _ *Weather Prediction by Numerical Process*.
Cambridge-reprinted by Dover Publications in 1966.

ROSSBY, C.G. (1938) _ J.Marine Res.1, 239-263.

TITCHMARSH, E.C. (1948) _ *Introduction to the Theory of Fourier integrals*. Oxford, Clarendon Press.

VAN DYKE, M. (1962) _ *Perturbation methods in Fluid Mechanics*. Academic Press, N-Y.

VERONIS, G. (1956) _ Deep- Sea Res., 3, n°3, 157.

ZEYTOUNIAN, R.Kh. (1984) _ C.R.Acad.Sci., Paris, t.299, I, n°20, 1033-36.

ZEYTOUNIAN, R.Kh. (1990) _ *Asymptotic modeling of Atmospheric Flows*. Springer-Verlag, Heidelberg.

# CHAPTER VI

## LEE WAVE
## LOCAL DYNAMIC PROBLEMS

When $\text{Re} \equiv \infty$ we obtain, instead of $f_0$-plane equations $(5,5)$-$(5,9)$, the Euler equations for the adiabatic, non-viscous, atmosphere and we have only possibility to impose, on the *flat ground*, the following slip condition:

$$z=0 \; : \; w=0.$$

For adiabatic, non-viscous, atmospheric phenomena at local scales (when $\varepsilon_0 \simeq 1$) we set, instead of $w_{|z=0} = 0$ ,

$$\text{on } z = \sigma h\left(\lambda \frac{x-x_0}{\varepsilon_0} \; , \; \mu \frac{y-y_0}{\varepsilon_0}\right): \qquad w = \sigma \vec{v} . \vec{\mathrm{D}} h\left(\lambda \frac{x-x_0}{\varepsilon_0} \; , \; \mu \frac{y-y_0}{\varepsilon_0}\right) \; ,$$

where

$$\sigma = \frac{h_0}{H_0} \; , \quad \lambda = \frac{H_0}{\ell_0} \; , \quad \mu = \frac{H_0}{m_0} \quad \text{and} \quad \varepsilon_0 = \frac{H_0}{L_0} \; .$$

The local ground h is characterized by the length scales $h_0, \ell_0$ and $m_0$; $x=x_0$, $y=y_0$ is a local origin and $h(0,0) \equiv 1$, but $h(\infty, \infty) \equiv 0$.

Here we start with the Euler equations in dimensionless form, i.e. the equations $(5,5)$-$(5,9)$, where $\text{Re} \equiv \infty$.

## 22 . EULER'S LOCAL DYNAMIC
### MODEL EQUATIONS

In connection with the Euler equations (equations $(5,5)$-$(5,9)$, where $\text{Re} \equiv \infty$) we can formally consider the following *two* local limiting processes:

$$(22,1) \qquad \qquad \varepsilon_0 \to 0 \; , \text{ with } t, \tilde{x}, \tilde{y} \text{ and } z \text{ fixed,}$$

and

$$(22,2) \qquad \qquad \varepsilon_0 \to 0 \; , \text{ with } \tilde{t}, \tilde{x}, \tilde{y} \text{ and } z \text{ fixed,}$$

where

(22,3) $\qquad \tilde{t} = \dfrac{t}{\varepsilon}$ , $\quad \tilde{x} = \dfrac{x-x_0}{\varepsilon_0}$ , $\quad \tilde{y} = \dfrac{y-y_0}{\varepsilon_0}$ .

Considering the Euler equations with the boundary condition:

(22,4) $\qquad w = \sigma \vec{v} . \vec{D} h(\lambda \tilde{x}, \mu \tilde{y})$, on $z = \sigma h(\lambda \tilde{x}, \mu \tilde{y})$

we are led to the limiting process (22,1), which is closely related to the model of local steady dynamic prediction. For this model we obtain the following set of limiting equations:

(22,5a) $\qquad (\vec{\tilde{v}}_0 . \vec{\tilde{D}}) \vec{\tilde{v}}_0 + \tilde{w}_0 \dfrac{\partial \vec{\tilde{v}}_0}{\partial z} + \dfrac{1}{\tilde{\rho}_0} \dfrac{1}{\gamma M_0^2} \vec{\tilde{D}} \tilde{p}_0 = 0 $ ;

(22,5b) $\qquad \vec{\tilde{v}}_0 . \vec{\tilde{D}} \, \tilde{w}_0 + \tilde{w}_0 \dfrac{\partial \tilde{w}_0}{\partial z} + \dfrac{1}{\gamma M_0^2} \left[ \dfrac{1}{\tilde{\rho}_0} \dfrac{\partial \tilde{p}_0}{\partial z} + Bo \right] = 0 $;

(22,5c) $\qquad \tilde{p}_0 = \tilde{\rho}_0 \tilde{T}_0 $ ;

(22,5d) $\qquad \vec{\tilde{D}} . (\tilde{\rho}_0 \vec{\tilde{v}}_0) + \dfrac{\partial \tilde{\rho}_0 \tilde{w}_0}{\partial z} = 0 $ ;

(22,5e) $\qquad \tilde{\rho}_0 \vec{\tilde{v}}_0 . \vec{\tilde{D}} \tilde{T}_0 - \dfrac{\gamma-1}{\gamma} \vec{\tilde{v}}_0 . \vec{\tilde{D}} \tilde{p}_0 + \tilde{w}_0 \left[ \tilde{\rho}_0 \dfrac{\partial \tilde{T}_0}{\partial z} - \dfrac{\gamma-1}{\gamma} \dfrac{\partial \tilde{p}_0}{\partial z} \right] = 0 $ ,

where $\quad \vec{\tilde{D}} = \dfrac{\partial}{\partial \tilde{x}} \vec{i} + \dfrac{\partial}{\partial \tilde{y}} \vec{j} = \varepsilon_0 \vec{D} $ ,

and

(22,6) $\qquad (\vec{\tilde{v}}_0, \tilde{w}_0, \tilde{p}_0, \tilde{\rho}_0, \tilde{T}_0) \equiv \lim_{\varepsilon_0 \to 0} (\vec{v}, \varepsilon_0 w, p, \rho, T)$ .

$\qquad\qquad\qquad\qquad\qquad$ $t, \tilde{x}, \tilde{y}, z$
$\qquad\qquad\qquad\qquad\qquad$ fixed

At this point, formal matching with the primitive model equations (7,1):

$$(22,7) \qquad \lim_{\substack{|\tilde{x}|\to\infty \\ |\tilde{y}|\to\infty}} \begin{Bmatrix} \tilde{\vec{v}}_0 \\ \tilde{w}_0 \\ \tilde{p}_0 \\ \tilde{\rho}_0 \\ \tilde{T}_0 \end{Bmatrix} = \begin{Bmatrix} \vec{v}(t,x_0,y_0,z) \\ 0 \\ p(t,x_0,y_0,z) \\ \rho(t,x_0,y_0,z) \\ T(t,x_0,y_0,z) \end{Bmatrix} \quad ,$$

and

$$(22,8) \qquad \left(\frac{\partial p}{\partial z} + Bo\rho\right)_{x_0,y_0} = 0 \quad , \quad (p-\rho T)_{x_0,y_0} = 0 \quad ,$$

may be interpreted as providing lateral boundary conditions at infinity for the local steady dynamic model (lee waves model), which take into account the prediction at $x=x_0$, $y=y_0$ according to the primitive equations $(7,1)$.

Of course, it is necessary to resolve the vertical structure problem for the internal lee waves of same type as the one considered in section 11 of the Chapter III.

For the model equations $(22,5)$ we get as condition on the ground:

$$(22,9) \qquad \tilde{w}_0 = \sigma\tilde{\vec{v}}_0 \cdot \vec{D}h(\lambda\tilde{x},\mu\tilde{y}), \text{ on } z = \sigma h(\lambda\tilde{x},\mu\tilde{y}).$$

However, if the initial conditions for the full Euler equations contain $\tilde{x}$ and $\tilde{y}$ :

$$(22,10) \qquad t=0: \vec{v}=\vec{V}^0, \; \varepsilon w=W^0, \; p=P^0, \; \rho=R^0,$$

where $\vec{V}^0$, $W^0$, $P^0$ and $R^0$ are given functions of $z$ and of the horizontal positions: $\vec{x}=x\vec{i}+y\vec{j}$ and $\tilde{\vec{x}}=\tilde{x}\vec{i}+\tilde{y}\vec{j}$, it is necessary to consider also, in the Euler equations, limiting process $(22,2)$, which lead to the local unsteady dynamic evolution model in lieu of the equations of adjustment to hydrostatic balance $(19,7)$, $(19,8)$. This last model (local nonlinear adjustment equations) is the most complete one, but it is coupled to the primitive equations model!

Therefore, if we introduce a fast time, $\tilde{t}=\dfrac{t-t^0}{\varepsilon_0}$, and the fast horizontal variables, $\tilde{x}=\dfrac{x-x_0}{\varepsilon_0}$ and $\tilde{y}=\dfrac{y-y_0}{\varepsilon_0}$ , and if we use the limiting process $(22,2)$, setting:

(22,11)
$$(\tilde{\vec{v}}_0^*, \tilde{w}_0^*, \tilde{p}_0^*, \tilde{\rho}_0^*, \tilde{T}_0^*) \equiv \lim_{\substack{\varepsilon_0 \to 0 \\ \tilde{t}, \tilde{x}, \tilde{y}, z \\ \text{fixed}}} (\vec{v}, \varepsilon_0 w, p, \rho, T) \ .$$

it is straightforward to derive, from the full Euler atmospheric equations, the following set of limiting unsteady equations for $\tilde{\vec{v}}_0^*, \tilde{w}_0^*, \tilde{p}_0^*, \tilde{\rho}_0^*$ and $\tilde{T}_0^*$:

(22,12)
$$\left\{ \begin{aligned}
&S \frac{\partial \tilde{\vec{v}}_0^*}{\partial \tilde{t}} + (\tilde{\vec{v}}_0^* . \tilde{D})\tilde{\vec{v}}_0^* + \tilde{w}_0^* \frac{\partial \tilde{\vec{v}}_0^*}{\partial z} + \frac{1}{\tilde{\rho}_0^*}\frac{1}{\gamma M_0^2} \tilde{D}\tilde{p}_0^* = 0 \ ; \\[2mm]
&S \frac{\partial \tilde{w}_0^*}{\partial \tilde{t}} + (\tilde{\vec{v}}_0^* . \tilde{D})\tilde{w}_0^* + \tilde{w}_0^* \frac{\partial \tilde{w}_0^*}{\partial z} + \frac{1}{\gamma M_0^2}\left[\frac{1}{\tilde{\rho}_0^*}\frac{\partial \tilde{p}_0^*}{\partial z} + Bo\right] = 0; \\[2mm]
&\tilde{p}_0^* = \tilde{\rho}_0^* \tilde{T}_0^* \ ; \\[2mm]
&S \frac{\partial \tilde{\rho}_0^*}{\partial \tilde{t}} + \tilde{D}.(\tilde{\rho}_0^* \tilde{\vec{v}}_0^*) + \frac{\partial \tilde{\rho}_0^* \tilde{w}_0^*}{\partial z} = 0 \ ; \\[2mm]
&\tilde{\rho}_0^*\left[S \frac{\partial \tilde{T}_0^*}{\partial \tilde{t}} + \tilde{\vec{v}}_0^* . \tilde{D}\tilde{T}_0^*\right] - \frac{\gamma-1}{\gamma}\left[S \frac{\partial \tilde{p}_0^*}{\partial \tilde{t}} + \tilde{\vec{v}}_0^* . \tilde{D}\tilde{p}_0^*\right] + \tilde{w}_0^*\left[\tilde{\rho}_0^*\frac{\partial \tilde{T}_0^*}{\partial z} - \frac{\gamma-1}{\gamma}\frac{\partial \tilde{p}_0^*}{\partial z}\right] = 0.
\end{aligned} \right.$$

We note that, if at the initial time $t = t^0$ we have a set of initial values, functions of $z$ and $\tilde{x}$, $\tilde{y}$, then in limiting process (22,2) we pose $\tilde{t} = \frac{t - t^0}{\varepsilon_0}$ ; then $\tilde{t} \to \infty$, i.e., when $t \to t^0$, we obtain the local steady dynamic model equations (22,5), corresponding to limiting process (22,1), where $t = t^0$ is a parameter.
For the unsteady equations (22,12) we get as condition on the ground (22,9) and corresponding initial conditions for $\tilde{\vec{v}}_0^*$, $\tilde{w}_0^*$, $\tilde{p}_0^*$, and $\tilde{\rho}_0^*$ at $\tilde{t} = 0$, where the initial values (see, (22,10)) are given functions of $z$ and $\tilde{x}$, $\tilde{y}$.
The equations (22,12), with the condition (22,9) and corresponding initial conditions, describe the *local nonlinear adjustment problem* and show how a local situation (corresponding to a fixed time $t^0$) changes into another local situation (with another time $t^0$), under the influence of the initial *local* conditions!

*CONSISTENCY OF THE MODEL EQUATIONS (22,5) AND (22,12)*

If we are sure that our singular perturbation problem (related to the limiting process $\varepsilon_0 \to 0$) can be resolved by the method of matched asymptotic expansions (see Van Dyke (1975)), then we have the following matching conditions, between the limiting processes:

$$(22,13) \qquad \lim_{\substack{t \to 0 \\ |\tilde{x}| \to \infty \\ |\tilde{y}| \to \infty}} \lim_{\substack{\varepsilon_0 \to 0 \\ t, \tilde{x}, \tilde{y}, z \\ \text{fixed}}} \equiv \lim_{\substack{\tilde{t} \to \infty}} \lim_{\substack{\varepsilon_0 \to 0 \\ \tilde{t}, x, y, z \\ \text{fixed}}} \qquad , \tilde{t} = \frac{t}{\varepsilon} \; ;$$

$$(22,14) \qquad \lim_{\substack{\tilde{t} \to \infty}} \lim_{\substack{\varepsilon_0 \to 0 \\ \tilde{t}, \tilde{x}, \tilde{y}, z \\ \text{fixed}}} \equiv \left. \lim_{\substack{\varepsilon_0 \to 0 \\ t, \tilde{x}, \tilde{y}, z \\ \text{fixed}}} \right|_{\text{for } t = t^0} \qquad , \tilde{t} = \frac{t - t^0}{\varepsilon} \; .$$

Naturally the nature of matching conditions depends vitally of the behaviour of solutions to the local problems when either one of $\tilde{t}, |\tilde{x}|$, $|\tilde{y}|$ tends to infinity.

If, in particular, we consider the behaviour of the steady solutions of equations (22,5), when $|\tilde{x}^2 + \tilde{y}^2| \to \infty$, we can suspect that the variations with respect to $\tilde{x}$ and $\tilde{y}$ occurs through two different scales. One of them grows with

$$\sqrt{|\tilde{x}^2 + \tilde{y}^2|}$$

but the other one corresponds to internal waves, as discussed in Chapter III , and its scale remains of order one, when $|\tilde{x}^2 + \tilde{y}^2| \to \infty$. The reason for the existence of such waves is that, when $|\tilde{x}^2 + \tilde{y}^2| \to \infty$, the relief is flat and the perturbations obey the linear equations with slip on a flat ground; this is precisely the situation discussed in Chapter III.

When $|\tilde{x}^2 + \tilde{y}^2| \to \infty$ the horizontal wavelength of these waves becomes very small in comparaison to the distance to the relief and they appear, locally, as plane waves radiated away. Finally, we note that it is possible to show that once excited, trapped lee waves travel along the rays and that their amplitude variation may, at least in principle, be computed; it is almost evident that this amplitude decays when travelling away from the relief but a formal proof is difficult.

## 23 . MODEL EQUATIONS FOR THE TWO-DIMENSIONAL STEADY LEE WAVES

Here we start from the local steady Euler's dynamic model equations (22,5), which has been obtained in section 22. But we consider only the two-dimensional case :

(23,1) $\qquad \dfrac{\partial}{\partial \tilde{y}} \equiv 0 \Leftarrow \mu \rightarrow 0$, with $\tilde{x}$, $\tilde{y}$ and z fixed.

Consequently we can write (the marks have been dropped from the nondimensional quantities):

(23,2)
$$
\begin{cases}
\rho\left[u\dfrac{\partial w}{\partial x} + w\dfrac{\partial w}{\partial z}\right] + \dfrac{1}{\gamma M_0^2}\dfrac{\partial p}{\partial x} = 0 \; ; \\[3mm]
\rho\left[u\dfrac{\partial u}{\partial x} + w\dfrac{\partial u}{\partial z}\right] + \dfrac{1}{\gamma M_0^2}\left[\dfrac{\partial p}{\partial z} + Bo\rho\right] = 0 \; ; \\[3mm]
\dfrac{\partial \rho u}{\partial x} + \dfrac{\partial \rho w}{\partial z} = 0 \; ; \; p = \rho T \; ; \\[3mm]
\rho\left[u\dfrac{\partial T}{\partial x} + w\dfrac{\partial T}{\partial y}\right] - \dfrac{\gamma-1}{\gamma}\left[u\dfrac{\partial p}{\partial x} + w\dfrac{\partial p}{\partial y}\right] = 0 \; ,
\end{cases}
$$

with the following boundary slip condition:

(23,3) $\qquad w = \sigma u \dfrac{dh_p(x)}{dx}$ , on $z = \sigma h_p(\lambda x)$.

The function $h_p(\lambda x)$ is of order unity being identically zero for $|x| > 1$. In (23,2) and (23,3) four length scales ratios enter, namely:

(23,4)
$$
\begin{cases}
\sigma = \dfrac{h_0}{H_0}, \; \lambda = \dfrac{H_0}{\ell_0}, \; Bo = \dfrac{H_0}{\dfrac{RT_\infty(0)}{g}} \; , \\[5mm]
\text{and } M_0^2 = \dfrac{U_0^2/\gamma g}{\dfrac{RT_\infty(0)}{g}} \; .
\end{cases}
$$

The lee wave two-dimensional steady problem is considered within the framework of the *nonlinear* equations (23,2), with (23,3).

Perturbations are assumed to be confined to the troposphere with vanishing vertical velocity w on the top. We assume also that far ahead from the mountain $z = \sigma h_p(\lambda x)$ there is a uniform flow with velocity (in dimensionless form):

(23,5)           u=1, w=0, when x→-∞,

and we set $z_\infty$ for the altitude far a head (where x≡-∞).
The equations (23,2) are reduced as usual by introducing the stream function
$\psi(x,z)$, such that:

(23,6)           $\rho u = -\dfrac{\partial \psi}{\partial z}$ and $\rho w = +\dfrac{\partial \psi}{\partial x}$ ,

but for convenience we follow the common technique in the theory of lee waves
(see, Long (1953) and Zeytounian (1979)) which amounts to replacing the stream
function by the *vertical displacement of the streamline* $\delta(x,z)$ (in
dimensionless form), in such a way that:

(23,7)           $z_\infty = Bo(z-\sigma\delta)$.

For the system of equations (23,2) we have, first, the Bernoulli's equation:

(23,8)           $\dfrac{u^2+w^2}{2} + \dfrac{1}{M_0^2(\gamma-1)} \rho^{\gamma-1}\, \pi(\psi) + \dfrac{Bo}{\gamma M_0^2}z = I(\psi)$,

where

(23,9)           $\pi(\psi) = \dfrac{p}{\rho^\gamma} \Longleftarrow \left[u\dfrac{\partial}{\partial x} + w\dfrac{\partial}{\partial z}\right]\left(\dfrac{p}{\rho^\gamma}\right) = 0$,

and, secondly, the vorticity theorem (see Zeytounian (1974)):

(23,10)           $\dfrac{\partial u}{\partial z} - \dfrac{\partial w}{\partial x} = -\rho\left\{\dfrac{dI}{d\psi} - \dfrac{1}{M_0^2(\gamma-1)\gamma}\dfrac{p}{\rho}\dfrac{d\text{Log}\pi}{d\psi}\right\}$ .

Equations (23,8)-(23,10) contain the arbitrary functions $I(\psi)$ and $\pi(\psi)$ of the
stream function $\psi$. The right-hand side of equation (23,10) can be determined
from the conditions in the unperturbed flow (see, (23,5)), where the vertical
distributions of all elements are known; in particular:

(23,11)           $p=p_\infty(z_\infty)$, $\rho=\rho_\infty(z_\infty)$ and $T=T_\infty(z)$, when x→-∞.

Through some lengthy but quite straightforward computations one may derive a
quasi-linear elliptic differential equation for $\delta(x,z)$, from (23,10) with
(23,6), (23,7) and (23,8). It is convenient to write this equation by singling
out the nondimensional density perturbation $\omega=\dfrac{\rho-\rho_\infty}{\rho_\infty}$ as an (implicit) function

of $\delta$ and its first order derivatives, namely:

(23,12)
$$(1+\omega)^{\gamma-1} = 1 - \left\{ \frac{\gamma-1}{2} M_0^2 \sigma^2 \frac{1}{(1+\omega)^2} \left[ \left(\frac{\partial\delta}{\partial x}\right)^2 + \left(\frac{\partial\delta}{\partial z}\right)^2 - \frac{2}{\sigma}\frac{\partial\delta}{\partial z} + \frac{1}{\sigma^2} \right] \right.$$

$$\left. + \frac{\gamma-1}{\gamma} Bo\sigma\delta - \frac{\gamma-1}{2} M_0^2 \right\} \frac{1}{T_\infty(z_\infty)} ,$$

where $\qquad T_\infty(z_\infty) \equiv T_\infty(Boz - Bo\sigma\delta)$.

Then the equation for $\delta(x,z)$ is rather awkward-looking one (Zeytounian(1979)):

(23,13)
$$\frac{\partial^2\delta}{\partial x^2} + \frac{\partial^2\delta}{\partial z^2} + \frac{Bo^2}{\gamma M_0^2} \frac{1}{T_\infty(z_\infty)} \left( \frac{\gamma-1}{\gamma} + \frac{dT_\infty}{dz} \right)(1+\omega)^2\delta$$

$$= \frac{1}{1+\omega} \left[ \frac{\partial\delta}{\partial x}\frac{\partial\omega}{\partial x} + \frac{\partial\delta}{\partial z}\frac{\partial\omega}{\partial z} - \frac{1}{\sigma}\frac{\partial\omega}{\partial z} \right]$$

$$+ \frac{1}{T_\infty(z_\infty)} \left( \frac{\gamma-1}{\gamma} + \frac{dT_\infty}{dz} \right) \left\{ \frac{Bo}{2\sigma}(\omega+2)\omega - \frac{\sigma}{2}Bo \left[ \left(\frac{\partial\delta}{\partial x}\right)^2 + \left(\frac{\partial\delta}{\partial z}\right)^2 - \frac{2}{\sigma}\frac{\partial\delta}{\partial z} \right] \right\} .$$

As is well known (Long (1953), Zeytounian (1969,1979)) the appropriate bounadary conditions are:

(23,14)
$$\begin{cases} \delta(x,\sigma h_p(\lambda x)) = h_p(\lambda x), \text{ where } |x|<1; \\[2mm] \delta(-\infty, z_\infty) = 0 \text{ , } \omega(-\infty, z_\infty) = 0; \\[2mm] \delta(x,\frac{1}{Bo}) = 0; \\[2mm] \lim_{x\to+\infty} \left[ \left|\frac{\partial\delta}{\partial x}\right| + \left|\frac{\partial\delta}{\partial z}\right| \right] < +\infty. \end{cases}$$

Concerning the second condition of (23,14) (for $x=-\infty$) it is merely the consequence of Long's hypothesis of *no* upstream influence; the role of this hypothesis was clarified by Mc Intyre (1972) who showed (on a linear version) that it emerges from a careful examination of the transient beheviour. The third condition of (23,14) is the consequence of the vanishing vertical velocity w on the top of the troposphere. The proper way to formulate the upper boundary condition relating to the top of the troposphere would be to match the previous model of lee waves in the troposphere with another model for waves in the upper atmosphere, taking into account the dissipation of such waves by viscosity at very high altitudes. The works of Yanowitch (1967),

Thomas and Stevenson (1972) and Bois (1984) would be helpful in formulating this matching process.

Needless to say, the problem (23,12)-(23,14) is *untractable* and the benifit of reducing the system of equations (23,2) to a single equation for $\delta(x,z)$ is lost through the awkward nonlinearity. The interest of the present formulation lies in the fact that it allows to work with asymptotic techniques in a very convenient way and we shall proceed with this now.

## 24 . BOUSSINESQ'S INNER SOLUTION

If we assume that:

$$(24,1) \qquad Bo = \hat{B}M_0, \ M_0 \to 0, \text{ with } \hat{B}, x, z, \sigma \text{ and } \lambda \text{ fixed},$$

and we suppose that, according to (23,12),

$$(24,2) \qquad \omega = O(M_0),$$

we find that $\lim\limits_{\substack{M_0 \to 0 \\ \hat{B} \text{ fixed}}} \delta \equiv \delta_0(x,z)$ is a solution of the Helmholtz equation:

$$(24,3) \qquad \frac{\partial^2 \delta_0}{\partial x^2} + \frac{\partial^2 \delta_0}{\partial z^2} + K_{00}^2 \delta_0 = 0,$$

where the constant $K_{00}^2$ is given by:

$$(24,4) \qquad K_{00}^2 = \frac{\hat{B}^2}{\gamma}\left[\frac{\gamma-1}{\gamma} + \left(\frac{dT_\infty}{dz_\infty}\right)_{z_\infty=0}\right]$$

and is assumed to be positive. In the limiting process (24,1) we assume that $\alpha_\infty^0 = O(1)$, where $\alpha_\infty^0$ is a dimensionless measure of stability of the standard atmosphere; we recall that:

$$\alpha_\infty^0 N_\infty^2(z_\infty) \equiv \frac{Bo}{T_\infty(z_\infty)}\left[\frac{\gamma-1}{\gamma} + \frac{dT_\infty(z_\infty)}{dz_\infty}\right].$$

On the other hand, (23,14) leads to:

$$(24,5) \quad \begin{cases} \delta_0(x, \sigma h_p(\lambda x)) = h_p(\lambda x), \text{ when } |x| < 1; \\[2ex] \delta(-\infty, z) = 0 \\[2ex] \lim_{x \to +\infty} \left( \left| \frac{\partial \delta}{\partial x} \right| + \left| \frac{\partial \delta}{\partial z} \right| \right) < +\infty, \end{cases}$$

but the third condition of (23,14) is no longer valid (in the framework of the Boussinesq inner approximation ) and a matching condition must be used instead (see, the following section 25).

The problem (24,3)-(24,5) for a semi-infinite half plane $z \geq \sigma h_p(\lambda x)$ must be complemented by a condition relating to $r \to \infty$, in order to achieve uniqueness (here we use polar coordinate $r$ and $\vartheta$ defined by: $x = r\cos\vartheta$, $z = r\sin\vartheta$, $r^2 = x^2 + z^2$). From the general results relating to the Helmholtz equation (see, Wilcox (1959)), we known that such a condition, assuming that *no waves* are radiated inwards, reads:

$$(24,6) \qquad \delta_0 \sim \left( \frac{2K_{00}}{\pi r} \right)^{1/2} \sin\vartheta \; \mathcal{R}eal \left\{ G(\cos\vartheta) \; \exp[i(K_{00}r - \pi/4)] \right\},$$

where the function $G(\cos\vartheta)$ is arbitrary and must depend on the form of the mountain through the function $h_p(\lambda x)$. As a reminiscence of the second condition of (24,5) we impose : $G(\cos\vartheta) \equiv 0$, for $\cos\vartheta < 0$.

In section 25 our goal was to show that Long's model[†] is the first inner approximation of an outer-inner approximation scheme.

*SEMICIRCULAR MOUNTAIN*

We consider now the case of a mountain ridge with a semicircular cross section of radius $r_0$

$$(24,7) \qquad h(x) = \begin{cases} \sqrt{r_0^2 - x^2} \;, & |x| \leq r_0, \\[2ex] 0 & , \; |x| > r_0, \end{cases}$$

---

[†] Long's classical model is obtained if we add to (24,3)-(24,5) the following *upper* condition : $\delta_0(x,1) = 0$.

We consider this model in the section 26.

and stream function $\psi_0(x,z)$ satisfies the equation[†]

(24,8)
$$\frac{\partial^2 \psi_0}{\partial x^2} + \frac{\partial^2 \psi_0}{\partial z^2} + k_0^2(\psi_0 - U_0 z) = 0.$$

For (24,8) the boundary conditions are:

(24,9)     $\psi_0 = 0$ ,   on $z = h(x)$;

(24,10)     $\psi_0 = U_0 z$ ,   for $x \rightarrow -\infty$.

In order to satisfy the condition (24,9) for the selected shape of the ridge we transform to polar coordinates

$$x = r\cos\vartheta, \quad z = r\sin\vartheta, \quad r \geq r_0, \quad 0 \leq \vartheta \leq \pi$$

and set

$$\Delta_0 = \psi_0 - U_0 z.$$

Equation (24,8) for $\Delta_0$, in the new coordinates, is:

(24,11)
$$\frac{\partial^2 \Delta_0}{\partial r^2} + \frac{1}{r} \frac{\partial \Delta_0}{\partial r} + \frac{1}{r^2} \frac{\partial^2 \Delta_0}{\partial \vartheta^2} + k_0^2 \Delta_0 = 0.$$

conditions (24,9) then takes the form of two conditions:

(24,12)     $\Delta_0(r,0) = \Delta_0(r,\pi) = 0$;

(24,13)     $\Delta_0(r_0,\vartheta) = U_0 r_0 \sin\vartheta.$

For a unique solution of the problem (24,11)-(24,13) it is sufficient that the solution of equation (24,11) should satisfy two conditions:

    (1) for some constant $C_0$, which is independent of position,

(24,14)     $|r^{1/2}\Delta_0| \leq C_0$;

    (2) for all angles within some range, smaller than $\pi$,

(24,15)     $\lim_{r\to\infty} (r^{1/2}\Delta_0) = 0.$

---

[†] $k_0^2 = \dfrac{\hat{B}^2}{\gamma U_0^2} \left[ \dfrac{\gamma-1}{\gamma} + \dfrac{dT_\infty}{dz_\infty} \bigg|_{z_\infty = 0} \right].$

The first of these requirements is actually the condition of generalized boundedness of the solution of equation (24,11).

The second amounts to requirements of particulary rapid damping of the solution along some selected directions, with the damping occuring faster than the reduction in $r^{-1/2}$ with increasing r. In the problem at hand this direction is apparently toward the incoming flow. It is clear from physical consideration that solution of (24,11) should damp out most rapidly in the direction toward the oncoming flow.

Hence in order to obtain a unique solution of the problem (24,11)-(24,13), the second condition (24,15) must be satisfed in the direction $\vartheta=\pi$ (before the study carried out by Magnus (1949), such a condition was used by Lyra (1943) to single out the unique solution of the Helmholtz equation).

Since $\Delta_0|_{\vartheta=\pi}=0$ for all $r \geq r_0$, and solution of (24,11) is continuous everywhere, following Merbt (1959), the second condition (24,15) can be used in the form:

(24,16)
$$\lim_{r\to\infty} (r^{1/2} \frac{\partial \Delta_0}{\partial \vartheta}) = 0.$$

Separation of variables is possible by seeking the solution of equation (24,11) in the form:

$$\Delta_0 = F(r)G(\vartheta)$$

and the equations governing $F(r)$ and $G(\vartheta)$ are then:

(24,17)
$$\frac{d^2F}{dr^2} + \frac{1}{r} \frac{dF}{dr} + \left[k_0^2 - \frac{n^2}{r^2}\right]F = 0;$$

(24,18)
$$\frac{d^2G}{d\vartheta^2} + n^2G = 0 ,$$

where n is a constant. By virtue of boundary conditions (24,12) the solution of equation (24,18) has the form:

$$G(\vartheta) = \sin(n\vartheta),$$

with n assuming integral values (n=1,2,3,...).

Equation (24,17) is an n-th order Bessel equation. For real $k_0$ (stable stratification) its solution are Bessel functions with real argument; for imaginary $k_0$ (unstable stratification) they are Bessel functions of imaginary argument. The former are oscillating and the latter exponential functions. It follows directly that wave-type disturbances may arise only in those air masses possessing stable stratification in the undisturbed state.

In general, the assumption that a large vertical extent of the atmosphere is stratified unstably is doubtful in the majority of problems. Hence it is assumed here and in all subsequent problems that the incoming flow as stable stratification:

$$-\left.\frac{dT_\infty}{dz_\infty}\right|_{z_\infty=0} < \frac{\gamma-1}{\gamma} .$$

The solution of equation (24,11) can be expressed in the form:

$$(24,19) \qquad \Delta_0(r,\vartheta) = \sum_{n=1}^\infty \left\{A_n J_n(k_0 r) + B_n Y_n(k_0 r)\right\}\sin(n\vartheta),$$

where $J_n(k_0 r)$ is a Bessel function of the first kind and $Y_n(k_0 r)$ is a Weber function, while $A_n$ and $B_n$ are constants to be defined.

Boundary condition (24,13) can be used to relate $A_n$ and $B_n$:

$$(24,20) \qquad A_1 = \frac{U_0 r_0}{J_1(k_0 r_0)} - x_1 B_1;$$

$$(24,21) \qquad A_n = -x_n B_n, \quad n\geq 2,$$

where $x_n = Y_n(k_0 r_0)/J_n(k_0 r_0)$, $n=1,2,\dots$ .

Coefficients $B_n$ remain undefined and requirement that the disturbances of $\psi$ damp out at high z *cannot* be used to eliminate this indeterminacy, because $J_n(k_0 r)\to 0$, and $Y_n(k_0 r) \to 0$ as $r\to\infty$!

The condition (24,16) is equivalent to the requirement that:

$$(24,22) \qquad \lim_{x\to-\infty}\left[|x|^{1/2}u'\,|_{z=0}\right] = 0.$$

Here u' denotes the disturbance of the horizontal velocity. Now substitute solution (24,19), with (24,20) and (24,21) into condition (24,16). Since this condition is stated at $r\to\infty$, functions $J_n$ and $Y_n$ are replaced by their asymptotic expansions, in which

$$\sin(k_0 r + \frac{\pi n}{2} - \frac{\pi}{4}) \text{ and } \cos(k_0 r + \frac{\pi n}{2} - \frac{\pi}{4})$$

are first transformed into functions having $(k_0 r + \frac{\pi}{4})$ as their argument. In the equation thus obtained, the coefficients of terms such as

$$r^{-n}\sin(k_0 r + \frac{\pi}{4}) \text{ and } r^{-n}\cos(k_0 r + \frac{\pi}{4})$$

are equaled to zero. This leads to a system of algebraic equations for coefficients $B_n$. Without analyzing this system of algebraic equations in detail, we note that Merbt, in the previously cited work, points out that, for pratical calculations, only a few of the first coefficients need be retained. Following this approach, only two harmonics in solution (24,19) will be considered, assuming all $B_n \equiv 0$ for $n \geq 3$.

The final form of the approximate expression for calculating the stream function is (Kozhevnikov (1963)):

$$(24,23) \quad \Delta_0 = U_0 r_0 \left\{ C_0 \left[ J_1(mR) + x_2 Y_1(mR) \right] - R \right\} \sin \vartheta - C_0 \left[ Y_2(mR) - x_2 J_2(mR) \right] \sin \vartheta \cos \vartheta,$$

$$R = \frac{r}{r_0} \,, \quad C_0 = \frac{1}{J_1(m)(1 + x_1 x_2)} \,, \quad m = k_0 r_0.$$

The stream function field in the vicinity of a mountain was calculated using the solution (24,23) with: $m = 2 \Rightarrow r_0 = 1\text{km}$, $U_0 = 6$ m/s and $\hat{B} \left( \frac{\gamma - 1}{\gamma} + \frac{dT_\infty}{dz_\infty} \Big|_{z_\infty = 0} \right) = 4°\text{C/km}$.

The results are plotted in dimensionless form in Fig. 9 below.

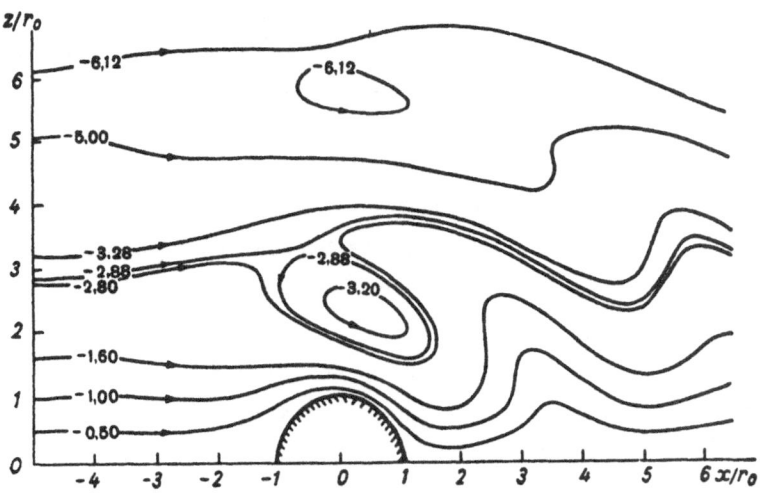

Fig. 9: Streamlines with m=2 (Kozhevnikov (1963))

It is clear that the disturbances produced by the mountain ridge are quite substantial. Very strong waves appear in the *lee* part of the flow. Their amplitude is damped as ones moves away horizontally from the mountain and, in addition, depends markedly on altitude. Disturbances in the windward part of the flow are practically imperceptible at distances exceeding $r_0$ by a factor of 4-5. It follows from solution (24,23) that disturbances disappear also at $z \rightarrow \infty$ and $x \rightarrow \infty$. The above Fig. 9 shows that closed streamlines (so-called *rotors*) with a horizontal axis form above the ridge. First we note that the rotors are shifted somewhat downstream and have a streamlined shape, similar to an aircraft wing. The air within the rotor moves in the same direction as that of the outside airflow, with the exception only of a small leading part. The air in the center of the rotor is at rest, while in the peripheral layers it moves at velocity close to that of the free flow. In the case at hand *two* rotors form above the mountain, with the center of the lower rotor situated at an altitude of about 2.5 km and that of the upper located at 5.7 km. The lower rotor is inclined noticeably to the horizontal. It, so to speak, forces the entire mass of air enclosed in the free flow between the ground and the altitude 2.8 km, to pass through a narrow (only about 800 m wide) passage between the mountain and the rotor. Here the wind speeds increase approximately fourfold, attaining (in the case under study) 24 m/s. Air particles located sufficiently far ahead of the mountain ridge at altitudes of from 1 to 2.8 km, rise upward after passing around the rotor from the bottom. Velocity reversal regions also exit here. The flow pattern between the rotors is similar to the one just examined, with the role of the mountain ridge-rotor pair played by the streamline moving in the immediate vicinity of the bottom and trailing part of the rotor. Above it, as above the mountain, the wind speed increases three-or fourfold downstream. This point is very dangerous for aircraft. As a whole the generation of rotors above a mountain ridge results, on the one hand, in increasing wind speeds between them and in the vinicity of the ridge, and on the other, in the formation of high-amplitude waves downstream from them.

We note that, the smaller m (smaller $r_0$ and larger $U_0$), the smaller is the number of rotors. For example, not more than one rotor can form when m=1.8, while when m=1.5 *no* rotors form at all.

Finally the reader interested in other computed examples is reffered to the original sources (Kozhevnikov (1968) and Miles (1968)).

## 25 . OUTER, GUIRAUD'S AND ZEYTOUNIAN'S SOLUTION

The condition $\delta(x, \frac{1}{Bo})=0$ (see,(23,14)), at the top of the troposphere is outside the domain of validity $\delta_0$ as an approximation of $\delta$ which is obtained through the inner limiting process (24,1).

In order to obtain an outer approximation to which the condition $\delta(x, \frac{1}{Bo})=0$ may be applied, we set:

$$(25,1) \qquad \xi=M_0 x, \quad \zeta=M_0 z, \quad \delta=M_0^{1/2} \tilde{\delta}(\xi,\zeta;M_0), \quad \omega=M_0^{3/2} \tilde{\omega}(\xi,\zeta;M_0).$$

The choise of scaling in (25,1) is dictated for $\delta$, by matching with (24,6) and for $\omega$, by equation (23,12).

In this case, instead of $\delta(x, \frac{1}{Bo})=0$, we have

$$(25,2) \qquad \tilde{\delta}(\xi, \frac{1}{\hat{B}}; M_0)=0$$

and otherwise $h_p(\lambda\xi/M_0)\to 0$, with $M_0\to 0$, except at $\xi\equiv 0$.

From the inner approximation we except that the perturbations in the outer region consist of lee waves with a wavelength of the order of $M_0$ (see, the relation (24,6)), at the scale of the outer region. As a consequence, there is a *double scale* built into the solution and we must take care of it. For the particular case

$$- \frac{dT_\infty}{dz_\infty} \equiv \Gamma_\infty^0 = \text{constant}$$

we have (in dimensionless form):

$$(25,3) \qquad T_\infty(z_\infty) = 1 - \frac{R}{g} \Gamma_\infty^0 z_\infty$$

$$= 1 - \frac{R}{g} \Gamma_\infty^0 \hat{B}\zeta + M_0^{3/2} \frac{R}{g} \Gamma_\infty^0 \hat{B} \sigma \tilde{\delta},$$

if we take into account (23,7) and (25,1).

As consequence of (25,1), (25,3) and

$$(25,4) \qquad z_\infty = \hat{B}(\zeta - \sigma M_0^{3/2}\tilde{\delta})$$

we find, from (23,13), the following *outer* equation for $\tilde{\delta}(\xi,\zeta;M_0)$:

$$(25,5) \qquad \frac{\partial^2 \tilde{\delta}}{\partial \xi^2} + \frac{\partial^2 \tilde{\delta}}{\partial \zeta^2} + \frac{1}{\sigma}\frac{\partial \tilde{\omega}}{\partial \zeta} - \frac{\gamma}{\hat{B}}\phi_\infty(\zeta)\frac{\partial \tilde{\delta}}{\partial \zeta} - \frac{\gamma}{\sigma}\frac{1}{\hat{B}}\phi_\infty(\zeta)\tilde{\omega}$$

$$+ \phi_\infty(\zeta)\frac{1}{M_0^2}\tilde{\delta} = 0\left[\frac{1}{M_0^{1/2}}\right],$$

with, from (23,12),

$$(25,6) \qquad \tilde{\omega} = -\frac{\sigma}{\gamma}\hat{B}\phi_\infty(\zeta)\frac{1}{K_{00}^2}\tilde{\delta} + \cdots .$$

In (25,5) and (25,6) we have

$$(25,7) \qquad 0 < \phi_\infty(\zeta) \equiv \frac{K_{00}^2}{1 - \frac{R}{g}\Gamma_\infty^0 \hat{B}\zeta} \ ; \ \phi_\infty(0) \equiv K_{00}^2 .$$

If we take into account (25,6) in (25,5) we see that the term

$$-\frac{\gamma}{\sigma}\frac{1}{\hat{B}}\phi_\infty(\zeta)\tilde{\omega} \cong \frac{\phi_\infty^2(\zeta)}{K_{00}^2}\tilde{\delta}$$

is *not* dominant, and strictly speaking, for the *outer* approximation we have the following *dominant equation*:

$$(25,8) \qquad \frac{\partial^2 \tilde{\delta}}{\partial \xi^2} + \frac{\partial^2 \tilde{\delta}}{\partial \zeta^2} + \frac{\phi_\infty(\zeta)}{M_0^2}\tilde{\delta} = \frac{\hat{B}}{\gamma}\frac{\phi_\infty(\zeta)}{K_{00}^2}\left\{\frac{d\text{Log}\phi_\infty(\zeta)}{d\zeta}\tilde{\delta} + \left[1 + \left(\frac{\gamma K_{00}}{\hat{B}}\right)^2\right]\frac{\partial \tilde{\delta}}{\partial \zeta}\right\} .$$

It is possible to consider (in an asymptotic sense ) the equation (25,8) as outer ones and the Helmholtz model equation (24,3), for $\delta_0(x,z)$, as inner equation. Therefore, it is shown that the upper condition

$$w=0 \ , \ \text{on } z=1/Bo$$

belongs to an *outer limiting* process:

$$(25,9) \qquad Bo=\hat{B}M_0 \ , \ M_0 \to 0 \ , \text{ with } \xi, \ \zeta, \ \hat{B} \text{ fixed}.$$

Such an outer asymptotic approximation is worked out by Guiraud and Zeytounian (1979) and it is shown that the upper and lower boundaries of the troposphere alternately reflect internal short gravity waves excited by the lee waves of the inner (Boussinesq) approximation, with a wavelength of the order of $M_0$, to the scale of the outer region. As a consequence, there is a double scale built into the solution and we must take care of it. The important point is that these short gravity excited waves propagate downstream and that *no* *feedback* occurs on the inner flow close to the mountain (to lowest order). We should understand the upper boundary condition as an artificial one, having asymptotically *no effects* on the inner flow (Boussinesq flow) which is the only really interesting one.

The equation (25,8) with the boundary conditions:

$$(25,10) \qquad \tilde{\delta}(\xi,0)=0 \quad , \quad \tilde{\delta}(\xi,1/\hat{B})=0 \quad ,$$

as well as the matching condition, at the origin, $\xi=0$, $\zeta=0$, with (24,6), was solved asymptotically for $M_0 \to 0$ and a double scale structure emerged from the solution (on account of the term $(\phi_\infty(\zeta)/M_0^2)\tilde{\delta}$ in equation (25,8)).

As a matter of fact, according to the technique of multiple scales (see, the Appendix 2), we set:

$$(25,11) \qquad \tilde{\delta} = \tilde{\delta}_0(\xi,\zeta;\chi) + M_0\tilde{\delta}_1(\xi,\zeta;\chi)+ \cdots \quad ,$$

where $\qquad \chi = \dfrac{\theta(\xi,\zeta)}{M_0}$ ,

and from (25,8), we find for $\tilde{\delta}_0$ the following equation:

$$(25,12) \qquad \left[\left(\frac{\partial\theta}{\partial\xi}\right)^2 + \left(\frac{\partial\theta}{\partial\zeta}\right)^2\right]\frac{\partial^2\tilde{\delta}_0}{\partial\chi^2} + \phi_\infty(\zeta)\tilde{\delta}_0 = 0 \quad .$$

We choose $\theta(\xi,\zeta)$ to be a solution of

$$(25,13) \qquad \left(\frac{\partial\theta}{\partial\xi}\right)^2 + \left(\frac{\partial\theta}{\partial\zeta}\right)^2 = \frac{\phi_\infty(\zeta)}{K_{00}^2} \quad ,$$

and an obvious solution to

$$(25,14) \qquad \frac{\partial^2\tilde{\delta}_0}{\partial\chi^2} + K_{00}^2\tilde{\delta}_0 = 0$$

is then

$$(25,15) \qquad \tilde{\delta}_0(\xi,\zeta;\chi) = \mathcal{Real}\left\{A(\xi,\zeta)\exp[i(K_{00}\chi - \pi/4)]\right\} ,$$

where the phase difference $\pi/4$ is used for effecting latter matching with (24,6).

In order to obtain an equation for the amplitude $A(\xi, \zeta)$ we go to the next order and we find, from (25,8), for $\tilde{\delta}_1(\xi, \zeta, \chi)$ the following nonhomogeneous equation:

(25,16)
$$\frac{\partial^2 \tilde{\delta}_1}{\partial \chi^2} + K_{00}^2 \tilde{\delta}_1 = - \frac{K_{00}^2}{\phi_\infty(\zeta)} \left\{ 2 \left[ \frac{\partial \theta}{\partial \xi} \frac{\partial^2 \tilde{\delta}_0}{\partial \xi \partial \chi} + \frac{\partial \theta}{\partial \zeta} \frac{\partial^2 \tilde{\delta}_0}{\partial \zeta \partial \chi} \right] \right.$$
$$\left. + \left[ \frac{\partial^2 \theta}{\partial \xi^2} + \frac{\partial^2 \theta}{\partial \zeta^2} - \mathcal{G}_\infty(\zeta) \frac{\partial \theta}{\partial \zeta} \right] \frac{\partial \tilde{\delta}_0}{\partial \chi} \right\} ,$$

where

(25,17)
$$\mathcal{G}_\infty(\zeta) \equiv \frac{\gamma}{\hat{B}} \left[ 1 + \left[ \frac{\hat{B}}{\gamma K_{00}} \right]^2 \right] \phi_\infty(\zeta) ,$$

and *secular* terms appears in the solution of equation (25,16) for $\tilde{\delta}_1(\xi, \zeta; \chi)$ which we must remove.

If we take into accont the solution (25,15), for $\tilde{\delta}_0(\xi, \zeta; \chi)$, we obtain, instead of (25,16):

(25,18)
$$\phi_\infty(\zeta) \left[ \frac{1}{K_{00}^2} \frac{\partial^2 \tilde{\delta}_1}{\partial \chi^2} + \tilde{\delta}_1 \right] + i K_{00} \hat{E} \left\{ 2 \left[ \frac{\partial \theta}{\partial \xi} \frac{\partial A}{\partial \xi} + \frac{\partial \theta}{\partial \zeta} \frac{\partial A}{\partial \zeta} \right] \right.$$
$$\left. + \left[ \frac{\partial^2 \theta}{\partial \xi^2} + \frac{\partial^2 \theta}{\partial \zeta^2} - \mathcal{G}_\infty(\zeta) \frac{\partial \theta}{\partial \zeta} \right] A \right\} = 0 ,$$

where
$$\hat{E} = \exp[i(K_{00}\chi - \pi/4)].$$

Accordingly, by elimination of the secular terms, we find that $A(\xi, \zeta)$ is determined by:

(25,19)
$$2 \left[ \frac{\partial \theta}{\partial \xi} \frac{\partial A}{\partial \xi} + \frac{\partial \theta}{\partial \zeta} \frac{\partial A}{\partial \zeta} \right] + \left[ \frac{\partial^2 \theta}{\partial \xi^2} + \frac{\partial^2 \theta}{\partial \zeta^2} - \mathcal{G}_\infty(\zeta) \frac{\partial \theta}{\partial \zeta} \right] A = 0 .$$

We observe that the equation (25,19) is an ordinary differential equation *along characteristic rays* of the equation (25,13), namely:

(25,20)
$$2 \frac{\partial \theta}{\partial \zeta} \frac{d \text{Log} A}{d \zeta} = \mathcal{G}_\infty(\zeta) \frac{\partial \theta}{\partial \zeta} - \left( \frac{\partial^2 \theta}{\partial \xi^2} + \frac{\partial^2 \theta}{\partial \zeta^2} \right) .$$

We must consider those characteristic rays which eventually reach a given point $\mathcal{P}(\xi,\zeta)$, having come from the origin ($\xi=0,\zeta=0$) and having been reflected many times on $\zeta=0$ and $\zeta=1/\hat{B}$ .

Let s be the distance along the ray from the origin to the point $\mathcal{P}$; through a straightforward computation[†], we find as a parametrization of the ray:

$$(25,21) \qquad \xi=s\lambda, \text{ with } s = 2\alpha(n)\int_0^{1/\hat{B}} \frac{dt}{\left[\frac{\phi_\infty(t)}{K_{00}^2} - \lambda^2\right]^{1/2}}$$

$$+ (-1)^n \int_0^\zeta \frac{dt}{\left[\frac{\phi_\infty(t)}{K_{00}^2} - \lambda^2\right]^{1/2}}$$

where n is the number of reflections experienced by the ray before reaching the point $\mathcal{P}$, while $\alpha(n)$ is the smallest integer greather than or equal to n/2. We observe that $\lambda$ is constant all along the ray, even after any number of reflections, and that, near the origin, (25,21) reduces to: $x=r\cos\vartheta$, $z=r\sin\vartheta$, $s=r$, which shows that $\lambda$ is related to the angle between the ray and the horizontal at the origin ($\lambda=\cos\vartheta$).

Let us denote by $\theta_n(\xi,\zeta)$ the value of $\theta$ for the solution which corresponds to the ray which reaches the point $\mathcal{P}$ after n reflections, we obtain:

$$(25,22) \qquad \theta = \theta_n = \xi\lambda + 2\alpha(n)\int_0^{1/\hat{B}} \left[\frac{\phi_\infty(t)}{K_{00}^2} - \lambda^2\right]^{1/2} dt$$

$$+ (-1)^n \int_0^\zeta \left[\frac{\phi_\infty(t)}{K_{00}^2} - \lambda^2\right]^{1/2} dt,$$

and near the origin we have: $\theta \approx \rho$ , where $\rho = M_0 r = (\xi^2 + \zeta^2)^{1/2}$. Hence

$$\theta_n(\xi,\zeta) \equiv \psi_n(\xi,\zeta;\lambda_n(\xi,\zeta)),$$

and

$$\left.\frac{\partial\psi_n}{\partial\lambda}\right|_{\lambda=\lambda_n} = 0, \text{ which is the definition of } \lambda_n.$$

---

[†] See, Zeytounian (1990; Sect.13.2.4), for the further details.

Finally, we obtain the following relation for the function $A_n(\xi,\zeta)^\dagger$:

(25,23)
$$\frac{d}{d\zeta}\left\{\text{Log}\left[A_n^2\left(\frac{\phi_\infty(\zeta)}{K_{00}^2} - \lambda_n^2\right)^{1/2}\Delta_n\right]\right\}_{\lambda_n = \text{const}} = \mathcal{G}_\infty(\zeta) \ ,$$

where

(25,24)
$$\Delta_n = 2\alpha(n)\int_0^{1/\hat{B}} \frac{\phi_\infty(t)/K_{00}^2}{\left[\frac{\phi_\infty(t)}{K_{00}^2} - \lambda_n^2\right]^{1/2}}\,dt + (-1)^n\int_0^\zeta \frac{\phi_\infty(t)/K_{00}^2}{\left[\frac{\phi_\infty(t)}{K_{00}^2} - \lambda_n^2\right]^{1/2}}\,dt.$$

Then, writing

(25,25)
$$\mathcal{F}_\infty(\zeta) = \exp\left\{\frac{1}{2}\int_0^\zeta \mathcal{G}_\infty(t)dt\right\}$$

we obtain the following relation for the coefficient $A_n(\xi,\zeta)$:

(25,26)
$$A_n = \left(\frac{2K_{00}}{\pi}\right)^{1/2} \frac{(1-\lambda^2)^{1/4}\mathcal{F}_\infty(\zeta)G(\lambda)}{\Delta_n^{1/2}[\phi_\infty(\zeta)/K_{00}^2 - \lambda^2]^{1/4}} \ ,$$

where the same function $G(\lambda)$ with $\lambda=\cos\vartheta$ as appears in (24,6). It is then easly checked that (25,26) matches with (24,6) and

$$\tilde{\delta}_0(\xi,\zeta,\chi) = \mathcal{R}eal\left\{\sum_{n\in W} A_n(\xi,\zeta)\exp\{i[K_{00}M_0^{-1}\theta_n(\xi,\zeta) + n\pi - \frac{\pi}{4}]\}\right\} \ ,$$

is the proper *outer* asymptotic approximation, where the $A_n$ are defined by (25,26) and $\theta_n$ by (25,22).

---

† From the solution (25,15) of (25,14) we can construct another which is able to satisfy the boundary condition $\zeta=0$ and $\zeta=1/\hat{B}$. That solution is

$$\tilde{\delta}_0(\xi,\zeta;\chi) = \mathcal{R}eal\left\{\sum_{n\in W} A_n(\xi,\zeta)\exp\{i[K_{00}M_0^{-1}\theta_n(\xi,\zeta) + n\pi - \frac{\pi}{4}]\}\right\} \ ,$$

where $W$ stands for the set of numbers of all possible reflections for rays which reach the given point.

At the origin the matching with (24,6) is to be used, namely:

$$A_n(\rho\cos\vartheta, \rho\sin\vartheta) \sim \left(\frac{2K_{00}}{\pi\rho}\right)^{1/2}\sin\vartheta\ G(\cos\vartheta),$$

when $\rho\equiv M_0 r\rightarrow 0$, for each n.

The main purpose of the present section 25 was twofold. First we wanted to understand the rôle of Kozhevnikov-Miles's model of lee waves in an unbounded two-dimensional atmosphere (see the section 24) and examine the relation between the linear Helmholtz equation (24,3) and the full set of equations describing lee waves in compressible troposphere. The answer to this question has been given in section 24 and it may be stated that, Long's equation (24,1) and Kozhevnikov-Miles's model (for the semicircular mountain) belong to an inner approximation for small mach numbers $M_0$ of incoming flow, a length scale of the mountain being of the order of $M_0$ in comparaison to the height of the troposphere.

The second point that we wanted to investigate was the rôle of the upper boundary condition, which is lost in the Kozhevnikov-Miles's model. The answer to this question is provided by the present section 25 where it is shown that the outward going waves which are generated at infinity in Kozhevnikov-Miles's model, according to the radiation condition of Lyra (1943) and Merbt (1959), excite internal waves in the main part of the troposphere. The internal waves are short waves and may be approximated by multiple scaling. They are found to propagate along rays of equation (25,13), being reflected alternately from the upper and lower boundaries of the troposphere. The important point is that these short waves propagate downstream and that no feedback occurs on the inner flow close to the mountain. We should understand the upper boundary condition as an artificial one, having asymptotically no effects on the inner flow which is the only really interesting one.

## 26 . LONG'S CLASSICAL PROBLEM

In the particular case of the two-dimensional isochoric flow (see, the section 14) in a duct with curvilinear bottom $z=h_p(x)$, and a rigid wall $z=H_0$, one obtains, in the case $U_\infty^2 \rho_\infty =$ constant[†], an equation of Helmholtz for the streamline altitude variation, $\delta=z-z_\infty(\psi_p)$, according to the section 14:

$$(26,1) \qquad \frac{\partial^2 \delta}{\partial x^2} + \frac{\partial^2 \delta}{\partial z^2} + \sigma_0^2 \delta = 0$$

[†] $U_\infty$ being the nonperturbed flow velocity at infinity upstream for $X \rightarrow -\infty$, and $\rho_\infty$ the fluid density, supposed to be a linear function of $Z_\infty$ ; $Z_\infty$ is the streamline altitude at infinity upstream. We note that for the isochoric flow the density $\rho$ is only function of $Z_\infty \equiv Z_\infty(\psi_p)$ and therefore $\rho \equiv \rho_\infty(Z_\infty)$. Here we follow the work of Zeytounian (1969).

with
$$\sigma_0^2 = \frac{g}{U_\infty^2 \rho_\infty} \left| \frac{d\rho_\infty}{dz_\infty} \right| \ ,$$

and the boundary conditions:

(26,2)
$$\begin{cases} z = h_p(x) \ : \ \delta = h_p(x) \ , \\[2mm] z = H_0 \qquad : \ \delta = 0 \ , \\[2mm] x \to -\infty \quad : \ \delta \to 0 \ , \\[2mm] |\delta| \ \text{uniformly bounded for } x \to +\infty \ . \end{cases}$$

In order to build a numerical algorithm leading to the solution of the Long's problem (26,1), (26,2) it is more convenient to use homogeneous boundary conditions; therefore new variables are introduced:

(26,3)
$$\xi = x/H_0 \ , \quad \zeta = \frac{z/H_0 - \eta(\xi)}{1 - \eta(\xi)} \ ,$$

with
$$\eta(\xi) = \frac{h_p(H_0\xi)}{H_0} \ .$$

Beginning with equation (26,1) one obtains, for the function

(26,4)
$$\psi(\xi,\zeta) = -\frac{\delta}{H_0} + \eta + (1-\eta)\zeta$$

a partial derivatives equation with variable coefficients of the form :

(26,5)
$$\frac{\partial^2 \psi}{\partial \xi^2} + 2\frac{\eta}{1-\eta}(\zeta-1)\frac{\partial^2 \psi}{\partial \xi \partial \zeta} + \frac{1+\dot{\eta}^2(\zeta-1)^2}{(1-\eta)^2}\frac{\partial^2 \psi}{\partial \zeta^2}$$
$$+ (\zeta-1)\frac{\ddot{\eta}(1-\eta)+2\dot{\eta}^2}{(1-\eta)^2}\frac{\partial \psi}{\partial \zeta} + K_0^2\psi = K_0^2\left[\eta+(1-\eta)\zeta\right] \ ,$$

where $K_0^2 = H_0^2\sigma_0^2$ and $\dot{\eta} \equiv d\eta/d\xi$ , with which must be associated the boundary conditions

(26,6)
$$\psi(\xi,0) = 0, \ \psi(\xi,1) = 1, \ \psi(-\infty,\zeta) = \zeta;$$

the function $\psi(\xi,\zeta)$ being uniformly bounded in the whole infinite domain

$$\mathbb{R} : \left\{-\infty \le \xi \le \infty, \ 0 \le \zeta \le 1\right\}.$$

It is obvious that the problem (26,5)-(26,6) *can* only be solved numerically and the main difficulty is that we cannot see how to take into account the conditions:

$\psi(-\infty,\zeta)\equiv\zeta$ and $\psi(\xi,\zeta)$ uniformly bounded in the $\mathbb{R}$?

First, we can try to bring the point $\xi=-\infty$ back to a point $\xi=\xi_0$, far enough upstream of the obstacle; but the computations carried out[†] show that the solutions obtained in this manner are only valid if the problem (relative to $\xi$), for equation (26,5) does not admit any increasing exponential components.

Let us remark that the exact solution of problem (26,5)-(26,6) does not admit such components. This is obvious if the linearized problem (26,1), (26,2) is examined with the assumption that the condition $\delta=h_p$ is written, not on the profile $z=h_p$, but approximately for $z=0$.

In this case the solution of problem, uniformly bounded in the domain $\mathbb{R}$, is easily written in the following form[††]

(26,7)
$$\delta(\xi,\zeta) = H_0\left\{(\zeta-1)\eta(\xi) + \sum_{n=1}^{\infty}\delta_n(\xi)\sin(n\pi\zeta)\right\},$$

where:

(26,8a)
$$\delta_n(\xi) = \frac{2}{n\pi\left[K_0^2-n^2\pi^2\right]^{1/2}}\int_{-\infty}^{\xi}\sin\left[\left[K_0^2-n^2\pi^2\right]^{1/2}(\xi-\xi')\right]\left[K_0^2\eta(\xi')\right.$$
$$\left.+ \ddot{\eta}(\xi')\right]d\xi', \text{ when } n<\frac{K_0}{\pi} \ ;$$

(26,8b)
$$\delta_n(\xi) = -\frac{1}{n\pi\left[n^2\pi^2-K_0^2\right]^{1/2}}\int_{-\infty}^{+\infty}\exp\left[-\left[n^2\pi^2-K_0^2\right]^{1/2}|\xi-\xi'|\right]\left[K_0^2\eta(\xi')\right.$$
$$\left.+ \ddot{\eta}(\xi')\right]d\xi', \text{ when } n>\frac{K_0}{\pi} \ .$$

We note that the function $\delta_n(\xi)$ is the solution of the following equation

(26,9)
$$\ddot{\delta}_n + \left[K_0^2-n^2\pi^2\right]\delta_n = \frac{2}{\pi n}\left[K_0^2\eta+\ddot{\eta}\right],$$

† According to Zeytounian (1963 and 1964).

†† See, for instance, Kotschin, Kibel and Rosé (1963).

with the boundary conditions

(26,10)                    $\delta_n(-\infty)=0$ and $|\delta_n|$ uniformly bounded for $-\infty<\xi<+\infty$ .

*PARASITE SOLUTIONS*

Let us again consider the equation (26,9) for $\delta_n(\xi)$; the solutions of the homogeneous equation corresponding to this equation are of the form:

(26,11a)          $A_0\sin\left[\left(K_0^2-n^2\pi^2\right)^{1/2}\xi\right] + B_0\cos\left[\left(K_0^2-n^2\pi^2\right)^{1/2}\xi\right]$

for all $n<K_0/\pi$, and

(26,11b)          $C_0\exp\left[\xi\left(n^2\pi^2-K_0^2\right)^{1/2}\right] + D_0\exp\left[-\xi\left(n^2\pi^2-K_0^2\right)^{1/2}\right]$

for all $n>K_0/\pi$.

When we wrote the solutions (26,8), we have *rejected* the solutions (26,11), for the $n<K_0/\pi$ as well as for the $n>K_0/\pi$; similarly when $\xi\longrightarrow-\infty$, the first of these solutions will not tend toward zero, and the second will not be bounded on the whole infinite straight line $-\infty \leq \xi \leq +\infty$.

Let us now suppose that we want to numerically solve this same problem and, for this, we transform it (relative to $\xi$), into a Cauchy problem; that is, we associate with the equation for $\delta_n(\xi)$, the initial conditions

(26,12)          $\delta_n(\xi_0) = \dot{\delta}_n(\xi_0) = 0,$

$\xi_0$ being a point far enough upstream of the obstacle.

However, from solution (26,8) it can be seen that the exact solution of our problem at the point $\xi=\xi_0$ is different from zero and hence (the equation being linear) the error $\mathcal{E}(\xi,\zeta)$ introduced into the solution through the transfer of the boundary conditions (relative to $\xi$) will be given by the solution of the following problem:

(26,13)    $\begin{cases} \dfrac{\partial^2\mathcal{E}}{\partial\xi^2} + \dfrac{\partial^2\mathcal{E}}{\partial\zeta^2} + K_0^2\mathcal{E} = 0, \\[2mm] \mathcal{E}=0 \text{ for } \zeta=0 \text{ and } \zeta=1, \\[2mm] \mathcal{E}=\alpha_1, \ \dfrac{\partial\mathcal{E}}{\partial\xi}=\alpha_2 \text{ for } \xi=\xi_0, \end{cases}$

where $\alpha_1(\xi_0,0)\equiv\alpha_1(\xi_0,1)=0$ and $\alpha_2(\xi_0,0)\equiv\alpha_2(\xi_0,1)=0$. Hence one can write

$$(26,14) \qquad \alpha_i(\xi_0,\zeta) = \sum_{n=1}^{\infty}\alpha_{in}(\xi_0)\sin(n\pi\zeta), \quad i=1,2,$$

and we look for the solution of the problem (26,13) a form similar to (26,7):

$$(26,15) \qquad \mathscr{E}(\xi,\zeta) = \sum_{n=1}^{\infty}\mathscr{E}_n(\xi)\sin(n\pi\zeta).$$

The $\mathscr{E}_n(\xi)$ satisfying the homogeneous equation

$$\ddot{\mathscr{E}}_n + \left[K_0^2 - n^2\pi^2\right]\mathscr{E}_n = 0$$

with the boundary conditions:

$$\mathscr{E}_n(\xi_0)=\alpha_{1n}(\xi_0) \text{ and } \dot{\mathscr{E}}_n(\xi_0)=\alpha_{2n}(\xi_0).$$

One can see that for $n>K_0/\pi$ the solution of this problem will include parasite increasing compenents of the exponential type which will completely modify the flow downstream of the obstacle, and which can even, in certain cases, *saturate* the computer memory.

The calculations that we have carried out (Zeytounian (1964)) as well as those of Pekelis (1966) concerning the numerical solution of the nonlinear problem (26,5), (26,6) show that the situation that has been pointed out for the linear problem is still valid in the nonlinear case.

Let us now see if the problem can be transformed into a boundary-value problem (still relative $\xi$) if we apply the conditions:

$$(26,16) \qquad \delta_n(\xi_0) = 0 \text{ and } \delta_n(\xi_N)=0,$$

to the equation (26,9); $\xi_N$ being a point for enough downstream of the obstacle. Reasoning in a manner similar to that used before, one can see that, in this case, and for $n<K_0/\pi$ the solutions are not zero at $\xi=\xi_N$; i.e., we shall artificially add to the solution of the equation the solution of the corresponding homogeneous equation with the boundary conditions: $\delta_n(\xi_0)=0$ but $\delta_n(\xi_N)=\beta_{2n}$ which is in fact, a linear combination of sines and cosines of type (26,11a), whose amplitude is of the same order as that of the exact solution. The solution thus obtained, even though it stays uniformly bounded at all $\xi$ points, will *not tend* toward zero for $\xi \rightarrow \infty$ and therefore will completely perturbed the flow being studied.

*THE NONLINEAR PROBLEM*

In a numerical computation of equation (26,5) satisfying the boundary conditions (26,6), it is convenient to effect first a discretization relative to the $\zeta$ variable, which gives, instead of (26,5) a system of ordinary differential equations in $\xi$. For that we rewrite the equation (26,5) in a *divergent* form:

$$(26,17) \quad \frac{\partial}{\partial \zeta}\left\{\frac{1+(1-\zeta)^2\dot{\eta}^2}{1-\eta}\frac{\partial\psi}{\partial\zeta} + (\zeta-1)\ \dot{\eta}\ \frac{\partial\psi}{\partial\xi}\right\} + \frac{\partial}{\partial\xi}\left\{(1-\eta)\frac{\partial\psi}{\partial\xi} + (\zeta-1)\ \dot{\eta}\ \frac{\partial\psi}{\partial\zeta}\right\}$$

$$= (1-\eta)K_0^2\left[(1-\zeta)\eta - \psi + \zeta\right].$$

To resolve equation (26,17) with the boundary conditions:

$$(26,18) \quad \begin{cases} \psi(\xi,0)=0,\ \ \psi(\xi,1)=1, \\[1mm] \lim_{\xi\to-\infty}\psi(\xi,\eta)=0, \\[1mm] \lim_{\xi\to+\infty}\psi(\xi,\eta)<\infty, \end{cases}$$

we are going to make use of the method of "*integral relations*" according to the vertical coordinate $\zeta$. This method, which is due to Dorodnitsyn (1958), is applied to the resolution of differential equations of the type:

$$(26,19) \quad \frac{\partial\mathcal{L}}{\partial\zeta} = \mu(\zeta).$$

Integrating equation (26,19) with respect to $\zeta$ from 0 to $\zeta_k=\frac{k}{n}$ (k=1,2,...,n) gives

$$\mathcal{L}_k - \mathcal{L}_0 = \int_0^{\zeta_k}\mu(\zeta)d\zeta \equiv \bar{\bar{\mu}}_k, \quad (k=1,2,\ldots,n).$$

We represent $\mu(\zeta)$ in the form of a polynomial taking the values $\mu_k = \mu(\zeta_k)$:

$$(26,20) \quad \mu(\zeta) = \mu_0 + \sum_{s=1}^{n}A_s\zeta^s;$$

in this case we may write that

$$(26,21) \quad \bar{\bar{\mu}}_k = \frac{k}{n}\mu_0 + \sum_{s=1}^{n}\frac{A_s}{s+1}\left(\frac{k}{n}\right)^{s+1}, \quad k=1,2,\ldots,n.$$

In writing the relation (26,21) for all the k values we obtain a system of equation to determine the n parameters

$$A_1,\ A_2,\ \ldots,\ A_n$$

as a function of $\mu_0$ and $\bar{\bar{\mu}}_k$. Finally by putting the expressions for these $A_s$ into (26,20) we will obtain the relation

$$(26,22) \qquad \mu_k = P^k_0 \mu_0 + \sum_{s=1}^{n} P^k_s \bar{\bar{\mu}}_s \ , \quad k=1,2,\ldots,n.$$

The coefficients $P^k_0$ and $P^k_s$, being function of n, are determined once and for all. For example, the calculations give (see, Zeytounian (1968)):

$$n=1 \ : \ P^1_0 = -1 \text{ and } P^1_1 = 2 \ ;$$

$$n=2 \ : \ P^1_0 = -\frac{1}{2}, \ P^1_1 = 2, \ P^1_2 = \frac{1}{2} \ ;$$

$$P^2_0 = 1, \ P^2_1 = -8, \ P^2_2 = 4 \ ;$$

$$n=3 \ : \ P^1_0 = -\frac{1}{3}, \ P^1_1 = \frac{3}{2}, \ P^1_2 = \frac{3}{2}, \ P^1_3 = -\frac{1}{6} \ ;$$

$$P^2_0 = \frac{1}{3}, \ P^2_1 = -6 \ P^2_2 = 3, \ P^2_3 = \frac{2}{3} \ ;$$

$$P^3_0 = -1, \ P^3_1 = \frac{27}{2}, \ P^3_2 = -\frac{27}{3}, \ P^3_3 = \frac{13}{2} \ ;$$

. . . . . . . . . . . . . . . . . . . . . . . . . . . . . . . . . .

For n=m we will have m(m+1) coefficients $P^k_s$ with k=1,2,...,m and s=1,2,...,m. Using (26,22) and from the fact that $\bar{\bar{\mu}}_k = \mathcal{L}_k - \mathcal{L}_0$ we will obtain, in place of equation (26,19), a system of algebraic equations of the form:

$$(26,23) \qquad \mu_k = P^k_0 \mu_0 - \mathcal{L}_0 \sum_{s=1}^{n} P^k_s + \sum_{s=1}^{n} P^k_s \mathcal{L}_s, \quad \text{with } k=1,2,\ldots,n.$$

Let us $\psi_k = \psi(\xi, \frac{k}{n})$ be the value of the stream function at the level $\xi_k = k/n$, with the purpose of constructing, in place of the partial derivative equation in the divergent form (26,17), a *system of ordinary differential equations to determine $\psi_k$, with* k=1,2,...,n.

To do this we apply the relation (26,23), by writing equation (26,17) in the form of a system of two equations:

$$(26,24) \qquad \frac{\partial \psi}{\partial \zeta} = (1-\eta)\omega \ ;$$

$$(26,25) \qquad \frac{\partial}{\partial \zeta}\left\{\left[1+(1-\zeta)^2\dot{\eta}^2\right]\omega + (\zeta-1)\left[2\dot{\eta}\,\frac{\partial \psi}{\partial \xi} + \ddot{\eta}\,\psi\right]\right\} = -(1-\eta)\frac{\partial^2 \psi}{\partial \xi^2}$$

$$+ 2\dot{\eta}\,\frac{\partial \psi}{\partial \xi} + \ddot{\eta}\,\psi + (1-\eta)K^2_0\left[(1-\zeta)\eta - \psi + \zeta\right];$$

we obtain then, taking into account that $\psi_0 \equiv 0$ and $\psi_n \equiv 1$:

(26,26)
$$(1-\eta)\omega_k = P_0^k(1-\eta)\omega_0 + \sum_{s=1}^{n-1} P_s^k \psi_k + P_n^k \; ;$$

$$-(1-\eta)\frac{d^2\psi_k}{d\xi^2} + 2\dot\eta\frac{d\psi_k}{d\xi} + \ddot\eta\psi_k + (1-\eta)K_0^2\left[(1-\frac{k}{n})\eta - \psi_k + \frac{k}{n}\right]$$

$$= P_0^k(1-\eta)\eta K_0^2 - (1+\dot\eta^2)\omega_0 \sum_{s=1}^{n} P_s^k + \sum_{s=1}^{n} P_s^k\left[1+\left(1-\frac{s}{n}\right)^2\dot\eta^2\right]\omega_s$$

(26,27)
$$+ \sum_{s=1}^{n-1} P_s^k(\frac{s}{n} - 1)\left[2\dot\eta\frac{d\psi_s}{d\xi} + \ddot\eta\psi_s\right],$$

$$k=1,2,\ldots,n.$$

With the aid of equation (26,26) we can eliminate from equation (26,27) all the $\omega_k$ (k=1,2,...,n); we then obtain one single equation comprising $\psi_k$ and $\omega_0$) namely:

(26,28)
$$(1-\eta)\frac{d^2\psi_k}{d\xi^2} - 2\dot\eta\frac{d\psi_k}{d\xi} - \ddot\eta\psi_k - (1-\eta)K_0^2\left[(1-\frac{k}{n})\eta - \psi_k + \frac{k}{n}\right] + D_n^k$$

$$+ \sum_{s=1}^{n-1}\left\{E_s^k\frac{d\psi_s}{d\xi} + F_s^k\psi_s\right\} = G_0^k\omega_0 \; ,$$

(26,29)
$$\begin{cases}
D_n^k = \frac{1}{1-\eta}\sum_{s=1}^{n} P_s^k P_n^s\left[1+\left(1-\frac{s}{n}\right)^2\dot\eta^2\right] + K_0^2 P_0^k(1-\eta)\eta; \\[2ex]
E_s^k = 2P_s^k\dot\eta(\frac{s}{n} - 1) \; ; \\[2ex]
F_s^k = \frac{1}{1-\eta}\sum_{m=1}^{n} P_m^k P_s^m\left[1+\left(1-\frac{m}{n}\right)^2\dot\eta^2\right] + P_s^k\ddot\eta(\frac{s}{n} - 1) \; ; \\[2ex]
G_0^k = (1+\dot\eta^2)\sum_{s=1}^{n} P_s^k - \sum_{s=1}^{n} P_s^k P_0^s\left[1+\left(1-\frac{s}{n}\right)^2\dot\eta^2\right] \; , \\[2ex]
k=1,2,\ldots,n.
\end{cases}$$

To eliminate $\omega_0$ from (26,28), we remark that for k=n when $\psi_k=1$ we obtain

(26,30)
$$\omega_0 = -\frac{\ddot{\eta}}{G_0^n} + \frac{D_n^n}{G_0^n} + \sum_{s=1}^{n-1}\left\{\frac{E_s^n}{G_0^n}\frac{d\psi_s}{d\xi} + \frac{F_s^n}{G_0^n}\psi_s\right\}.$$

From (26,28) and (26,30) we can eliminate $\omega_0$, which gives us a single equation for $\psi_k$ with k=1,2,...,n, strictly speaking we obtain a differential system of the following type[†]:

(26,31)
$$\frac{d^2\psi_k}{d\xi^2} - \frac{2\dot{\eta}}{1-\eta}\frac{d\psi_k}{d\xi} - \frac{\ddot{\eta}}{1-\eta}\psi_k - K_0^2\left[(1-\frac{k}{n})\eta - \psi_k + \frac{k}{n}\right]$$

$$+ \sum_{s=1}^{n-1}\left\{\mathcal{E}_{s,n}^k\frac{d\psi_s}{d\xi} + \mu_{s,n}^k\psi_s\right\} = \lambda_n^k.$$

In this last equation (26,31) the coefficients $\mathcal{E}_{s,n}^k$, $\mu_{s,n}^k$ and $\lambda_n^k$ are of the form:

(26,32)
$$\begin{cases} \mathcal{E}_{s,n}^k = \frac{1}{1-\eta}\left\{E_s^k - \frac{G_0^k}{G_0^n}E_s^n\right\}; \\[2em] \mu_{s,n}^k = \frac{1}{1-\eta}\left\{F_s^k - \frac{G_0^k}{G_0^n}F_s^n\right\}; \\[2em] \lambda_n^k = \frac{1}{1-\eta}\left\{\frac{G_0^k}{G_0^n}D_n^n - D_n^k - \ddot{\eta}\frac{G_0^k}{G_0^n}\right\}. \end{cases}$$

## NONLINEAR MODEL WITH ONE INTERMEDIARY LEVEL

If we consider the more simple case of a model with only a *single* intermediary level $\zeta_1=1/2$, between $\zeta_0=0$ and $\zeta_2=1$ (then n≡2 and k admits only one value, k=1, in the system (26,31)), for the system (26,31) we shall obtain only *one* equation for the function

$$\psi_1(\xi) \equiv \psi(\xi,\zeta=\frac{1}{2}).$$

---

[†] According to Zeytounian (1964,1969).

This equation may be written in the form:

$$(26,33) \qquad \frac{d^2\psi_1}{d\xi^2} + p(\xi)\frac{d\psi_1}{d\xi} + q(\xi)\psi_1 = r(\xi),$$

with:

$$(26,34)\quad
\begin{cases}
p(\xi) = \dfrac{2\dot\eta}{\alpha(\xi)}\left[\dfrac{1}{12}\dot\eta^2 - 1\right] \ ; \\[3mm]
q(\xi) = \dfrac{1}{\alpha(\xi)}\left\{\ddot\eta\left[\dfrac{1}{12}\dot\eta^2 - 1\right] + K_0^2(1-\eta)\left[1+\dfrac{5}{12}\dot\eta^2\right]\right. \\[3mm]
\qquad\qquad \left. - \dfrac{12}{1-\eta}\left[1+\dfrac{11}{12}\dot\eta^2+\dfrac{1}{24}\dot\eta^4\right]\right\} \ ; \\[3mm]
r(\xi) = \dfrac{1}{\alpha(\xi)}\left\{\dfrac{1}{4}\dddot\eta(1+11\dot\eta^2) - \dfrac{3}{1-\eta}\left[2+\dfrac{4}{3}\dot\eta^2-\dfrac{1}{24}\dot\eta^4\right]\right. \\[3mm]
\qquad\qquad \left. + K_0^2(1-\eta)\left[\dfrac{1}{2}+\dfrac{3}{4}\eta+\dfrac{5}{24}\dot\eta^2+\dfrac{3}{16}\eta\dot\eta^2\right]\right\} \ ,
\end{cases}$$

where
$$\alpha(\xi) = (1-\eta)\left[1 + \frac{5}{12}\dot\eta^2\right] .$$

Equation (26,33) must be solved with the boundary conditions:

$$(26,35)\quad
\begin{cases}
\text{for } \xi\longrightarrow-\infty:\ \psi_1\rightarrow\dfrac{1}{2};\ \psi_1 - \text{ must be uniformly bounded in the} \\[2mm]
\text{whole internal } \xi\in[-\infty,+\infty].
\end{cases}$$

The calculations carried out have shown that, when the coefficient $q(\xi)\le0$ (is negative), at least for one value of $\xi$, in the internal $[-\infty,+\infty]$, then the process by which the conditions (26,35) are transformed into a condition of Cauchy:

$$(26,36) \qquad \psi_1(\xi_0) = \frac{1}{2} \ , \qquad \frac{d\psi_1}{d\xi}\bigg|_{\xi=\xi_0} = 0 \ ,$$

with $\xi=\xi_0$ being a point sufficiently distant upstream from the obstacle, *is not correct*. This procedure (for a model with a single level) remains correct if the parameter $K_0^2$ satisfies the relation

$$(26,37) \qquad K_0^2 > \min_{-\infty<\xi<+\infty} \left\{\frac{1}{(1-\eta)\left[1+\dfrac{5}{12}\dot\eta^2\right]}\left[\dfrac{12}{1-\eta}\left(1+\dfrac{11}{12}\dot\eta^2+\dfrac{1}{24}\dot\eta^4\right) - \ddot\eta\left[\dfrac{1}{12}\dot\eta^2+1\right]\right]\right\} .$$

Once $\psi_1(\xi)$ is calculated by finite differences method, then $\psi_m(\xi)$ will be calculated approximately at any level $\zeta_m = \frac{m}{n}$ (m=1,2,.. ,n-1) by means of the following polynomial of the second degree

(26,38)          $\psi_m(\xi) = [\ 4\psi_1(\xi) - 1\ ]\zeta_m - [\ 4\psi_1(\xi) - 2\ ]\zeta_m^2\ .$

Figure 10-14 illustrate the method presented above.
Figure 10 shows the flow above and downstream of a typical isolated obstacle, calculated for $K_0^2 = 25$ .

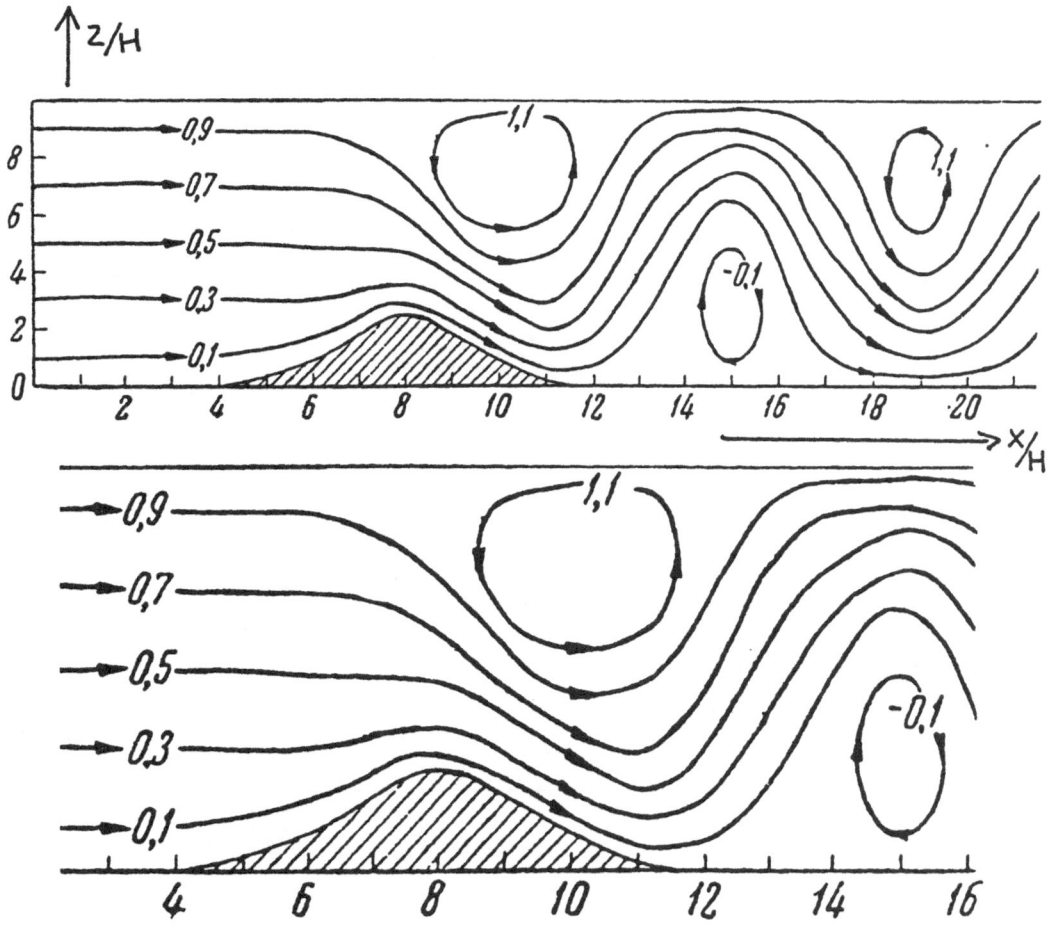

$\mathcal{F}ig.$ 10: $\mathcal{F}low$ $above$ $and$ $downstream$ $of$ $an$ $isolated$ $obstacle$ $for$ $K_0^2 = 25$.

We obtain as seen downstream of the obstacle, a regime of rotors during which the eddy vortices (closed streamlines) are periodically liberated by the crest of the obstacle, leading to a configuration analogous to that of the Von Karman vortex.

Figure 11 represents the flow downstream of an asymmetric obstacle. We have put in our numerical result (case b) along with that of Long (case a) obtained analytically in investigation. We see in this example that the model with a single intermediary level already gives a satisfactory approximation to the exact solution of Long. This example clearly shows the influence of the curvature of an obstacle upstream on the formation of the regime of rotors downstream of the obstacle (see Fig. 10 for comparaison).

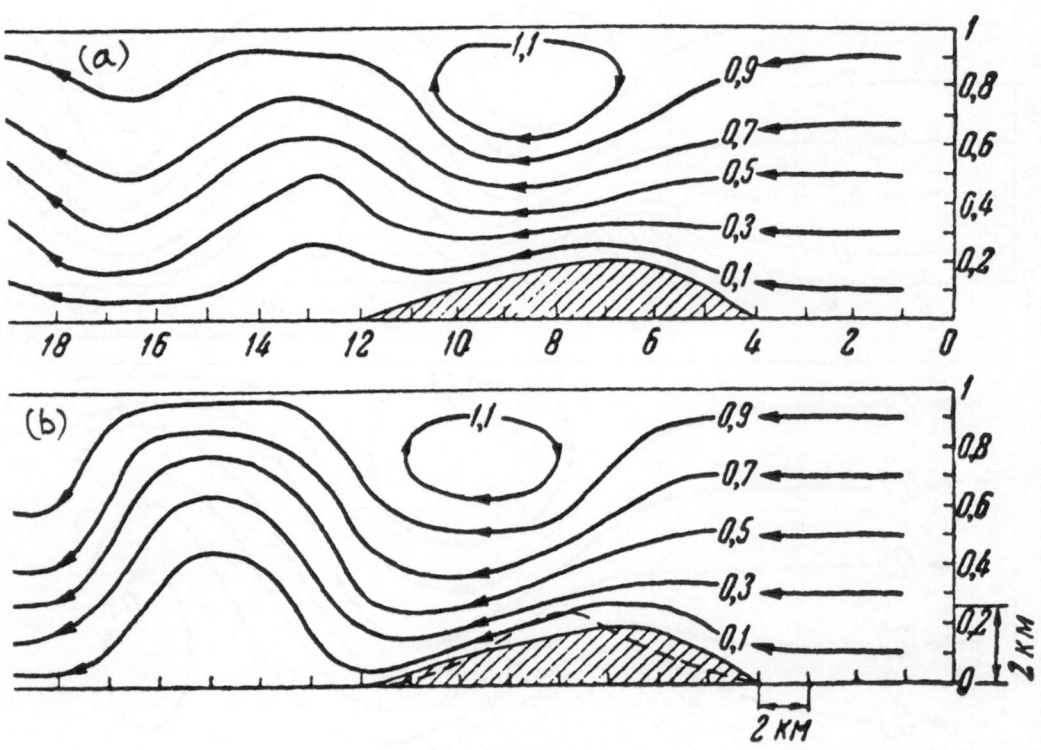

Fig. 11: Flows downstream of an obstacle:
(a)- from Long's investigation ($K_0^2$=24);
(b)- from the model for this investigation ($K_0^2$=24).

In Figure.12 we have represented the calculation (a),(b) and (d) of the relief waves above Sierra-Nevada for different values of $K_0^2$ and also the flows (c) and (e), constructed from known observations (Long (1959)), for comparaison with real flows. We point out that the values of $K_0^2$ in the corresponding flows are different; it is partly due to differences in the speed profile at infinity upstream and also to the fact that our theoretical models define the flow by a plane solid surface.

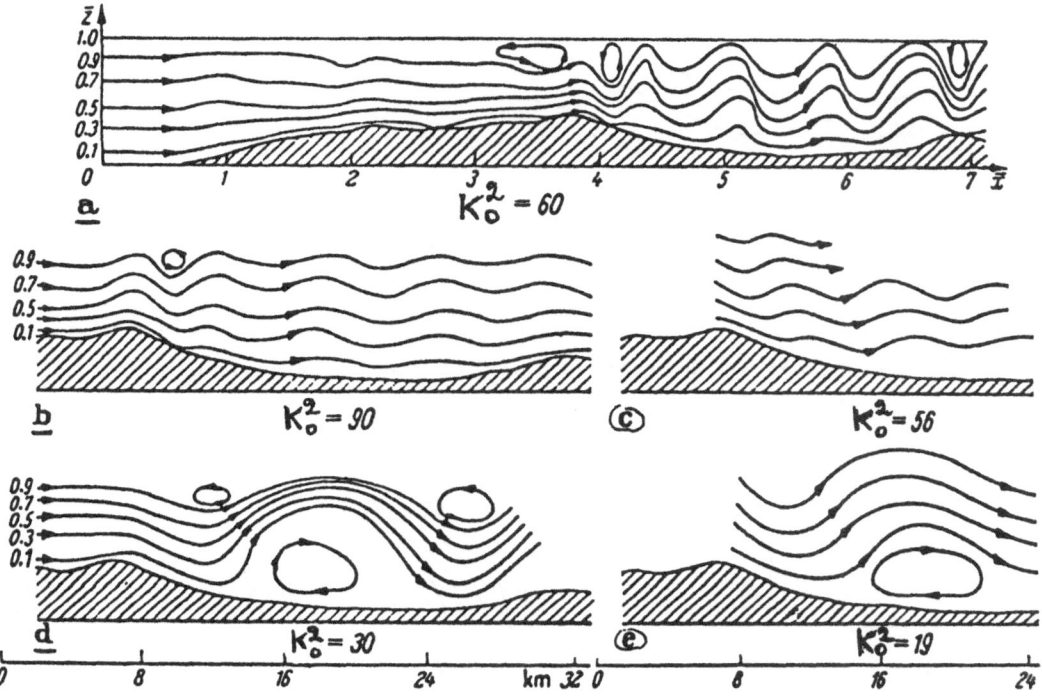

Fig. 12: Flows above Sierra-Nevada:
(a)- for $K_0^2=60$ , (b)- $K_0^2=90$ ,
(c)- from observation when $K_0^2=56$ ,
(d)- for $K_0^2=30$ , (e)- from observation
when $K_0^2=19$.

148

Figures 13 and 14 illustrate the method of Long (see below the remarks concerning this method). Long could have obtained the flow of Fig.14 if he had been able to resolve the problem correctly by applying the *exact* slip condition $\psi=0$ to the profile: z=0.175(1+cosπx); since he did not take this condition into account but wrote a linearized condition he obtained the flow of Fig.13 which is the flow above an asymmetric obstacle and has nothing in common with the flow sought in Fig.14

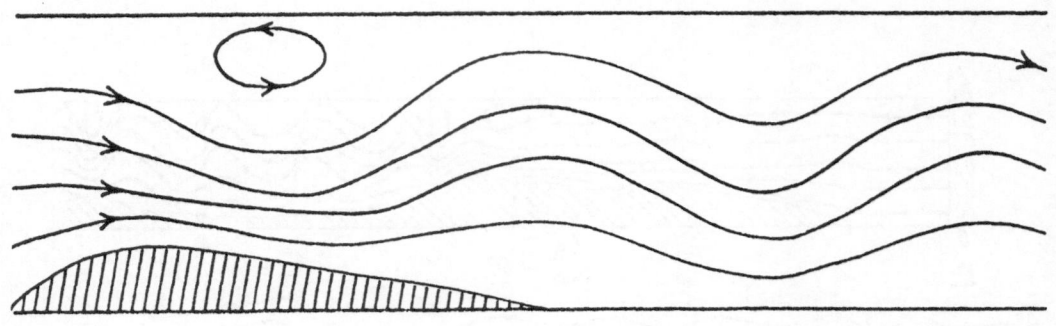

Fig. 13: Flow for an asymmetric obstacle with $K_0^2$=25. This solution of Long corresponds to an exact solution of the nonlinear problem for the profile in Fig. 13 (unknown apriori!).

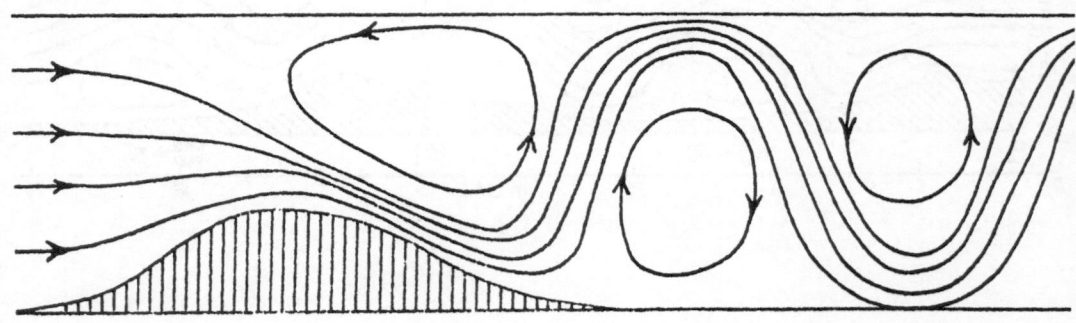

Fig. 14: Flow for a symmetric obstacle z=0,175(1+cosπx) for $K_0^2$= 25 (nonlinear solution).

Long resolves the linear equation (26,1) with a linear slip condition: $\delta(x,0)=h_p(x)$ instead of the nonlinear condition (26,2). We then obtain an analytical solution of the problem; but this solution is, in fact, only valid for values of $h_p(x)$ which are intrinsically small. However, Long takes for $h_p(x)$ a profile with a height of relative importance: $h_p(x)=0,175(1+\cos\pi x)$ and calculates the flow for $K_0^2=25$, which gives him the flow in Fig.13, where the obstacle is represented by the stream line $\psi_p=0$. The real flow that should be obtained for the obstacle $h_p(x)=0,175(1+\cos\pi x)$ with $K_0^2=25$ by taking account of condition (26,2): $\delta(x,h_p(x))=h_p(x)$ is represented in Fig.14.

The method of Long enables the nonlinear flows to be uniquely found for unknown obstacles apriori-for example, in the case of Fig.13, Long obtain a flow (nonlinear) for an asymmetric profil whose equation is $h^*(x)=?$; the condition $\psi_p = 0$ being uniquely fulfilled on this unknown profile. As for the flow for the symmetric profile of Fig.14, Long resolved nothing. It would be necessary for him to choose a certain profile such that, using his calculation, he had $\psi_p=0$ for $z=0.175\ (1+\cos\pi x)$ which appeared troublesome!!

## 27 . MODELS FOR LEE WAVES THROUGHOUT THE TROPOSPHERE

Let us consider the case of lee waves involving the entire thickness of the troposphere: $H_0 \approx RT_\infty(0)/g$, and in this case Bo $\approx$ 1.

It can be assumed for the troposphere that: $- \dfrac{dT_\infty}{dz}=\Gamma_\infty^0$ , with $\Gamma_\infty^0$=constant, i.e.,

$T_\infty(z_\infty) = 1 - \dfrac{R}{g}\Gamma_\infty^0 z_\infty$ (with the dimensionless variables ).

In this case of constant velocity upstream (for unperturbed mean flow) and linear temperature profile, we have that, in the equation (23,13), the term proportional to $\delta$

$$\frac{Bo^2}{\gamma M_0^2}\frac{1}{T_\infty(z_\infty)}\left[\frac{\gamma-1}{\gamma}+\frac{dT_\infty}{dz_\infty}\right](1+\omega)^2\delta \equiv (1-\mu_0 z+\mu_0\sigma\ \delta)^{-1}(1+\omega)^2\chi_{00}^2\delta\ ,$$

where

(27,1) $$\chi_{00}^2= \frac{Bo}{\gamma M_0^2}\left[\frac{\gamma-1}{\gamma}Bo-\mu_0\right]\ ,\quad \mu_0= \frac{H_0}{T_\infty(0)/\Gamma_\infty^0}\ .$$

As Bo $\approx$ 1 we assume that:

(27,2)       $$\frac{\frac{\gamma-1}{\gamma}Bo-\mu_0}{\gamma M_0^2} = O(1) \implies \chi_{00}^2 = O(1) \ , \quad \text{when } M_0 \to 0.$$

In this case we obtain that

(27,3)       $$\mu_0 \approx \frac{\gamma-1}{\gamma} Bo \ , \quad \text{when } M_0 \to 0.$$

*NONLINEAR MODEL*

We assume here that $\sigma = O(1)$ and in this case with (27,2) and (27,3), we find the asymptotic solution of problem (23,12)-(23,14) in the form:

(27,4)       $$\delta = \delta_0 + o(M_0) \ , \quad \omega = \omega_0 + o(M_0).$$

We then obtain, from (23,12),

$$(1+\omega_0)^{\gamma-1} = 1 - \frac{\frac{\gamma-1}{\gamma}Bo\sigma\delta_0}{1-\frac{\gamma-1}{\gamma}Bo(z-\sigma\delta_0)}$$

or

(27,5)       $$\omega_0 = \left\{ \frac{1-\frac{\gamma-1}{\gamma}Boz}{1-\frac{\gamma-1}{\gamma}Bo(z-\sigma\delta_0)} \right\}^{\frac{1}{\gamma-1}} \ ,$$

where $\delta_0$ satisfies the following limiting equation[†]

(27,6)       $$\left[1-\frac{\gamma-1}{\gamma}Bo(z-\sigma\delta_0)\right]\left\{\frac{\partial^2\delta_0}{\partial x^2} + \frac{\partial^2\delta_0}{\partial z^2} - \frac{1}{1+\omega_0}\left[\frac{\partial\delta_0}{\partial x}\frac{\partial\omega_0}{\partial x} + \frac{\partial\delta_0}{\partial z}\frac{\partial\omega_0}{\partial z}\right.\right.$$

$$\left.\left. - \frac{1}{\sigma}\frac{\partial\omega_0}{\partial z}\right]\right\} + (1+\omega_0)^2 BoA_{00}\delta_0 = 0 \ ,$$

where       $$A_{00} \equiv (\frac{\gamma-1}{\gamma} Bo - \mu_0)/\gamma M_0^2 \ .$$

Combining (27,5) and (27,6), we obtain a single equation for the single function $\delta_0$, which is very complex and nonlinear. The boundary conditions for $\delta_0$ have the form (see (23,14)):

---

[†] From the exact equation (23,13), take into account the relations (27,2) and (27,3).

$$\begin{cases} \delta_0(x, \sigma h_p(\lambda x)) = h_p(\lambda x), \text{ when } |x| < 1; \\ \\ \delta_0(-\infty, z_\infty) = 0, \quad \delta_0(x, 1/Bo) = 0; \\ \\ \lim_{x \to +\infty} \left( \left| \dfrac{\partial \delta_0}{\partial x} \right| + \left| \dfrac{\partial \delta_0}{\partial z} \right| \right) < \infty. \end{cases}$$

(27,7)

The nonlinear problem (27,5)-(27,7) describes the mesoscale steady two-dimensional perturbations throughout the troposphere outside of the boundary layers above a locally curved surface of arbitrary form and elevation.

*LINEAR MODEL*

Equation (27,6) is linearized on the assumption that the obstacle height is small:

(27,8)   $\sigma \ll 1 \implies \sigma = \hat{\sigma} M_0^2 \implies h_0 \simeq U_0^2 / \gamma g$ ,

since Bo $\approx$ 1.

When $\sigma \to 0$, we obtain the following linear problem, in place of (27,5)-(27,7),

(27,9)

$$\begin{cases} \dfrac{\partial^2 \varphi_0}{\partial x^2} + \dfrac{\partial^2 \varphi_0}{\partial z^2} + \mathcal{D}(Boz)\varphi_0 = 0, \\ \\ \varphi_0(x,0) = h_p(\lambda x), \quad |x| < 1; \\ \\ \varphi_0(x, 1/Bo) = 0; \\ \\ \varphi_0(-\infty, z_\infty) = 0; \\ \\ \lim_{x \to +\infty} \left( \left| \dfrac{\partial \varphi_0}{\partial x} \right| + \left| \dfrac{\partial \varphi_0}{\partial z} \right| \right) = 0, \end{cases}$$

where

(27,10)

$$\begin{cases} \varphi_0 = \left\{ 1 - \dfrac{\gamma-1}{\gamma} Boz \right\}^{\frac{1}{2(\gamma-1)}} \delta_0; \\ \\ \mathcal{D}(Boz) = A_{00} Bo \left[ 1 - \dfrac{\gamma-1}{\gamma} Boz \right]^{-1} - Bo^2 \dfrac{2\gamma-1}{4\gamma^2} \left[ 1 - \dfrac{\gamma-1}{\gamma} Boz \right]^{-2}. \end{cases}$$

Under ordinary meteorological conditions, $\mathcal{D}(Boz) > 0$. We note that the linear problem (27,9) duplicates that of Dorodnitsyn (1950).

The solution of (27,9) is of the following form:

$$(27,11) \qquad \varphi_0(x,z) = \sum_{n=1}^{\infty} \left[\frac{d\chi_n}{dz}\right]_{z=0} \psi_n(x)\chi_n(z)$$

where

$$(27,12a) \qquad \psi_n(x) = -\frac{1}{2\mu_n}\int_{-\infty}^{+\infty} \exp[-\mu_n|x-x'|]\, h_p(\lambda x')dx', \qquad \text{when } \mu_n^2 > 0$$

and

$$(27,12b) \qquad \psi_n(x) = \frac{1}{2\gamma_n}\int_{-\infty}^{x} \sin[\gamma_n(x-x')]h_p(\lambda x')dx', \qquad \text{when } \mu_n^2 = -\gamma_n^2 < 0.$$

The eigenfunctions $\chi_n(z)$ and eigenvalues $\mu_n^2$ satisfy the following Sturm-Liouville problem:

$$(27,13) \qquad \begin{cases} \dfrac{d^2\chi_n}{dz^2} + \left[\mu_n^2 + \mathcal{D}(Boz)\right]\chi_n = 0, \\[2ex] \chi_n(0) = \chi_n\left[\dfrac{1}{Bo}\right] = 0. \end{cases}$$

When $Bo \to 0$, both the height z and the slow height $\zeta = Boz$ appear in this problem (27,9). Therefore the perturbed solution is in fact singular (when $Bo \to 0$), since it incorporates two different length scales for height.

It would be interesting analyze the behaviour of nonlinear problem (27,5)-(27,7) in detail as $Bo \to 0$ and $A_{00} \to \infty$ so that
$$A_{00}Bo = K_{00}^2 = O(1).$$
This would enable us better to understand the essential of the classical Boussinesq approximation.

The numerical solution of the nonlinear problem (27,5)-(27,7) should bring out interesting nonlinear effects for the lee waves that have thus far been developed only partially by Pekelis (1976).

*REFERENCES TO WORKS CITED IN THE TEXT*

BOIS, P.A.(1984) _ Geophys.Astrophys.Fluid Dynamics, 29, 267-303.

DORODNITSYN, A.A.(1958) _ in Proceed.of the Third All_Union Math.Congr.,Vol.3, 447, (Akad.Nauk SSSR, Moscow).

DORODNITSYN, A.A. (1950) _ Trudy Ts.I.P, 21 (48), 3-25 (in Russian).

GUIRAUD, J.P. and ZEYTOUNIAN, R.Kh. (1979) _ Geophys.Astrophys. Fluid Dynamics, 12, 61.

KOTSCHIN, N.E., KIBEL, I.A. and ROSE, N.V. (1963) _ Theoretical Hydromechanics Vol.1, Moscow (in Russian).

KOZHEVNIKOV, V.N. (1963) _ Izvestiya AN SSSR, Seriya Geofizicheskaya, n°7, 1108-1116.

KOZHEVNIKOV, V.N. (1968) _ Atmospheric and Oceanic physics, Vol.4, n°1, 16-27.

LONG, R.R. (1953) _ Tellus, 5, 42-57.

LONG, R.R. (1959) _ in the "*Rossby Memorial Volume*"; Rockefeller Institute Press, New-York.

LYRA, G. (1943) _ Z.Angew.Math.Mech., 23,1-28.

MAGNUS, W.(1949) _ Seminar der Universität Hamburg. Vol.16, n°6, 1/2 may.

MC.INTYRE, M.E. (1972) _ J.Fluid Mech.,52, 209-43.

MERBT, H. (1959) _ Beitr.Phys.Atmos., 31, 152-161.

MILES, J.W.[†] (1968) _ J.Fluid Mech., 33, part4, 803-814.

PEKELIS, E. (1966) _ Bull.Acad.Sci.USSR.Atmos.Ocean Phys., 2, 689.

PEKELIS, E. (1976) _ Bull.Acad.Sci.USSR, Atmosph. and Oceanic Physics, 12, n°5, 470-477.

THOMAS,N.H., and STEVENSON, T.N. (1972) _ J.Fluid Mech. 54, 495-506.

WILCOX, C.H. (1959) _ Arch.Rat.Mech.3, 133.

YANOWITCH, N. (1967) _ J.Fluid Mech. 29, 203.

ZEYTOUNIAN, R.Kh.(1963) _ Trudy of the Centre of Meteorological Calculations, Moscow, Note 1, 72.

ZEYTOUNIAN, R.Kh. (1964)_ Bull.Acad.of.Sci.USSR, geophysics series, sept. 865.

ZEYTOUNIAN, R.Kh. (1968) _ La Recherche Aerospatiale, N°127, 59-61.

ZEYTOUNIAN, R.Kh. (1969) _ The Physics of Fluids, Supplement II, Vol.12, n°12, part II, 46-50.

---

[†] See also: MILES,J.W.(1969) _ *Waves and wave drag in stratified flows.*
Proced. twelfth Int. Congr. Appl. Mech., Stanford. Eds.M.Hetenyi and
W.G.Vincenti, Berlin, Springer-Verlag, pp.50-76.

ZEYTOUNIAN, R.Kh. (1974) _ *Notes sur les Ecoulements Rotationnels de Fluides Parfaits*. Lecture Notes in Physics, Vol.27, Springer-Verlag, Heidelberg.

ZEYTOUNIAN, R.Kh. (1979) _ Izvestiya of Acad.of Sci.USSR, Atmospheric and Oceanic physics, Vol.15, n°5, 498-507.

ZEYTOUNIAN, R.Kh. (1990) _ *Asymptotic modeling of Atmospheric Flows*. Springer-Verlag, Heidelberg; Chapter 13.

# CHAPTER VII

# BOUNDARY LAYER
# PROBLEMS

## 28 . THE EKMAN LAYER

We start from the so-called hydrostatic equations for the *tangent* viscous and non-adiabatic atmospheric motions[†]. These equations are the non adiabatic, viscous, equivalent of the equations (9,1) written at the section 9, using the $(x,y,p)$ independent variables, namely:

(28,1)
$$Ki\left\{\frac{\partial\vec{v}}{\partial t} + \frac{1}{S}\left[\vec{v}.\vec{D}\ \vec{v} + \omega\frac{\partial\vec{v}}{\partial p}\right]\right\} + \left[1 + \frac{\beta}{S}\ Kiy\right]\left[\vec{k}\wedge\vec{v}\right] + \frac{\lambda_0 Bo}{Ki}\ \vec{D}\mathcal{H}$$

$$= Bo^2 x_0 Ki^2\ \frac{\partial}{\partial p}\left[\rho\ \frac{\partial\vec{v}}{\partial p}\right]\ ;$$

(28,2)
$$\vec{D}.\vec{v} + \frac{\partial\omega}{\partial p} = 0;$$

(28,3)
$$T = -Bo\ p\ \frac{\partial\mathcal{H}}{\partial p};$$

(28,4)
$$Ki\left\{\frac{\partial T}{\partial t} + \frac{1}{S}\left[\vec{v}.\vec{D}T + \omega\left[\frac{\partial T}{\partial p} - \frac{\gamma-1}{\gamma}\ \frac{T}{p}\right]\right]\right\}$$

$$= \frac{Bo^2 x_0 Ki^2}{Pr}\left\{\frac{\partial}{\partial p}\left[\ \rho\frac{\partial T}{\partial p}\right] + Pr\ \frac{\gamma-1}{\gamma}\ \frac{Ki^2}{S\lambda_0}\ \rho\left|\frac{\partial\vec{v}}{\partial p}\right|^2 - \sigma_0\frac{d\hat{R}_\infty}{dp}\right\}.$$

The similarity relation

(28,5)
$$\frac{Ki}{SRe_\perp} \equiv Ek_\perp = x_0 Ki^2\ ,\ \text{with}\ x_0 = O(1),$$

is motivated by the fact that it corresponds to the least degeneracy of hydrostatic-Navier-stokes equations; this similarity relation has been obtained by Zeytounian (1976).

---

† See the section 6 (Chapter II). For *tangent* atmospheric motions, in (6,2)-(6,7) we carry out the limiting process: $\delta_0\to 0\ (\varphi\to\varphi_0)$.

We note that

$$(28,6) \qquad Ek_\perp = \frac{\mu_0/\rho_\infty(0)}{f_0 H_0^2} \; ,$$

is the vertical Ekman number.

The complete derivation of the quasi-geostrophic model (when Ki→0) in the Guiraud and Zeytounian (1980) asymptotic theory, uses concurrently with the principal expansion (see, the section 9, Chapter II), *two* local ones: the first of these leads to the problem of adjustment to geostrophy (see, the section 21, Chapter V) and gives the initial condition that must be supplied for the equation (9,14), the second one leads to the Ekman steady layer and classical Ackerblom's problem which gives the boundary condition at the ground that must be supplied for the main equation (9,14).

The second inner (local) region corresponds to

$$|p-1| = O(Ki)$$

and we call it the *Ekman (steady)* region (see, the sketch below).

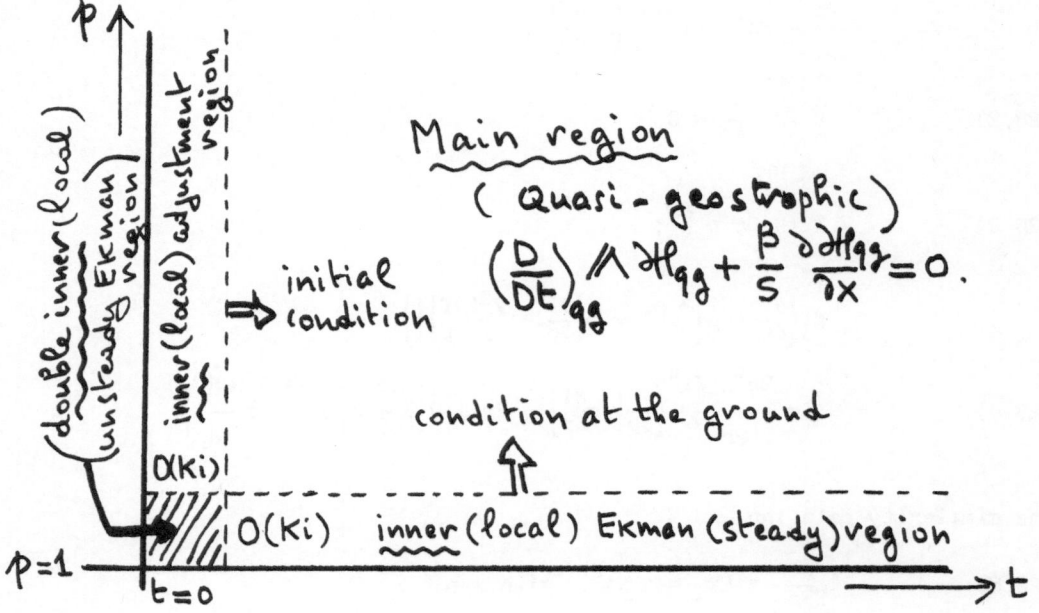

Within the inner steady Ekman region the independent variables are: $t, x, y$, $\hat{p} = \frac{1-p}{Ki}$ and we have:

$$(28,7) \qquad Ki \to 0, \text{ with } t, \ x, \ y \text{ and } \hat{p} \text{ fixed.}$$

If we take into account that: $\dfrac{\partial}{\partial p} = -\dfrac{1}{Ki} \dfrac{1}{\partial \hat{p}}$ , and we assume that:

$$(28,8) \qquad \begin{bmatrix} \vec{v} \\ \omega \\ \mathcal{H} \\ T \\ \rho \end{bmatrix} = \begin{bmatrix} \hat{\vec{v}}_0 \\ \hat{\omega}_0 \\ \hat{\mathcal{R}}_0 \\ \hat{T}_0 \\ \hat{\rho}_0 \end{bmatrix} + Ki \begin{bmatrix} \hat{\vec{v}}_1 \\ \hat{\omega}_1 \\ \hat{\mathcal{R}}_1 \\ \hat{T}_1 \\ \hat{\rho}_1 \end{bmatrix} + \ldots ,$$

then, from $(28,1)$-$(28,4)$, we obtain, as first approximation, the following relations:

$$(28,9) \quad \begin{cases} \vec{D}\hat{\mathcal{R}}_0 = 0 \ , \quad \dfrac{\partial \hat{\omega}_0}{\partial \hat{p}} = 0 \ , \quad \dfrac{\partial \hat{\mathcal{R}}_0}{\partial \hat{p}} = 0, \\[3mm] \hat{\rho}_0 \hat{T}_0 = 1 \ , \quad \dfrac{Bo^2}{Pr} x_0 \dfrac{\partial}{\partial \hat{p}} \left( \hat{\rho}_0 \dfrac{\partial \hat{T}_0}{\partial \hat{p}} \right) = \hat{\omega}_0 \dfrac{\partial \hat{T}_0}{\partial \hat{p}} \ . \end{cases}$$

But for the full equations $(28,1)$-$(28,4)$ we have on the flat ground the following boundary conditions:

$$(28,10) \qquad \vec{v} = 0 \ , \quad \omega = Bo \ \rho \ S \ \dfrac{\partial \mathcal{H}}{\partial t} \ , \quad \rho \dfrac{\partial T}{\partial p} = \sigma_0 \hat{R}_\infty(p), \ \mathrm{on} \ \mathcal{H} = 0.$$

As consequence of $(28,10)$, we have

$$(28,11) \qquad \dfrac{\partial \hat{T}_0}{\partial \hat{p}} = 0 \ , \ \mathrm{on} \ \hat{\mathcal{R}}_0 = 0.$$

From matching with the main region, when $\hat{p} \to \infty$, we get:

$$(28,12) \qquad \hat{\omega}_0 = 0, \ \underset{\hat{p}\to\infty}{\mathrm{Lim}} \ \hat{T}_0 = T_0(1), \ \underset{\hat{p}\to\infty}{\mathrm{Lim}} \ \hat{\mathcal{R}}_0 = \mathcal{H}_0(1),$$

but $\hat{\mathcal{R}}_0 \equiv 0$.

Hence, for the next order we find the following set of equations:

$$(28,13a) \quad \begin{cases} \lambda_0 Bo \vec{D}\hat{\mathcal{R}}_1 + (\vec{k} \wedge \hat{\vec{v}}_0) - Bo^2 x_0 \dfrac{\partial}{\partial \hat{p}} \left( \rho_0(1) \dfrac{\partial \hat{\vec{v}}_0}{\partial \hat{p}} \right) = 0; \\[3mm] \dfrac{Bo^2}{Pr} x_0 \dfrac{\partial}{\partial \hat{p}} \left( \rho_0(1) \dfrac{\partial \hat{T}_1}{\partial \hat{p}} \right) = 0; \end{cases}$$

$$(28,13b) \quad \begin{cases} \dfrac{\partial \hat{\omega}_1}{\partial \hat{p}} = \vec{D} \cdot \hat{\vec{v}}_0 ; \\[4mm] Bo \, \dfrac{\partial \hat{\mathcal{R}}_1}{\partial \hat{p}} = T_0(1), \end{cases}$$

and from (28,10) we have also

$$(28,14) \quad \hat{\vec{v}}_0 = 0, \quad \hat{\omega}_1 = \frac{Bo}{T_0(1)} S \frac{\partial \hat{\mathcal{R}}_1}{\partial t}, \quad \frac{1}{T_0(1)} \frac{\partial \hat{T}_1}{\partial \hat{p}} = \sigma_0 \hat{R}_\infty(1) \quad on \ \hat{\mathcal{R}}_1 = 0,$$

if we suppose that the main radiative transfer not have of Ekman boundary layer structure.

We note that the flat ground in the Ekman layer is characterized by

$$(28,15) \quad \hat{p} = \hat{p}_{g0} + Ki \, \hat{p}_{g1} + \dots \ .$$

From (28,13), (28,14) we obtain, after the matching with the main region,

$$(28,16) \quad \hat{T}_1 = T_{qg,1} + \left( \frac{dT_0}{dp} \right)_{p=1} \hat{p} \ ,$$

where $T_{qg,1} \equiv T_{qg}(t,x,y,1) = -B_0 \left( \dfrac{\partial \mathcal{R}_{qg}}{\partial p} \right)_{p=1}$ (see, the relation (9,11)).

The last equation of (28,13) gives:

$$(28,17) \quad \frac{\partial \hat{\mathcal{R}}_1}{\partial \hat{p}} = \frac{T_0(1)}{Bo} = \text{Const} \implies \hat{\mathcal{R}}_1 = \mathcal{H}_{qg,1} + \frac{T_0(1)}{Bo} \hat{p}$$

and $\hat{\mathcal{R}}_1 = 0$ imply

$$(28,18) \quad \hat{p}_{g0} = -\frac{Bo}{T_0(1)} \mathcal{H}_{qg,1},$$

where $\mathcal{H}_{qg,1} \equiv \mathcal{H}_{qg}(t,x,y,1)$.

*THE ACKERBLOM'S PROBLEM*

We consider now the first equation, for $\hat{\vec{v}}_0$, of (28,13).
We set

$$\hat{\vec{v}}_0 = \vec{v}_{qg,1} + \hat{\vec{v}}_0' \quad , \text{ with } \vec{v}_{qg,1} \equiv \vec{v}_{qg}(t,x,y,1).$$

From matching with the main region, when $\hat{p} \to \infty$, we have

$$\lim_{\hat{p}\to\infty} \hat{\vec{v}}' = 0 \quad \text{and} \quad \lambda_0 \text{Bo } \vec{D}\mathcal{H}_{qg,1} + \vec{k} \wedge \vec{v}_{qg,1} = 0,$$

and we obtain, from (28,17),

$$\lambda_0 B_0 \vec{D}\mathcal{H}_1 + \vec{k} \wedge \hat{\vec{v}}_0 \equiv \vec{k} \wedge \hat{\vec{v}}_0' \ .$$

As consequence, for $\hat{\vec{v}}_0'$ we get the Ackerblom's problem:

$$(28,19) \quad \begin{cases} \text{Bo}^2 x_0 \dfrac{\partial^2 \hat{\vec{v}}_0'}{\partial \hat{p}^2} - \vec{k} \wedge \hat{\vec{v}}_0' = 0, \\[2mm] \hat{\vec{v}}_0' = -\vec{v}_{qg,1} \text{ on } \hat{p} = -\dfrac{\text{Bo}}{T_0(1)}\mathcal{H}_{qg,1}; \\[2mm] \hat{\vec{v}}_0' \to 0, \text{ when } \hat{p} \to \infty. \end{cases}$$

The solution to (28,19) is obtained in a standard way:

$$(28,20) \quad \hat{\vec{v}}_0' - i\vec{k} \wedge \hat{\vec{v}}_0' = -(\vec{v}_{qg,1} - i\vec{k} \wedge \vec{v}_{qg,1})\hat{E},$$

where $i \equiv (-1)^{1/2}$ and

$$(28,21) \quad \hat{E} = \exp\left\{-\frac{1+i}{\sqrt{2\text{Bo}^2 x_0}}\left[\hat{p} + \frac{\text{Bo}}{T_0(1)}\mathcal{H}_{qg,1}\right]\right\}.$$

We consider now the equation of continuity, for the system (28,13), and we find for $\hat{\omega}_1$ the following relation:

$$\hat{\omega}_1 = \int\limits_{\hat{p}_{g0}}^{\hat{p}} (\vec{D}.\hat{\vec{v}}_0)d\hat{p} + S \frac{Bo}{T_0(1)}\left(\frac{\partial \hat{\mathcal{H}}_1}{\partial t}\right)_{\hat{p}=\hat{p}_{g0}},$$

where $\quad \hat{\vec{v}}_0 = \mathcal{R}eal\left\{\left[\vec{v}_{qg,1} - i\vec{k} \wedge \vec{v}_{qg,1}\right](1-\hat{E})\right\}.$

But $\quad \int\limits_{\hat{p}_{g0}}^{\hat{p}} \hat{\vec{v}}_0 d\hat{p} = (\hat{p}-\hat{p}_{g0})\vec{v}_{qg,1} + \sqrt{Bo^2 x_0}\ (\vec{k} \wedge \vec{v}_0),$

and for $\hat{p}\to\infty$, we get:

$$\int\limits_{\hat{p}_{g0}}^{\infty} (\vec{D}.\hat{\vec{v}}_0)d\hat{p} = \vec{D}.\int\limits_{\hat{p}_{g0}}^{\infty} \hat{\vec{v}}_0 d\hat{p} = \sqrt{Bo^2 x_0}\ \vec{D}.(\vec{k} \wedge \hat{\vec{v}}_0),$$

since
$$\vec{D}.\vec{v}_{qg,1}=0,$$

$$\vec{v}_{qg,1}.\vec{D}\hat{p}_{g0} \equiv -\frac{Bo}{T_0(1)}\vec{v}_{qg,1}.\vec{D}\mathcal{H}_{qg,1}=0.$$

Therefore

(28,22) $\qquad \hat{\omega}_1^\infty \equiv \underset{\hat{p}\to\infty}{Lim}\ \hat{\omega}_1 = S \frac{Bo}{T_0(1)} \frac{\partial \mathcal{H}_{qg,1}}{\partial t} - \lambda_0 B_0 \sqrt{Bo^2 x_0}\ \vec{D}^2\mathcal{H}_{qg,1},$

since $\quad \dfrac{\partial \mathcal{H}_1}{\partial t} = \dfrac{\partial \mathcal{H}_{qg,1}}{\partial t}$ is independent of $\hat{p}$.

But, from matching with the main region, we have:

(28,23) $\qquad \hat{\omega}_1^\infty \equiv \omega_{qg,1} = \omega_{qg}(t,x,y,1) = -Bo\ \frac{1}{K_0(1)}\left[S\ \frac{\partial}{\partial t}+\vec{v}_{qg,1}.\vec{D}\right]\left(\frac{\partial \mathcal{H}_{qg}}{\partial p}\right)_{p=1}.$

Finally, we derive the boundary condition at the flat ground which must be supplied to the main equation (9,14), namely (with $x_0 \equiv \mu_0 t_0/\rho_\infty(0)H_0^2$):

$$(28,24) \qquad \left\{\frac{S}{T_0(1)}\frac{\partial}{\partial t} + \frac{1}{K_0(1)}\left[S\frac{\partial}{\partial t} + \vec{v}_{qg}\cdot\vec{D}\right]\frac{\partial}{\partial p} - \lambda_0\sqrt{Bo^2 x_0}\ \vec{D}^2\right\}\mathcal{H}_{qg} = 0,$$

$$\mathit{on}\ p = 1,$$

where $\vec{v}_{qg} \equiv \lambda_0 B_0 (\vec{k} \wedge \vec{D}\mathcal{H}_{qg})$. This last boundary condition (28,24) is naturally different of the slip condition (9,16) and take into account of the *influence* of the Ekman boundary layer on the main quasi-geostrophic flow which is ruled by the equation (9,14).

## 29 . MODEL EQUATIONS FOR BREEZES[†]

Here we consider only the influence of a localized thermal non-homogeneity of the flat ground surface.
We assume on the ground, z=0, with the dimensional variables, that:

$$(29,1) \qquad T = T_\infty(0) + \Delta T_0 \Xi\left(\frac{t}{t^0}, \frac{x}{\ell^0}, \frac{y}{m^0}\right),$$

where $\Xi$ is a known function describing the temperature field on the ground in a localized region characterized by the length scales $\ell^0$ and $m^0$, at the proximity of some origin (0,0).
For the breeze phenomenon it is necessary of take into account the Coriolis terms in the dynamical equations and for that we must assume:

$$(29,2) \qquad \ell^0 = m^0 \cong 10^5 m.$$

On the other hand we have, in our breeze phenomenon, the following particular value of the vertical scale:

$$h^0 = \frac{R\Delta T_0}{g} \ll \frac{RT_\infty(0)}{g}$$

and

$$(29,3) \qquad \tau_0 \equiv \frac{\Delta T_0}{T_\infty(0)} \ll 1.$$

In this case:

[†] According to Zeytounian (1977).

(29,4) $$\varepsilon = \frac{h^0}{\ell^0} \ll 1,$$

since $\Delta T_0 \ll \ell^0 \frac{g}{R}$, and we may consider the main limiting process (see,(6,1)):

(29,5) $$\varepsilon \rightarrow 0, \quad Re \rightarrow \infty, \quad \text{with } \varepsilon^2 Re \equiv Re_\perp = O(1),$$

in the full equations (4,24), (4,27), (4,30), (4,31) and (4,33) (see the section 4, Chapter II), *assuming that* $\delta_0 \equiv 0$.[†]

We note, that from (29,5),

(29,6) $$h^0 = Re_\perp^{1/2} \left( \frac{\ell^0 \nu_0}{U_0} \right)^{1/2}, \quad \text{with } \nu_0 \equiv \frac{\mu_0}{\rho_\infty(0)},$$

and

(29,7) $$\ell^0 \cong \frac{1}{Re_\perp} \frac{U_0}{g^2} \frac{(R\Delta T_0)^2}{\nu_0}.$$

As a matter of fact corresponding to the typical values:

$$U_0 \sim 5 \text{ m/s}, \quad \nu_0 \sim 5 \text{ m}^2\text{/s} \quad \text{and} \quad \Delta T_0 \sim 10°C \text{ [††]},$$

we have $\ell^0 \sim 10^5 \text{m}$.

Finally, for $t^0$ we use $\Omega_0^{-1}$s and in this case:

$$S \equiv \frac{1}{2Ro \sin\varphi_0} \cong 1$$

if $\varphi_0$ is a mid-latitude and $Ro = U_0/2\Omega_0 \sin\varphi_0 \ell^0$ is the Rossby number corresponding to $\ell^0$.

But we need also to specify the value of main characteristic velocity $U_0$; we can suppose that, for the breeze phenomenon,

(29,8) $$U_0 = \sqrt{\gamma R T_\infty(0)}, \quad \tau_0 \equiv \Delta T_0 \sqrt{\frac{\gamma R}{T_\infty(0)}}$$

and we introduce a Grashof number $Gr_\perp$, related to the Reynolds number $Re_\perp$ by the relation:

---

[†] In this case our dimensionless variables are : $t/t^0$, $x/\ell^0$, $y/m^0$ and $z/h^0$. Naturally $\delta_0 \equiv \ell^0/a_0 \ll 1$.

[††] We note that $\Delta T_0 \sim \frac{g}{R} \left( \frac{\ell^0 \nu_0}{U_0} \right)^{1/2}$ if $Re_\perp \cong 1$.

(29,9)
$$Gr_\perp \equiv Re_\perp^2 = \epsilon^2 \frac{\gamma R \Delta T_0^2 (\hbar_0^0)^2}{T_\infty(0)\nu_0^2}.$$

Finally, our Boussinesq number Bo, in this case, is given by the relation:

(29,10)
$$\tau_0 \equiv \frac{\hbar^0 g}{RT_\infty(0)} = Bo \ll 1,$$

and we can consider the Boussinesq-Zeytounian's, main limiting process:

(29,11)
$$\tau_0 \to 0, \quad M_0 \to 0; \quad \frac{\tau_0}{M_0} = \hat{\tau} = O(1),$$

with[†]
$$\hat{\tau} = \frac{\Delta T_0}{U_0} \sqrt{\frac{\gamma R}{T_\infty(0)}} \equiv 1.$$

Now, if we represent the solution of the full Navier-Stokes equations (4,24), (4,27), (4,30), (4,31) and (4,33), where $\delta_0 \equiv 0$, by asymptotic expansions of the form:

$$u' = u_0 + \ldots, \quad v' = v_0 + \ldots, \quad w' = w_0 + \ldots,$$

$$p' = p_\infty(z_\infty)\left[1 + M_0^2\pi_2 + \ldots\right],$$

(29,12)
$$\rho' = \rho_\infty(z_\infty)\left[1 + M_0\omega_1 + \ldots\right],$$

$$T' = T_\infty(z_\infty)\left[1 + M_0\theta_1 + \ldots\right],$$

with $z_\infty = \hat{\tau} M_0 z$, we can easily show that the function $u_0$, $v_0$, $w_0$, $\pi_2$, $\omega_1$ and $\theta_1$ (as functions of dimensionless variables t, x, y and z), under the hydrostatic main limiting process (29,5) satisfy the *hydrostatic Boussinesq equations*:

(29,13a)
$$\begin{cases} S\dfrac{\partial \vec{v}_0}{\partial t} + (\vec{v}_0 \cdot \vec{D})\vec{v}_0 + w_0\dfrac{\partial \vec{v}_0}{\partial z} + \left[\dfrac{1}{Ro} + \beta y\right](\vec{k} \wedge \vec{v}_0) + \dfrac{1}{\gamma}\vec{D}\pi_2 = Gr_\perp^{-1/2}\dfrac{\partial^2 \vec{v}_0}{\partial z^2}; \\[2ex] \dfrac{\partial \pi_2}{\partial z} = \hat{\tau}\,\theta_1; \end{cases}$$

---

[†] Since $M_0 = U_0 / \sqrt{\gamma RT_\infty(0)}$.

$$(29,13b) \quad \begin{cases} \vec{D}.\vec{v}_0 + \dfrac{\partial w_0}{\partial z} = 0; \\[2mm] S\dfrac{\partial \theta_1}{\partial t} + \vec{v}_0.\vec{D}\theta_1 + w_0\dfrac{\partial \theta_1}{\partial z} + \hat{\tau}\left[\dfrac{\gamma-1}{\gamma} + \left(\dfrac{dT_\infty}{dz_\infty}\right)_{z_\infty=0}\right]w_0 = Gr_\perp^{-1/2}\dfrac{\partial^2 \theta_1}{\partial z^2}; \end{cases}$$

and $\omega_1 = -\theta_1$, $\vec{v}_0 \equiv u_0\vec{i} + v_0\vec{j}$, $\vec{D} = \dfrac{\partial}{\partial x}\vec{i} + \dfrac{\partial}{\partial y}\vec{j}$, $\vec{D}.\vec{k} = 0$.

These equations (29,13) can be considered as an inner degeneracy of full Navier-Stokes equations (4,24), (4,27), (4,30), (4,31) and (4,33), when $\delta_0 \equiv 0$, under the conditions (29,11), (29,12). The outer degeneracy (see, for instance, Zeytounian (1977)) gives the trivial *zero solution* (there is *no outer* fields) which determines the *behaviour* of the solutions of the equations (29,13) *far from the wall* z=0.

Therefore, we must consider, for model equations (29,13), the following boundary conditions:

$$(29,14) \quad \begin{cases} z=0: \ \vec{v}_0 = w_0 = 0, \ \theta_1 = \hat{\tau}\ \Xi\ (t,x,y); \\[2mm] z \to +\infty: \ \vec{v}_0 = w_0 = \pi_2 = \theta_1 \to 0; \\[2mm] |x^2+y^2| \to \infty: \ \vec{v}_0 = w_0 = \pi_2 = \theta_1 \to 0. \end{cases}$$

It is clear that the order of equations (29,13) with respect to z makes it impossible to specify the condition for $w_0$ at z=∞. It would appear that, by analogy with problems of Prandtl's boundary layer theory, in the given problem $w_0$ may not become zero at z=∞! In many published solutions of the problem of breeze $w_0 \neq 0$ at z=∞ for all t>0. If one assumes a *neutrally* stratified atmosphere, i.e.,

$$(29,15) \quad -\left.\dfrac{dT_\infty}{dz_\infty}\right|_{z_\infty=0} \equiv \dfrac{\gamma-1}{\gamma},$$

the above cicumstance does not result in any contradictions in the equations. However, if $-\left.\dfrac{dT_\infty}{dz_\infty}\right|_{z_\infty=0} \neq \dfrac{\gamma-1}{\gamma}$, then by virtue of equations (29,13), for $\theta_1$, and condition $\theta_1 \to 0$ for $z \to \infty$ we should have

$$(29,16) \quad w_0 = 0 \text{ for } z=\infty.$$

Note that condition (29,16), not being a boundary condition, should be satisfied *automatically*. Integration of continuity equation of the system (29,13) with respect to z from 0 to $\infty$, while making use of (29,14), yields

(29,17)          $\vec{D} . \int_{0}^{\infty} \vec{v}_0 dz = 0.$

We notice that the condition (29,17) is not consistent with the "boundary-layer" equations (29,13) but it must be enforced as a consequence

of: $- \dfrac{dT_\infty}{dz_\infty}\Big|_{z_\infty=0} < \dfrac{\gamma-1}{\gamma}$, since $\theta_1 \to 0$, with $z \to \infty$.

The constraint (29,17) gives the possibility to obtain the formation of the so-called "antibreeze" over the main breeze. The model problem of the breeze (29,13)-(29,17), over the thermal spot simulated by the function $\Xi(t,x,y)$, where $(x,y) < \mathcal{D}_0$, has zero as a solution when $\Xi(0,x,y) \equiv 0$ and this solution plays the rôle of initial conditions for t=0.

The fact that the lower atmospheric layer is generally stably stratified leads to a new solution as compared with the case of the neutrally stable stratified atmosphere (see, for this case, for instance, Zeytounian (1964)) by revealing the origin of the antibreeze (see, Gutman (1972; §7,3)).

The breeze model problem (29,13) and (29,14), with the *rest* of t=0 and constraint (29,17), is different from the classical Prandtl boundary layer problem, since we have the constraint for the vertical component of the velocity, $w_0$, at $z=\infty$.

The presence in equation (29,13) for $\theta_1$ of the term $\hat{\tau}\left[\dfrac{\gamma-1}{\gamma} + \dfrac{dT_\infty}{dz_\infty}\Big|_{z_\infty=0}\right]w_0$ which,

in general is not small compared with the other terms, is responsible for the fact that in a stably stratified atmosphere a perceptible compensating flow (antibreeze), well known from observations in nature, should exist over the main breeze. At the same time, all published solutions of the problem either have no antibreeze, or show a very weak backflow, induced by other factors (such as the Coriolis force or unsteadiness of the process) and not serving to compensate the breeze.

This can be attributed to the fact that the methods of solution used are found

to be unsuitable at $- \dfrac{dT_\infty}{dz_\infty}\Big|_{z_\infty=0} < \dfrac{\gamma-1}{\gamma}$.

All these methods reduce , in one manner or another, to successive approximations; here the terme $\hat{\tau}\left[\dfrac{\gamma-1}{\gamma} + \dfrac{dT_\infty}{dz_\infty}\bigg|_{z_\infty=0}\right]w_0$, is always found in the second approximation. As a result the solution departs from the *no* proper branch and no matter how many approximations are sought, relations (29,16) and (29,17) cannot be satisfied.

The difficulties in solving the problem defined by equations (29,13) and conditions (29,14), with (29,17), are due to the fact that, as result of the presence of the term $\hat{\tau}\left[\dfrac{\gamma-1}{\gamma} + \dfrac{dT_\infty}{dz_\infty}\bigg|_{z_\infty=0}\right]w_0$, which must be taken into account already in the first approximation, the system is found to be coupled in a very complex manner and reduces to a six-order nonlinear equation with respect to z.

Several solutions are known in which damping of $w_0$ with altitude z is specified as a boundary condition. Such an approach was taken, for example, by Estoque (1961) and Magata (1965), who use an equation obtained by differentiating the continuity equation with respect to z.

It is clear that here the continuity equation proper may not be satisfied and thus relations (29,16) and (29,17) will not be satisfied. It is seen from the figures presented in articles by Estoque and Magata that the law of conservation of mass *is not* satisfied in their solutions.

*A SIMPLE SOLUTION*

Consider now the solution of the *model* simple problem, with which we shall attempt to gain some insight into the complex interaction of meteorological fields during breeze. The statement of the problem will be simplified as much as possible. Noting that the most characteristic factor in the breeze mechanism is the existence of a temperature difference between the land and sea surface varying periodically during the day, we consider the simplest case of linear horizontal variation of surface temperature with a periodic cycle in time. The coordinate origin is taken at the shore line on the assumption of a straight and infinite shore. The x-axis is directed normal to the shore and the y-axis along the shore. Then the entire process does not depend on y.

FOr simplicity, no initial allowance is made for the Coriolis force, setting Ro≡∞ and β≡0. There is very little information on the turbulence coefficient during breezes, and it will hence be assumed that $\nu=\nu_0$=Const. By virtue of the above assumptions system of equations (29,13) assumes the form:

$$(29,18) \quad \begin{cases} S\dfrac{\partial u_0}{\partial t} + u_0\dfrac{\partial u_0}{\partial x} + w_0\dfrac{\partial u_0}{\partial z} = -\dfrac{\partial \cap_2}{\partial x} + Gr_\perp^{-1/2}\dfrac{\partial^2 u_0}{\partial z^2} \; ; \\[2mm] \dfrac{\partial \cap_2}{\partial z} = \lambda_0\theta_1 ; \\[2mm] \dfrac{\partial u_0}{\partial x} + \dfrac{\partial w_0}{\partial z} = 0; \\[2mm] S\dfrac{\partial \theta_1}{\partial t} + u_0\dfrac{\partial \theta_1}{\partial x} + w_0\dfrac{\partial \theta_1}{\partial z} + S_0 w_0 = Gr_\perp^{-1/2}\dfrac{\partial^2 \theta_1}{\partial z^2} \; , \end{cases}$$

where $\dfrac{\pi_2}{\gamma} \equiv \cap_2$, $\lambda_0 \equiv \dfrac{\hat\tau}{\gamma}$ and $S_0 = \hat\tau\left[\dfrac{\gamma-1}{\gamma} + \dfrac{dT_\infty}{dz_\infty}\Big|_{z_\infty=0}\right].$

Also, in accordance with the assumptions mades we set

$$(29,19) \qquad \Xi(t,x,y) = (a_0 + a_1 x)\sin t^\dagger,$$

where $a_0$ and $a_1$ are specified constants. Quantity $a_1$ can be interpreted as some characteristic gradient of the underlying-surface temperature, i.e., $a_1$ is the maximum difference between the temperature of land and sea, divided by the characteristic length of the phenomenon. Condition (29,19) is satisfied best in some sufficiently small region in the vicinity of the shore, of the order of several kilometers in both directions. It can hence also be expected that the results of the solution will be most valid in this region.

The picture of the phenomenon should be somewhat distorted, since no allowance is made for the effects of regions far from the shore, where the temperature no longer varies in the x direction. However, since the statement of the problem takes into account factors most characteristic for the breeze mechanism, one may except that the main features of the phenomenon have been correctly obtained.

Thus a periodic solution of the problem is required, described by equations (29,18), the conditions:

$$(29,20) \qquad u_0 = w_0 = 0, \text{ for } z=0,$$

and

$$(29,21) \qquad \theta_1 = \hat\tau(a_0 + a_1 x)\sin t, \text{ for } z=0, \ t>0.$$

Initial conditions are not needed in this case.

---

† With the dimensions we have for the time variable, $t/\Omega_0$, where $\Omega_0$ is the angular (constant) speed of rotation of the Earth.

Using (29,21), the solution of system (29,18) is sought in the form

(29,22a)
$$\begin{cases} u_0 = u(t,z), \quad \theta_1 = \vartheta(t,z) + x\sigma(t,z), \\ \cap_2 = \pi(t,z) + x\varphi(t,z). \end{cases}$$

This solution can have physical meaning only at moderate x. Hence boundary conditions with respect to x are disregarded.

By virtue of equation of continuity of the system (29,18) and (29,20), (29,22),

(29,22b)      $w_0 \equiv 0.$

Substitution of the solution (29,22) into equations (29,18) yields a system in which the variables are not a function of x:

(29,23)
$$\begin{cases} S\dfrac{\partial u}{\partial t} = -\varphi + Gr_\perp^{-1/2} \dfrac{\partial^2 u}{\partial z^2}; \\[2mm] S\dfrac{\partial \vartheta}{\partial t} + u\sigma = Gr_\perp^{-1/2} \dfrac{\partial^2 \vartheta}{\partial z^2}; \\[2mm] S\dfrac{\partial \sigma}{\partial t} = Gr_\perp^{-1/2} \dfrac{\partial^2 \sigma}{\partial z^2}; \\[2mm] \dfrac{\partial \pi}{\partial z} = \lambda_0 \vartheta, \quad \dfrac{\partial \varphi}{\partial z} = \lambda_0 \sigma. \end{cases}$$

This system of equations should be solved subject to the following boundary conditions:

(29,24)
$$\begin{cases} u = 0, \quad \vartheta = \hat{\tau}\, a_0 \sin t, \quad \sigma = \hat{\tau}\, a_1 \sin t, \text{ for } z=0; \\ u = \vartheta = \sigma = \pi = \varphi = 0, \text{ for } z = \infty, \end{cases}$$

which follow from conditions (29,20), (29,21) and of the behaviour of the solutions far from the wall z=0.

To simplify mathematical manipulations we convert to news variables (symbols with bars) using the expressions:

(29,25)
$$\begin{cases} z = \sqrt{2}\, Gr_\perp^{-1/4} \bar{z}, \quad \sigma = \hat{\tau} a_1 \bar{\sigma}, \quad u = \sqrt{2}\, \lambda_0 a_1 Gr_\perp^{-1/4} \hat{\tau}\bar{u}, \\[2mm] \varphi = \sqrt{2}\, \lambda_0 a_1 Gr_\perp^{-1/4} \hat{\tau}\bar{\varphi}, \quad \vartheta = \sqrt{2}\, \lambda_0 a_1^2 Gr_\perp^{-1/4} \hat{\tau}^2 \bar{\vartheta}, \\[2mm] \pi = 2Gr_\perp^{-1/2}(a_1\lambda_0)^2\, \hat{\tau}^2\bar{\pi}, \quad a_0 = \hat{\tau} a_1^2 \lambda_0 \sqrt{2}\, Gr^{-1/4} \bar{a}_0. \end{cases}$$

As a result equations (29,23) take the form (the bars are dropped) if S≡1:

(29,26a)    $$\frac{\partial \sigma}{\partial t} = \frac{1}{2} \frac{\partial^2 \sigma}{\partial z^2};$$

(29,26b)    $$\frac{\partial \varphi}{\partial z} = \sigma;$$

(29,26c)    $$\frac{\partial u}{\partial t} = -\varphi + \frac{1}{2} \frac{\partial^2 u}{\partial z^2};$$

(29,26d)    $$\frac{\partial \vartheta}{\partial t} + u\sigma = \frac{1}{2} \frac{\partial^2 \vartheta}{\partial z^2};$$

(29,26e)    $$\frac{\partial \pi}{\partial z} = \vartheta.$$

Of conditions (29,24) only those for the temperature at z=0 change their form:

(29,27)    $\sigma = \sin t, \ \vartheta = a_0 \sin t$   for z=0.

Solving equations (29,26a)-(29,26e) it is possible to find successively $\sigma$, $\varphi$, u, $\vartheta$ and $\pi$.

It is clear that system of equations (29,26a)-(29,26e) represents the chain of interactions between physical factors in the breeze mechanism. It follows from equations (29,26a) that the horizontal temperature gradient is produced in the atmosphere due to heating of air by conduction of heat from the underlying surface. Equation (29,26b) shows that the appearance of the horizontal temperature gradient should result in the appearance in the atmosphere of a horizontal pressure gradient. Equation (29,26c) indicates that the pressure gradient induces the onset of wind: here an important rôle is played by eddy diffusion. Equation (29,26d) demonstrates the opposite effect, exerted by the wind on the temperature field. Nonlinear term $u\sigma$ represents a negative heat source in the heat conduction equation. It is precisely this term which describes the wind transport of heat.

Finally, the above scheme is quite rough, since it is based on a model simplified to the utmost. All the regions of breeze and the corresponding field of meteorological variables interact, and here an important rôle is played by nonlinear terms (in the equations of motion), which in this model are found to be identically equal to zero. A particularly important rôle in breeze is played by the vertical velocity field.

The solution of equations (29,26a)-(29,26e) with boundary conditions (29,27) and (29,24) is elementary:

(29,28a)    $\sigma = \exp(-z) \sin(t-z);$

$$(29,28b) \qquad \varphi = \exp(-z)\cos(t-z+\tfrac{\pi}{4});$$

$$(29,28c) \qquad u = -\frac{z}{\sqrt{2}} \exp(-z)\cos(t-z);$$

$$(29,28d) \qquad \vartheta = a_0 \exp(-z)\sin(t-z) + \frac{z}{4\sqrt{2}} \exp(-2z)\cos[2(t-z)]$$

$$+ \frac{1}{4} \exp(-2z)\cos[2(t-z-\tfrac{\pi}{8})] - \exp(z\sqrt{2})\cos[2(t-\frac{z}{\sqrt{2}} - \tfrac{\pi}{8})].$$

The expression for $\pi$ is not given, since it is too cumbersome. The daily pressure variation at the underlying-surface level is:

$$(29,29) \qquad \pi\big|_{z=0} = a_0 \cos(t+\tfrac{\pi}{4}) + 0.02\, \sin 2t.$$

Solutions (29,18) show that the structure of this breeze model in vinicity of the shore is similar to the wind and temperature progressive wave damping out with altitude. These waves move from the ground upward, as confirmed by observations of breezes.

We now establish the instant when the wind appears at the ground on the onset of breeze. At this instant, $\frac{\partial u}{\partial z}\big|_{z=0} = 0$. From this it is found that the breeze lags behind the variation in the soil temperature by 6 hours; this is a rough result, since observations yield from 2 to 5 hours. It is important that the above solution should point to the cause of this lag, which is inertia moving air.

## 30 . MODEL EQUATIONS
## OF THE SLOPE WIND

Consider a local wind arising above a slope, the steepness of which is not less than several degrees and the deviation of the surface temperature of which from the temperature of the free atmosphere at the same altitude exceeds, in absolute value, several degrees centigrades and change little along the slope. This wind will be called *the slope wind* on the assumption that it develops in an atmosphere at rest.

The slope wind arises as a result of the difference in air temperature in the vicinity of the slope in the free atmosphere at the same altitude. Hence, as in the case of breeze, a major role in the slope wind mechanism should be played by eddy heat coduction.

As staring system of equations we consider the equations (29,13). The topography of the slope is assumed to vary according to the following equation:

(30,1)         $z = \alpha_0 \chi(x,y),$

where $\alpha_0 = \dfrac{\chi_0}{\hbar^0}$ with $\chi_0 \equiv \underset{x,y \in D_2}{\text{Max}} |\chi(x,y)|$ and $\hbar^0 = \dfrac{R\Delta T_0}{g}$ (see the section 29).

We now transform from variables x,y and z to new independent variables:

(30,2)         $\begin{cases} \xi \equiv x, \quad \eta \equiv y; \\ \zeta = z - \alpha_0 \chi(\xi,\eta). \end{cases}$

We have the following relations:

$$\frac{\partial}{\partial z} \equiv \frac{\partial}{\partial \zeta}, \quad \frac{\partial}{\partial x} = \frac{\partial}{\partial \xi} - \alpha_0 \frac{\partial \chi}{\partial \xi} \frac{\partial}{\partial \zeta}, \quad \frac{\partial}{\partial y} = \frac{\partial}{\partial \eta} - \alpha_0 \frac{\partial \chi}{\partial \eta} \frac{\partial}{\partial \zeta},$$

and

$$\vec{D}.\vec{v}_0 + \frac{\partial w_0}{\partial z} = \frac{\partial u_0}{\partial \xi} + \frac{\partial v_0}{\partial \eta} + \frac{\partial \omega_0}{\partial \zeta},$$

where

(30,3)         $\omega_0 \equiv w_0 - \alpha_0 \left[ \dfrac{\partial \chi}{\partial \xi} u_0 + \dfrac{\partial \chi}{\partial \eta} v_0 \right].$

Then elementary transformations yield, in place of (29,13), the following equations:

(30,4)
$$\begin{cases} S\dfrac{\partial \vec{v}_0}{\partial t} + (\vec{v}_0.\vec{D})\vec{v}_0 + \omega_0 \dfrac{\partial \vec{v}_0}{\partial \zeta} + \left[\dfrac{1}{Ro}+\beta\eta\right](\vec{k} \wedge \vec{v}_0) + \dfrac{1}{\gamma} \vec{D}\pi_2 - \dfrac{\alpha_0}{\gamma} \hat{\tau} \, \vec{D}\chi \, \theta_1 \\ \qquad\qquad\qquad = Gr_\perp^{-1/2} \dfrac{\partial^2 \vec{v}_0}{\partial \zeta^2}; \\[2mm] \dfrac{\partial \pi_2}{\partial \zeta} = \hat{\tau} \, \theta_1; \\[2mm] \vec{D}.\vec{v}_0 + \dfrac{\partial \omega_0}{\partial \zeta} = 0; \\[2mm] S\dfrac{\partial \theta_1}{\partial t} + \vec{v}_0.\vec{D}\theta_1 + \omega_0 \dfrac{\partial \theta_1}{\partial \zeta} + \hat{\tau}\mu_0 (\omega_0 + \alpha_0\vec{v}_0.\vec{D}\chi) = Gr_\perp^{-1/2} \dfrac{\partial^2 \theta_1}{\partial \zeta^2}, \end{cases}$$

where $\vec{D} = \left( \dfrac{\partial}{\partial \xi} , \dfrac{\partial}{\partial \eta} \right)$, $\omega_0 = w_0 - \alpha_0 \vec{v}_0 \cdot \vec{D} \chi$, $\vec{v}_0 = (u_0, v_0)$ and $\mu_0 = \dfrac{\gamma - 1}{\gamma} + \left. \dfrac{dT_\infty}{dz_\infty} \right|_{z_\infty = 0}$.

For equations (30,4) we must consider the following boundary conditions:

(30,5) $\qquad \vec{v}_0 = 0$, $\omega_0 = 0$, $\theta_1 = \hat{\tau} \Xi(t, \xi, \eta)$ for $\zeta = 0$, $t > 0$ and $\xi, \eta \in D_2$;

(30,6) $\qquad \vec{v}_0 \to 0$, $\omega_0 \to 0$, $\pi_2 \to 0$, $\theta_1 \to 0$ when $|\xi^2 + \eta^2| \to \infty$ and $\zeta \to +\infty$.

But for the slope wind we have that:

$$Gr_\perp^{-1/2} \ll 1,$$

and we consider the limiting process: $Gr_\perp^{-1/2} \to 0$.
When

$$Gr_\perp^{-1/2} \to 0, \text{ with } t, \xi, \eta, \zeta \text{ fixed}$$

we have only possibility to impose the conditions (30,6) and these conditions yield a trivial outer limiting solution:

$$\bar{\vec{v}}_0 = \bar{\omega}_0 = \bar{\pi}_2 = \bar{\theta}_1 \equiv 0.$$

Now we introduce the following inner variables:

(30,7) $\qquad \hat{\zeta} = \zeta / Gr_\perp^{-1/4}$ and $\hat{\omega}_0 = \dfrac{\omega_0}{Gr_\perp^{-1/4}}$

and we get the inner limiting process:

(30,8) $\qquad Gr_\perp^{-1/2} \to 0$, with $t, \xi, \eta$ and $\hat{\zeta}$ fixed.

If we associate to (30,8) the following asymptotic expansions:

(30,9) $\qquad \vec{v}_0 = \hat{\vec{v}} + \dots$, $\hat{\omega}_0 = \hat{\omega} + \dots$, $\pi_2 = \hat{\pi} + \dots$, $\theta_1 = \hat{\vartheta} + \dots$,

we obtain, for $\hat{\vec{v}}$, $\hat{\omega}$, $\hat{\pi}$ and $\hat{\vartheta}$, as functions of $t$, $\xi$, $\eta$ and $\hat{\zeta}$, the following *boundary layer* equations:

(30,10) $\qquad S \dfrac{\partial \hat{\vec{v}}}{\partial t} + (\hat{\vec{v}} \cdot \vec{D}) \hat{\vec{v}} + \hat{\omega} \dfrac{\partial \hat{\vec{v}}}{\partial \hat{\zeta}} + \left[ \dfrac{1}{Ro} + \beta \eta \right] (\vec{k} \wedge \hat{\vec{v}}) + \dfrac{1}{\gamma} \vec{D} \hat{\pi} - \dfrac{\alpha_0}{\gamma} \hat{\tau} \vec{D} \chi \, \hat{\vartheta} = \dfrac{\partial^2 \hat{\vec{v}}}{\partial \hat{\zeta}^2}$;

(30,11) $\qquad \dfrac{\partial \hat{\pi}}{\partial \hat{\zeta}} = 0$;

(30,12)
$$\vec{\mathbb{D}} \cdot \hat{\vec{v}} + \frac{\partial \hat{\omega}}{\partial \zeta} = 0;$$

(30,13)
$$S \frac{\partial \hat{\vartheta}}{\partial t} + \hat{\vec{v}} \cdot \vec{\mathbb{D}} \hat{\vartheta} + \hat{\omega} \frac{\partial \hat{\vartheta}}{\partial \zeta} + \hat{\tau} \, \mu_0 \, \alpha_0 \, \hat{\vec{v}} \cdot \vec{\mathbb{D}} \chi = \frac{\partial^2 \hat{\vartheta}}{\partial \zeta^2}.$$

But the equation (30,11) yield, with the matching,

$$\hat{\pi} \equiv \hat{\pi}(t, \xi, \eta) = \bar{\pi}_2(t, \xi, \eta, \zeta=0) \equiv 0.$$

On the other hand, the condition $Gr_\perp^{-1/2} \ll 1$, yield

(30,14)
$$\ell^0 \ll \frac{(h^0)^2 \Delta T_0}{\nu_0} \sqrt{\frac{\gamma R}{T_\infty(0)}}$$

and as consequence we have that: $Ro \gg 1$ and $\beta \ll 1$.

Finally, we obtain the following *model equations of the slope wind*:

(30,15)
$$\begin{cases} S \dfrac{\partial \hat{\vec{v}}}{\partial t} + (\hat{\vec{v}} \cdot \vec{\mathbb{D}}) \hat{\vec{v}} + \hat{\omega} \dfrac{\partial \hat{\vec{v}}}{\partial \zeta} - \dfrac{\alpha_0}{\gamma} \hat{\tau} \, \vec{\mathbb{D}} \chi \, \hat{\vartheta} = \dfrac{\partial^2 \hat{\vec{v}}}{\partial \zeta^2}; \\[3mm] \vec{\mathbb{D}} \cdot \hat{\vec{v}} + \dfrac{\partial \hat{\omega}}{\partial \zeta} = 0; \\[3mm] S \dfrac{\partial \hat{\vartheta}}{\partial t} + \hat{\vec{v}} \cdot \vec{\mathbb{D}} \hat{\vartheta} + \hat{\omega} \dfrac{\partial \hat{\vartheta}}{\partial \zeta} + \hat{\tau} \mu_0 \alpha_0 \hat{\vec{v}} \cdot \vec{\mathbb{D}} \chi = \dfrac{\partial^2 \hat{\vartheta}}{\partial \zeta^2}. \end{cases}$$

Specific features of this system are that, first $\hat{\vec{v}} = \hat{u} \vec{\xi} + \hat{v} \vec{\eta}$ and $\hat{\omega}$ are wind velocity components along "curvilinear coordinates" $\xi$, $\eta$ and $\zeta$[†]; second, the equations containing new terms makes allowance in explicit form for the buoyancy forces acting along the $\xi$ and $\eta$ axes; and third, new terms appear in the equation of heat conduction, and describe the wind transport of the heat flux component associated with stratification of the undisturbed atmosphere. The boundary conditions, as in the problem of a breeze, are taken to ensure contact of air at the surface level:

(30,16)
$$\hat{\vec{v}} = \hat{\omega} = 0 \text{ for } \zeta = 0, \ t > 0,$$

---

[†] According to the Kaplun's correlation theorem (see, Van Dyke (1975; section 7,12)).

specification of the slope temperature:

(30,17)     $\hat{\vartheta} = \hat{\tau} \ \Xi(t,\xi,\eta)$ for $\hat{\zeta} = 0$, t>0,

where $\Xi$ is assumed to be known function of time, and damping of disturbances of meteorological variables with increasing distance from the slope surface:

(30,18)     $\begin{cases} \hat{\vec{v}} = \hat{\omega} = \hat{\vartheta} \rightarrow 0 & \text{for } |\xi^2 + \eta^2| \rightarrow \infty, \\ \hat{\vec{v}} = \hat{\vartheta} \rightarrow 0 & \text{for } \hat{\zeta} \rightarrow \infty. ^\dagger \end{cases}$

The atmosphere is again assumed to be initially at rest:

(30,19)     $\hat{\vec{v}} = 0$ , $\hat{\vartheta} = 0$ for t=0.

*LINEAR SOLUTION OF PRANDTL (1944)*

Note that when the slope can be treated as an infinite plane and conditions (30,16)-(30,19) are satisfied, system (30,15) becomes linear, since all the unknown quantities cease to depend on $\xi$:

(30,20)     $\begin{cases} S \dfrac{\partial \hat{u}}{\partial t} - \dfrac{\alpha_0}{\gamma} \hat{\tau} \ \hat{\vartheta} \ \sin\beta_0 = \dfrac{\partial^2 \hat{u}}{\partial \hat{\zeta}^2} \ , \\[4mm] S \dfrac{\partial \hat{\vartheta}}{\partial t} + \hat{\tau} \ \mu_0 \alpha_0 \hat{u} \ \sin\beta_0 = \dfrac{\partial^2 \hat{\vartheta}}{\partial \hat{\zeta}^2} \ . \end{cases}$

Here it is assumed that $\dfrac{\partial \chi}{\partial \xi} \equiv \sin(\beta_0)$, $\dfrac{\partial \chi}{\partial \eta} \equiv 0$ and $\hat{V} \equiv 0$, which does not detract from the generality of the problem and the continuity equation together with condition (30,16) yield $\hat{\omega} \equiv 0$.

System of equations (30,20) is the case, infrequent in mesometeorology, when the interaction between the velocity and temperature fields is described by linear equations.

The steady-state solution of equation (30,20) satisfying conditions (30,16)-(30,18) at $\mu_0 > 0$ and $\Xi \equiv 1$ is

(30,21)     $\begin{cases} \hat{u} = \dfrac{\hat{\tau}}{\sqrt{\gamma \mu_0}} \ e^{-\psi} \sin\psi; \\[4mm] \hat{\vartheta} = \hat{\tau} e^{-\psi} \cos\psi, \end{cases}$

---

$\dagger$ Here we have only $\partial \hat{\omega}/\partial \hat{\zeta} \rightarrow 0$ with $\hat{\zeta} \rightarrow \infty$.

where
$$\psi = \hat{\zeta} \sqrt[4]{\frac{1}{4\gamma}\mu_0 \alpha_0^2 \hat{\tau}^2 \sin^2\beta_0} \ .$$

Thus for neutral or unstable stratification ($\mu_0 \leq 0$) of the undisturbed atmosphere, equations (30,20) do not have steady solutions which would satisfy conditions (30,16)-(30,18). As expected, the diurnal wind ($\hat{\tau} > 0$) is directed upslope ($\hat{u} > 0$), while the nocturnal wind ($\hat{\tau} < 0$) is directed downslope ($\hat{u} < 0$).

It is interesting that, according to solution (30,21), the maximum of $\hat{u}$ does not depend on $\beta_0$. The physical cause for this is as follows: it is more difficult for air to rise along a steeper slope, but then the buoyancy force component is larger.

It is seen from relation for $\psi$ that the boundary layer becomes narrower with increasing slope stepness. Conversely, for a shallower slope the boundary layer thickness becomes greater.

Finally we note that as a result of eddy friction, the rising heated-air particles set into motion also air layers, which themselves are not heated. As a result of their ascent these layers become cooler than the neighboring layers which are not set into motion. The latter, upon being cooled by eddy removal of heat, start to descend and form a weak descending flow. As consequence it is clear that the temperature deviations, as well as the wind speed at some distance from the slope, take on small negative values.

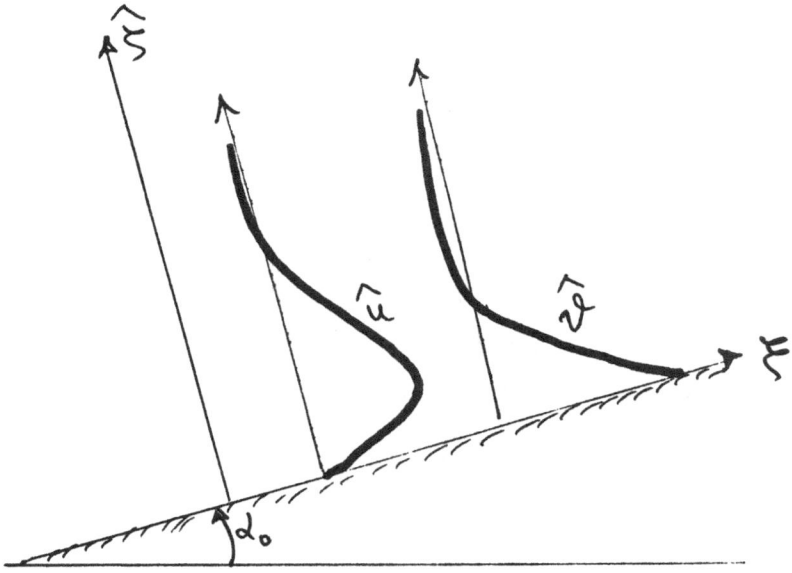

Fig. 15: Profiles of $\hat{u}$ and $\hat{\vartheta}$ during slope wind according to Prandtl solution (30,21).

## 31 . MODEL PROBLEM FOR THE
## LOCAL THERMAL PREDICTION
## (THE TRIPLE DECK VIEWPOINT)

Here we consider only problems which are two-dimensional and steady and we rewrite the thermal boundary condition (29,1) in the following form:

$$(31,1) \qquad \frac{T}{T_\infty(0)} = 1 + \tau_0 \Xi\left[\frac{x}{\ell^0}\right], \quad on \ z=0,$$

where $\Xi \neq 0$, if $\left|\dfrac{x}{\ell_0}\right| \leq 1$.

Far upstream, when $x \rightarrow -\infty$ and $\Xi \equiv 0$, we assume that we have a *basic* undisturbed flow which is characterized by an Ekman layer profile:

$$(31,2) \qquad U_{Ek}\left[\frac{x}{L_0}, \frac{z/\ell^0}{\alpha_0}\right] = U_{geost}\left[\frac{x}{L_0}\right]\left\{1-\exp\left[-\frac{z/\ell^0}{\alpha_0}\right]\cos\left[\frac{z/\ell^0}{\alpha_0}\right]\right\},$$

with

$$(31,3) \qquad \alpha_0 = \frac{\left[\dfrac{\Omega_0\sin\varphi_0}{\nu_0}\right]^{-1/2}}{\ell^0} \equiv \left[\frac{Re}{2Ro}\right]^{-1/2},$$

where:

$$(31,4) \qquad Re = \frac{\ell^0 U_0}{\nu_0} \text{ and } Ro = \frac{U_0/\ell^0}{2\Omega_0\sin\varphi_0}$$

are *local* Reynolds and Rossby numbers, based on local scale length $\ell^0$.

If we nondimensionalize the coordinates x and z with $\ell^0$, then we introduce, in the dimensionless problem, the following Boussinesq number:

$$(31,5) \qquad Bo = \frac{\ell^0 g}{RT_\infty(0)} \ .$$

If $\ell^0 \sim 10^3$m, then $Bo \ll 1$; but if $\ell^0 \sim 10^3$m then we have $2Ro \gg 1$ and also $2Ro \ll Re$. Therefore, in this case (when $\ell^0 \sim 10^3$m), we can assume that

$$(31,6) \qquad 2Ro = Re^{-1/a} \Longrightarrow \alpha_0 = Re^{-1/m},$$

with

$$(31,7) \qquad m = \frac{2a}{a-1} > 2.$$

For $U_0 \sim 10 \text{m/sec}$, $\nu_0 = 5 \text{m}^2/\text{sec}$ and $f_0 \equiv 2\Omega_0 \sin\varphi_0 \cong 10^{-4} 1/\text{sec}$, the case of $\ell^0 \sim 10^3 \text{m}$ leads to $m=5$. For this case we have the possibility of using:

(31,8)
$$\ell^0 \sim \frac{U_0}{g} \sqrt{\frac{RT_\infty(0)}{\gamma}} \implies \mathcal{B}o/M_0 \equiv \hat{\mathcal{B}} \cong 1,$$

and the Boussinesq approximation is correct.

The value m=5 is the same as the one used by Smith (1973) and Smith, Sykes and Brighton (1977) for the flow over an isolated two-dimensional short hump in boundary layers. For the Boussinesq stratified fluid see the work of Sykes (1978).

When m=5 we have that:

(31,9)
$$\frac{\ell^0}{L_0} \sim Re^{-3/8}, \quad \text{where } Re = \frac{U_0 L_0}{\nu_0}.$$

According to the Boussinesq approximation (see the section 8) and taking into account the relation (31,9) we have the possibility to formulate the following *dimensionless local* problem, if we impose that

$$\tau_0 \to 0, \text{ with } M_0 \to 0 \text{ and } \tau_0/M_0 = \lambda_0 \cong 1^\dagger:$$

(31,10a)
$$
\begin{cases}
u\dfrac{\partial u}{\partial x} + w\dfrac{\partial u}{\partial z} + \dfrac{1}{\gamma}\dfrac{\partial \pi}{\partial x} = \alpha_0^5 \left(\dfrac{\partial^2 u}{\partial x^2} + \dfrac{\partial^2 u}{\partial z^2}\right); \\[2mm]
u\dfrac{\partial w}{\partial x} + w\dfrac{\partial w}{\partial z} + \dfrac{1}{\gamma}\dfrac{\partial \pi}{\partial z} - \dfrac{\hat{B}}{\gamma}\theta = \alpha_0^5 \left(\dfrac{\partial^2 w}{\partial x^2} + \dfrac{\partial^2 w}{\partial z^2}\right); \\[2mm]
\dfrac{\partial u}{\partial x} + \dfrac{\partial w}{\partial z} = 0; \\[2mm]
u\dfrac{\partial \theta}{\partial x} + w\dfrac{\partial \theta}{\partial z} + \hat{B}\left[\dfrac{\gamma-1}{\gamma} + \dfrac{dT_\infty}{dz_\infty}\Big|_{z_\infty=0}\right]w = \dfrac{\alpha_0^5}{Pr}\left(\dfrac{\partial^2 \theta}{\partial x^2} + \dfrac{\partial^2 \theta}{\partial z^2}\right); \\[2mm]
\omega = -\theta, \\[2mm]
z = 0: \ u = w = 0, \ \theta = \lambda_0 \Xi(x), \ 0 \leq x \leq 1;
\end{cases}
$$

---

† In this case we obtain for $\Delta T_0$ the following valuation:

$$\Delta T_0 \sim U_0 \sqrt{\frac{T_\infty(0)}{\gamma R}},$$

since $M_0 = U_0/\sqrt{\gamma R T_\infty(0)}$ and $\tau_0 = \Delta T_0/T_\infty(0)$.

$$(31,10b) \qquad x \longrightarrow -\infty \begin{cases} u \longrightarrow 1-\exp(-z/\alpha_0)\cos(z/\alpha_0) \equiv U^\infty(z/\alpha_0)^\dagger, \quad w \longrightarrow 0, \\[2mm] \pi = \theta = -\omega \longrightarrow 0. \end{cases}$$

We note that:

I) if $\dfrac{z}{\alpha_0} \longrightarrow \infty$ , then $u \longrightarrow 1$, for $x \longrightarrow -\infty$,

II) if $\dfrac{z}{\alpha_0} \longrightarrow 0$ , then $u \sim \dfrac{z}{\alpha_0}$, for $x \longrightarrow -\infty$.

Now if we require to take into account the boundary conditions on the ground z=0, it is necessary to introduce the inner variable

$$(31,11) \qquad \hat{z} = \frac{z}{\alpha_0^\alpha}, \quad \alpha>1$$

and in this case

$$(31,12) \qquad u \sim \alpha_0^{\alpha-1}\, \hat{z}\ , \quad \text{for } x \longrightarrow -\infty.$$

From the first equation of (31,10) we verify that, if $u \sim \alpha_0^{\alpha-1}\hat{u}(x,\hat{z})$, with $\hat{z}=z/\alpha_0^\alpha$, then:

$$\alpha_0^{\alpha-1}\, \hat{u}\, \frac{\partial \hat{u}}{\partial x} + \ldots = \alpha_0^{5-2\alpha}\, \frac{\partial^2 \hat{u}}{\partial \hat{z}^2} + \ldots$$

and it is necessary to impose that

$$(31,13) \qquad \alpha-1 = 5-2\alpha \implies \alpha=2.$$

Finally we establish that three vertical variables are necessary for the asymptotic analysis of the system (31,10):

I)    z, for the *upper* region, where

$$u \sim \bar{u} \longrightarrow 1, \quad \text{when } x \longrightarrow -\infty,$$

II)   $\tilde{z} = z/\alpha_0$, for the *middle* region, where

$$u \sim \tilde{u} \longrightarrow 1-e^{-\tilde{z}}\cos\tilde{z} \equiv U^\infty(\tilde{z}), \quad \text{when } x \longrightarrow -\infty,$$

III)  $\hat{z} = z/\alpha_0^2$, for the *lower* wall viscous region, where

$$u \sim \alpha_0\hat{u}, \quad \text{and } \hat{u} \longrightarrow \hat{z}, \quad \text{when } x \longrightarrow -\infty.$$

---

† We note that $\ell^0/L_0 \sim \alpha_0^3$, if m=5 and as consequence we have that:

$U_{geost}\left[\dfrac{x}{L_0}\right] \sim 1+O(\alpha_0^3)$, when $\alpha_0 \to 0$, and it is sufficient to take into account,

in the boundary condition for $x \longrightarrow -\infty$, only the first term $U^\infty(z/\alpha_0)$.

For the other case, when $\ell^0 < \alpha_0^3 L_0$ and $\ell^0 > \alpha_0^3 L_0$, it is necessary to apply a different asymptotic analysis (see, for example, the work of Smith et al. (1981)). But the case m=6 and m=4 can be analysed from the problem (31,10). For the case m=3 it is necessary to start from another problem, where the Boussinesq approximation does not emerge. For m=3, we have $\ell^0 \sim 10^4$m and we may neglect the Coriolis terms in the local non Boussinesq equations.

On the below sketch (Figure 16) we have demonstrated the triple-deck structure for the analysis of boundary Ekman layer flow interaction with the termal non-homogeneity on the ground z=0 (with the dimensionless variables).

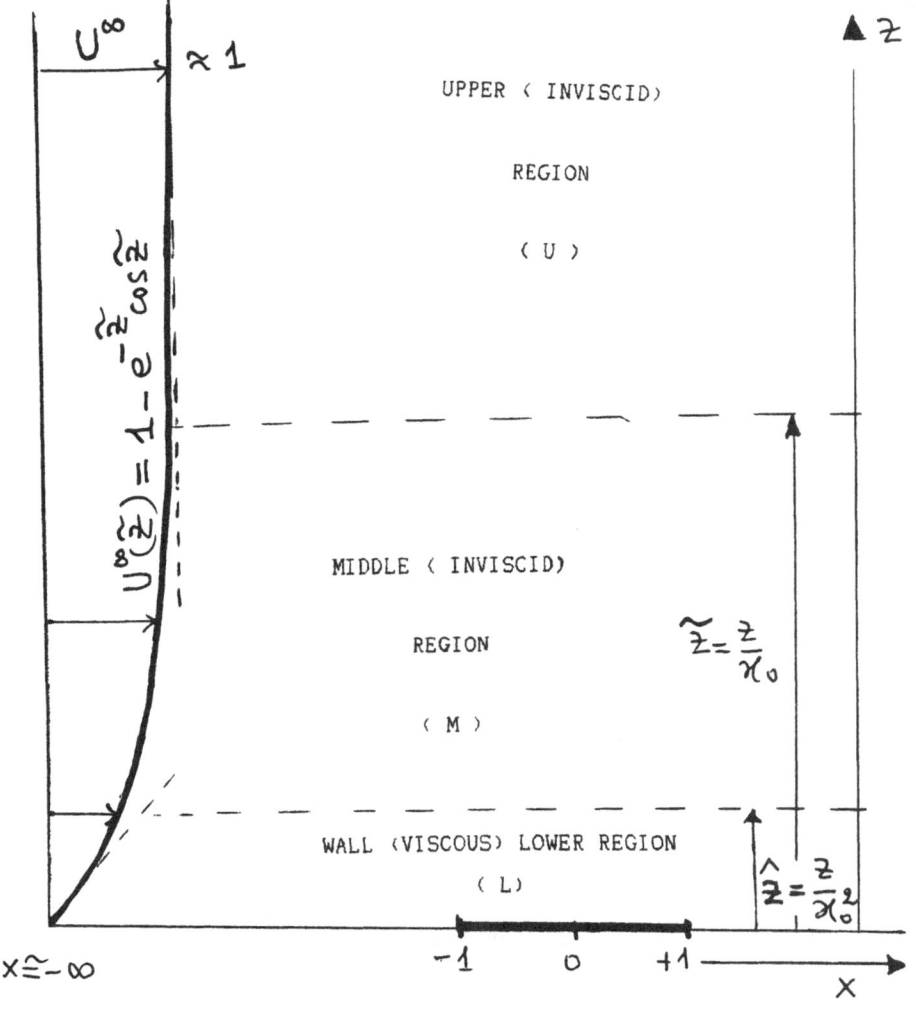

Fig. 16: Definition sketch of the asymptotic regions and stretched vertical coordinates for the triple-deck analysis (case of $\ell^0 \sim \alpha_0^3 L_0$).

*THE TRIPLE-DECK STRUCTURE*

We shall give the analysis for the three regions: $\mathcal{U}$, $\mathcal{M}$ and $\mathcal{L}$ of the triple-deck theory (see the Figure 16 above), with little discussion[†].
Beginning in the *middle deck* ($\mathcal{M}$), where x and $\tilde{z}=z/\alpha_0$ are the order one coordinates, we expand the flow variables as:

$$(31,14) \quad \begin{cases} u = U^\infty(\tilde{z}) + \alpha_0^\varphi\, \tilde{u} + \ldots\,; \\[2mm] w = \qquad\quad \alpha_0^\psi\, \tilde{w} + \ldots\,; \\[2mm] \pi = \qquad\quad \alpha_0^2\, \tilde{\pi} + \ldots\,; \\[2mm] \theta = \qquad\quad \alpha_0^\sigma\, \tilde{\theta} + \ldots\,, \end{cases}$$

and substitute in problem (31,10) to find for the first lowest order:

$$(31,15) \quad \begin{cases} U^\infty(\tilde{z})\dfrac{\partial\tilde{u}}{\partial x} + \dfrac{dU^\infty}{d\tilde{z}}\, \tilde{w} = 0\,; \\[4mm] \dfrac{\partial\tilde{u}}{\partial x} + \dfrac{\partial\tilde{w}}{\partial\tilde{z}} = 0\,; \\[4mm] \dfrac{\partial\tilde{\pi}}{\partial\tilde{z}} = \hat{B}\tilde{\theta}\,; \\[4mm] \dfrac{\partial\tilde{\theta}}{\partial x} = 0\,, \end{cases}$$

if we assume that:

$$(31,16) \qquad \varphi=1, \quad \psi=1+\varphi \text{ and } \sigma=1.$$

This choice (31,16) is necessary if we want to obtain a significant degeneracy of the problem (31,10) in the lower ($\mathcal{L}$) viscous region in the vicinity of $\hat{z}=0$, near the wall.
Notice that the effects of the expansion of the boundary layer are $O(\alpha_0)$ in u and $O(\alpha_0^3)$ in w. Furthermore in the boundary layer, if we take into account

---

[†] The reader being referred to Stewartson and Williams (1969) original work or Zeytounian (1987; see the "Leçon IX").

that $\alpha=2$ (see (31,13)), we have necessarily

(31,17)        $\pi = \alpha_0^2 \overset{\wedge}{\pi} + \ldots,$

and, by continuity, we obtain the form of the expansion for $\pi$ in (31,14). Solutions for $\tilde{u}(x,\tilde{z})$ and $\tilde{w}(x,\tilde{z})$ satisfying the upstream boundary conditions are:

(31,18)        $\tilde{u} = A(x)\dfrac{dU^\infty(\tilde{z})}{d\tilde{z}}$ and $\tilde{w} = -\dfrac{dA(x)}{dx} U^\infty(\tilde{z}),$

which represent simply a vertical displacement of the streamlines through a distance $-\alpha_0 A(x)$.

The flow in the *upper* ($\mathcal{U}$) *deck* is driven by an outflow from the middle deck. Far from (31,18) we have:

(31,19)        $\underset{\tilde{z}\to\infty}{\text{Lim}} \, \tilde{w}(x,\tilde{z}) = -\dfrac{dA(x)}{dx}.$

We introduce a new vertical coordinate z and we have

$z = \alpha_0 \tilde{z};$

in this case the flow expansions for the upper deck is:

(31,20)
$\begin{cases} u = 1 + \alpha_0^2 \, \bar{u} + \ldots \\ w = \alpha_0^2 \, \bar{w} + \ldots \\ \pi = \alpha_0^2 \, \bar{\pi} + \ldots \\ \theta = \alpha_0^2 \, \bar{\theta} + \ldots \end{cases}$

Substitution in the equations of the local problem (31,10) shows that the motion is inviscid:

(31,21a)        $\dfrac{\partial\bar{u}}{\partial x} + \dfrac{1}{\gamma}\dfrac{\partial\bar{\pi}}{\partial x} = 0;$

(31,21b)        $\dfrac{\partial\bar{w}}{\partial x} + \dfrac{1}{\gamma}\dfrac{\partial\bar{\pi}}{\partial z} = \dfrac{\hat{\mathcal{B}}}{\gamma}\bar{\theta};$

(31,21c)        $\dfrac{\partial\bar{u}}{\partial x} + \dfrac{\partial\bar{w}}{\partial z} = 0;$

(31,21d)
$$\frac{\partial \bar{\theta}}{\partial x} + \hat{\mathcal{B}}\left[\frac{\gamma-1}{\gamma} + \frac{dT_\infty}{dz_\infty}\bigg|_{z_\infty=0}\right]\bar{w} = 0.$$

From (31,21) we obtain, for $\bar{\pi}$, a Helmoltz's equation in a half space:

(31,22)
$$\left[\frac{\partial^2}{\partial x^2} + \frac{\partial^2}{\partial z^2} + K_0^2\right]\frac{\partial \bar{\pi}}{\partial x} = 0,$$

where

(31,23)
$$K_0^2 = \frac{\hat{\mathcal{B}}^2}{\gamma}\left[\frac{\gamma-1}{\gamma} + \left(\frac{dT_\infty}{dz_\infty}\right)_{z_\infty=0}\right].$$

Note that in (31,14) we have

$$\tilde{\theta}(x,\tilde{z}) \equiv \chi(\tilde{z})$$

according to the last of equations (31,15).

But

$$\frac{\partial^2 \tilde{\pi}}{\partial x \partial \tilde{z}} = 0 \implies \tilde{\pi} = \mathcal{P}(x) + Q(\tilde{z})$$

and consequently

(31,24)
$$\tilde{\theta} = \frac{1}{\tilde{\mathcal{B}}}\frac{dQ}{d\tilde{z}} \equiv \chi(\tilde{z}).$$

However the solutions for $\tilde{\theta}$ and $\tilde{\pi}$ satisfying the upstream boundary conditions (for $x \to -\infty$) are:

(31,25)
$$Q(\tilde{z}) \implies \begin{cases} \chi(\tilde{z}) \equiv 0 \implies \tilde{\theta} \equiv 0, \\ \tilde{\pi} \equiv \mathcal{P}(x). \end{cases}$$

Now, for the equation (31,22) we have, as a consequence of matching with the middle deck, the following condition:

$$\underset{z \to 0}{\text{Lim}}\,\bar{\pi} \equiv \bar{\pi}(x,0) = \mathcal{P}(x) \equiv \underset{\tilde{z} \to \infty}{\text{Lim}}\,\tilde{\pi}.$$

But

$$\bar{w}(x,0) = -\frac{dA(x)}{dx},$$

as a consequence of (31,19) and matching with the upper deck, and we can write the following relation between $\bar{\pi}$ and $A(x)$:

(31,26)
$$\frac{\partial}{\partial x}\left[\frac{\partial \bar{\pi}}{\partial z}\bigg|_{z=0}\right] = \gamma\left[K_0^2\frac{dA}{dx} + \frac{d^3A}{dx^3}\right].$$

This last relation (31,26) is a boundary condition for the equation (31,22) for $\bar{\pi}$.

It is obvious that the middle deck solution (31,18) do not satisfy the no-slip conditions on $\hat{z}=0$. A situation which is remedied by the analysis of the lower viscous deck $(\mathcal{L})$, where the stretched variable is

$$\hat{z} = \frac{z}{\alpha_0^2} \equiv \frac{\tilde{z}}{\alpha_0}.$$

Matching with the expansions (31,14), when $\tilde{z}=O(\alpha_0)$ implies the inner expansions

(31,27)
$$\begin{cases} u = \alpha_0 \hat{u} + \ldots \\[4pt] w = \alpha_0^3 \hat{w} + \ldots \\[4pt] \pi = \alpha_0^2 \hat{\pi} + \ldots \\[4pt] \theta = \hat{\theta} + \ldots \end{cases}$$

Substitution of (31,27) into the full equations of the local problem (31,10) yields the following, *nonlinear, viscous equations for* $\hat{u}$, $\hat{w}$ *and* $\hat{\theta}$:

(31,28a)
$$\hat{u}\,\frac{\partial \hat{u}}{\partial x} + \hat{w}\,\frac{\partial \hat{u}}{\partial \hat{z}} + \frac{\hat{\mathcal{B}}}{\gamma}\int_\infty^{\hat{z}} \frac{\partial \hat{\theta}}{\partial x}\,d\hat{z} + \frac{1}{\gamma}\frac{d\mathcal{P}(x)}{dx} = \frac{\partial^2 \hat{u}}{\partial \hat{z}^2};$$

(31,28b)
$$\frac{\partial \hat{u}}{\partial x} + \frac{\partial \hat{w}}{\partial \hat{z}} = 0;$$

(31,28c)
$$\hat{u}\,\frac{\partial \hat{\theta}}{\partial x} + \hat{w}\,\frac{\partial \hat{\theta}}{\partial \hat{z}} = \frac{1}{Pr}\frac{\partial^2 \hat{\theta}}{\partial \hat{z}^2},$$

with the boundary conditions:

(31,29)
$$\begin{cases} \hat{z} = 0:\ \hat{u} = \hat{w} = 0,\ \hat{\theta} = \lambda_0\Xi(x),\ 0\le x\le 1; \\[8pt] \hat{z}\to +\infty: \begin{cases} \hat{u}\to\hat{z},\ \hat{w}\to 0,\ \hat{\theta}\to 0 \\[4pt] \mathcal{P}(x)\to 0,\ A(x)\to 0,\ \dfrac{dA}{dx}\to 0; \end{cases} \\[12pt] x\to -\infty:\ \hat{u}\to\hat{z} + A(x),\ \hat{w}\to -\hat{z}\,\dfrac{dA}{dx},\ \hat{\theta}\to 0, \end{cases}$$

after matching with the middle deck $(\mathcal{M})$

$$(31,30) \qquad \lim_{\tilde{z}\to 0} \begin{bmatrix} U^\infty(\tilde{z}) + \alpha_0 \dfrac{dU^\infty}{d\tilde{z}} A(x) \\[2mm] -\alpha_0 U^\infty(\tilde{z})\dfrac{dA}{dx} \end{bmatrix} = \lim_{\hat{z}\to +\infty} \begin{bmatrix} \alpha_0 \hat{u} \\[2mm] \alpha_0^3 \hat{w} \end{bmatrix}$$

and taking into account that $(\tilde{z}=\alpha_0\hat{z})$:

$$(31,31) \qquad U^\infty(\tilde{z}) \sim \tilde{z}, \quad \frac{dU^\infty}{d\tilde{z}} \sim 1, \quad \text{when } \tilde{z}\to 0.$$

We note that for $\hat{\pi}$ we have the following expression

$$(31,32) \qquad \frac{\partial\hat{\pi}}{\partial\hat{z}} = \mathcal{B}\theta \implies \hat{\pi} = \mathcal{B}\int_\infty^{\hat{z}}\theta\, d\hat{z} + \mathcal{P}(x).$$

The specification of the problem $(31,28)$-$(31,32)$ is completed by the relation $(31,26)$ between $\mathcal{P}(x)$ and $A(x)$, since $\overline{\pi}(x,0)\equiv\mathcal{P}(x)$. The well-known interpretation of $(31,26)$ is that the pressure $\mathcal{P}(x)$ driving the flow in the lower deck $(\mathcal{L})$ is itself induced in the main stream, i.e. the upper deck $(\mathcal{U})$, by the displacement thickness of the lower deck transmitted through the middle deck by the passive effect of displacement of the streamlines.

The *strong singular self-induced coupling* arises because the problem $(31,28)$-$(31,29)$ to be solved in the lower viscous layer $(\mathcal{L})$ *does not accept* $\mathcal{P}(x)$ as data known prior to the resolution (as is the case in classical boundary layer problems). On the contrary, this pressure perturbation $\mathcal{P}(x)$ must be calculated at the same time as the velocity components $\hat{u}$ and $\hat{w}$, as well as the temperature perturbation $\hat{\theta}$.

Nevertheless, it must be emphasized that this function $\mathcal{P}(x)$ is not completely arbitrary and that it is connected to the function $A(x)$ through a relation. This last relation is obtained via the analysis of perfect fluid flow in the upper layer $(\mathcal{U})$ (see $(31,26)$).

If it is assumed that the parameter $\lambda_0$ in $(31,29)$ satisfies the condition:

$$(31,33) \qquad \lambda_0 \ll 1 \implies \tau_0 \ll M_0,$$

then the equations $(31,28)$ may be linearized about the undisturbed boundary layer profile by making a further expansion

$$(31,34) \quad \begin{cases} \hat{u} = \hat{2} + \lambda_0 u' + \ldots ; \\ \hat{w} = \quad \lambda_0 w' + \ldots ; \\ \hat{\theta} = \quad \lambda_0 \theta' + \ldots ; \\ \mathcal{P}(x) = \lambda_0 \mathcal{P}' + \ldots ; \\ A(x) = \lambda_0 A' + \ldots . \end{cases}$$

Finally we simply record here the solutions obtainable by using Fourier transform in x, defined for the function f(x) by:

$$\overline{\overline{f}}(k) = \int_{-\infty}^{+\infty} f(x) e^{-ikx} dx.$$

In particular, we find instead of (31,26), the relation:

$$(31,35) \quad \overline{\overline{A}}(k) = \frac{iN_0}{\gamma} \frac{\overline{\overline{\mathcal{P}}}(k)}{K_0^2 - k^2},$$

where

$$(31,36) \quad N_0 = \begin{cases} i \left[ k^2 - K_0^2 \right]^{1/2}, & \text{if } |k| > K_0 ; \\ \left[ K_0^2 - k^2 \right]^{1/2}, & \text{if } |k| < K_0. \end{cases}$$

Here we have applied the standard radiation condition for $z \to +\infty$, choosing the sign of $N_0$, for $|k| < K_0$, so that the wave modes carry energy only upwards.

*REFERENCES TO WORKS CITED IN THE TEXT*

ESTOQUE, M.A.  (1961) _ Q.J.Roy.Met.Soc.,vol.87, pp.136-146.

GUIRAUD, J.P. and ZEYTOUNIAN, R,Kh.  (1980) _ Geophys. Astrophys. Fluid Dynamics, 15, 283.

GUTMAN, L.M.  (1972) _ *Introduction to the nonlinear theory of mesoscale meteorological process.* Guidrometeorologicheskoe izdalel'stvo, Leningrad (1969). Israel Program for Sci. Transl. Jerusalem.

MAGATA, M. (1965) _ Pap. Met. a. Geophys., vol. 16, n°1.

PRANDTL, L. (1944) _ *Führer durch die Strömungslehre.* Braunschweig, Vieweg and Sohn.

SMITH, F.T. (1973) _ J.F.M, vol. 57, pt. 4, 803-824.

SMITH, F.T, SYKES, R.I and BRIGHTON, P.W.M.(1977) _ J.F.M, vol.83, part 1, 163-176.

SMITH, F.T, BRIGHTON, P.W.M, JACKSON, P.S. and HUNT, J.C.R. (1981) _ J.F.M. 113, 123.

STEWARTSON, K. and WILLIAMS, P.G. (1969) _ Proc. Roy.Soc., London A312, 181-206.

SYKES, R.I. (1978) _ Proc.Roy.Soc., London A361, 225-243.

VAN DYKE, M. (1975) _ *Perturbation methods in Fluid Mechanics.* Parabolic Press, Stanford, USA.

ZEYTOUNIAN, R.Kh. (1964) _ Trudy of the World Meteorological Center, Moscow, URSS; vol.3, pp.19-74 (in Russian).

ZEYTOUNIAN, R.Kh. (1976) _ La Météorologie du point de vue du Mécanicien des Fluides. Fluid Dynamics Transactions, 8, 289-352.

ZEYTOUNIAN, R.Kh. (1977) _ J. of Engineering Mathematics, vol.11, n°3, 241-247.

ZEYTOUNIAN, R.Kh. (1987) _ *Les Modèles Asymptotiques de la Mécanique des Fluides* II. Lecture Notes in Physics, vol.276; Springer-Verlag, Heidelberg.

# CHAPTER VIII

## 32 . WHAT IS STABILITY ?

We consider here the main equation of the quasi-geostrophic model (9,14), with (9,15):

(32,1)
$$S \frac{\partial \hat{\Lambda}\mathcal{H}_{qg}}{\partial t} + \frac{\partial (\mathcal{H}_{qg}, \hat{\Lambda}\mathcal{H}_{qg} + \beta y)}{\partial (x, y)} = 0,$$

where

(32,2)
$$\hat{\Lambda}\mathcal{H}_{qg} = \vec{D}^2 \mathcal{H}_{qg} + S \frac{\partial}{\partial p}\left[\frac{p^2}{K_0(p)} \frac{\partial \mathcal{H}_{qg}}{\partial p}\right]$$

and $Bo \equiv 1$, $\lambda_0 \equiv 1$[†].

In the adiabatic nonviscous atmosphere (see the formula (9,16)) the boundary condition at the Earth's surface (reduction of the vertical velocity to zero) reduces to the form:

(32,3)
$$S \frac{\partial \mathcal{H}_{qg}}{\partial t} + \frac{T_0(1)}{K_0(1)}\left[S \frac{\partial}{\partial t}\left(\frac{\partial \mathcal{H}_{qg}}{\partial p}\right) + \frac{\partial (\mathcal{H}_{qg}, \partial \mathcal{H}_{qg}/\partial p)}{\partial (x, y)}\right] = 0, \; \text{on } p=1.$$

Let us consider a *basic flow*, $U_B(y,p)$, having a purely zonal velocity (i.e., directed along the circles of latitude) which is expressed from the geostrophic stream function $\mathcal{H}_B(y,p)$:

(32,4)
$$U_B(y,p) = - \frac{\partial \mathcal{H}_B}{\partial y}.$$

This basic current is naturally assumed to be a solution to equation (32,1), with (32,2), and we consider now the evolution of a perturbation $h(t,x,y,p)$ of

---

† We note that:
$$\frac{\partial (a,b)}{\partial (x,y)} = \frac{\partial a}{\partial x} \frac{\partial b}{\partial y} - \frac{\partial a}{\partial y} \frac{\partial b}{\partial x}.$$

this basic flow; i.e.,

(32,5)
$$\mathcal{H}_{qg}(t,x,y,p) = \mathcal{H}_{B}(y,p) + h(t,x,y,p).$$

If (32,5) is inserted into (32,1), with (32,2), then the following equation is obtained for h:

(32,6)
$$\left[S \frac{\partial}{\partial t} + U_B \frac{\partial}{\partial x}\right]q + \frac{\partial \sqcap}{\partial y} \frac{\partial h}{\partial x} + \frac{\partial(h,q)}{\partial(x,y)} = 0,$$

where $q(t,x,y,p)$ is the potential vorticity (quasi-geostrophic and baroclinic) of perturbation defined by

(32,7)
$$q = \vec{D}^2 h + S \frac{\partial}{\partial p}\left[\frac{p^2}{K_0(p)} \frac{\partial h}{\partial p}\right].$$

The term $\partial \sqcap / \partial y$ in (32,6) is the gradient along the meridian of the potential vorticity of basic flow:

(32,8)
$$\sqcap = \beta y + \frac{\partial^2 \mathcal{H}_B}{\partial y^2} + S \frac{\partial}{\partial p}\left[\frac{p^2}{K_0(p)} \frac{\partial \mathcal{H}_B}{\partial p}\right],$$

and we note that:

(32,9)
$$\frac{\partial \sqcap}{\partial y} = \beta - \frac{\partial^2 U_B}{\partial y^2} - S \frac{\partial}{\partial p}\left[\frac{p^2}{K_0(p)} \frac{\partial U_B}{\partial p}\right].$$

One of the fundamental questions to be clarified is: *how the given structure of* $U_B(y,p)$ *determines the evolution of perturbations field* $h(t,x,y,p)$?
Stated more precisely, this means that given the basic flow $U_B(y,p)$, the behavior of $h(t,x,y,p)$, stemming from (32,5), must be studied in order to determine wheter it increases or decreases.
If it increases, then the *instability* of $U_B$ with respect to h is ascertained.
The basic flow $U_B$ can be said to be "truly" *stable* only when it is stable with respect to *all* h. On the contrary, *instability* takes places if $U_B$ is unstable for *even one* h.
The equation dealing with $h(t,x,y,p)$ is quasi-linear (on account of the term $\partial(h,q)/\partial(x,y)$), and generally speaking it is quite difficult to study the behavior of its solution under the boundary condition:

$$(32,10) \qquad \left[S\frac{\partial}{\partial t} + U_B\frac{\partial}{\partial x}\right]\left[\frac{\partial h}{\partial p} + \frac{K_0(1)}{T_0(1)}h\right] - \left(\frac{\partial U_B}{\partial p} + \frac{K_0(1)}{T_0(1)}U_B\right)\frac{\partial h}{\partial x}$$

$$+ \frac{\partial(h,\partial h/\partial p)}{\partial(x,y)} = 0 \quad \text{on } p=1,$$

according to (32,3) and (32,4), (32,5).

Thus the *linear case* is often adopted and it is assumed that $|h|<<1$. In this way, the following equation, which governs the *linear stability problem*, can replace (32,6):

$$(32,11) \qquad \left[S\frac{\partial}{\partial t} + U_B\frac{\partial}{\partial x}\right]\left[\bar{D}^2 h' + S\frac{\partial}{\partial p}\left(\frac{p^2}{K_0(p)}\frac{\partial h'}{\partial p}\right)\right]$$

$$+ \left[\beta - \frac{\partial^2 U_B}{\partial y^2} - S\frac{\partial}{\partial p}\left(\frac{p^2}{K_0(p)}\frac{\partial U_B}{\partial p}\right)\right]\frac{\partial h'}{\partial x} = 0,$$

and *for* p=1, we have the following boundary condition,

$$(32,12) \qquad \left[S\frac{\partial}{\partial t} + U_B\frac{\partial}{\partial x}\right]\left[\frac{\partial h'}{\partial p} + \frac{1}{T_0(p)}\frac{K_0(p)}{p}h'\right]$$

$$- \left(\frac{\partial U_B}{\partial p} + \frac{1}{T_0(p)}\frac{K_0(p)}{p}U_B\right)\frac{\partial h'}{\partial x} = 0.$$

In order to solve (32,11), under the condition (32,12), we can set the following

$$(32,13) \qquad h'(t,x,y,p) = \mathcal{Real}\left\{\tilde{h}(y,p)\exp\left[ik\left(x - \frac{c}{S}t\right)\right]\right\},$$

where the zonal wave number k must be real since h' must remain finite for all $x \rightarrow \pm\infty$. It can be assumed that k>0. On the other hand, the phase velocity c can be written in the form: $c=c_r+ic_i$, and therefore, the following replace (32,13):

$$(32,14) \qquad h'(t,x,y,p) = \mathcal{Real}\left\{\tilde{h}(y,p)\exp\left[k\frac{c_i}{S}t\right]\exp\left[ik\left(x - \frac{c_r}{S}t\right)\right]\right\}.$$

From (32,11) and (32,14) the following equation results for $\tilde{h}(y,p)$:

(32,15)
$$\left[U_B(y,p) - c\right]\left\{S \frac{\partial}{\partial p}\left[\frac{p^2}{K_0(p)} \frac{\partial \tilde{h}}{\partial p}\right] + \frac{\partial^2 \tilde{h}}{\partial y^2} - k^2 \tilde{h}\right\}$$

$$+ \left[\beta - \frac{\partial^2 U_B}{\partial y^2} - S \frac{\partial}{\partial p}\left[\frac{p^2}{K_0(p)} \frac{\partial U_B}{\partial p}\right]\right]\tilde{h} = 0.$$

The boundary condition (32,12) at the Earth's surface, p=1, reduces to the form:

(32,16)
$$\left[U_B(y,p) - c\right]\left[\frac{\partial \tilde{h}}{\partial p} + \frac{1}{T_0(p)} \frac{K_0(p)}{p} \tilde{h}\right]$$

$$- \left(\frac{\partial U_B}{\partial p} + \frac{1}{T_0(p)} \frac{K_0(p)}{p} U_B\right)\tilde{h} = 0, \quad \textit{for } p=1.$$

As a general rule, boundary conditions in y and p→0 must be superimposed on (32,15).

It turns out that those conditions are *homogeneous* and hence, the corresponding linear stability problem usually has only the trivial solution which is identically zero.

One exception is when k and c are linked by a relation depending on the profile of $U_B(y,p)$ which can be called by *dispersion relation* of the stability problem.

For a fixed profile $U_B(y,p)$, if k is fixed, the dispersion relation allows a sequence of complex roots in c. If $c_i < 0$ for all the roots (we remark that k is real and positive), then the perturbations (called *normal modes*) attenuate exponentially as a function of time and corresponding Rossby waves are stable for the type of perturbations considered.

If, however $c_i > 0$ for *at least one* normal mode, then the Rossby waves are unstable for the perturbations of wave number k fixed.

In the *barotropic case*, when:

$$U_B \equiv U_B^*(y),$$

the instability process is related essentially to the existence of the term $d^2 U_B^* / dy^2$ and the situation is then referred as a *barotropic instability*.

However, when

$$U_B = U_B(y,p),$$

*baroclinic case*, the vertical shearing, $\partial U_B/\partial p$, is an important cause of instability and the corresponding process gives us the *baroclinic instability*. The Eady (1949) model with:

$$U_B \equiv p, \quad \beta \equiv 0,$$

is a simple and very good example of baroclinic instability (see, the next section 33).

We note, finally, that for the *barotropic flow* the quasi-geostrophic model equation (32,1), with (32,2), reduces to the form:

(32,17)
$$S \frac{\partial \hat{\Lambda}^* \mathcal{H}^*_{qg}}{\partial t} + \frac{\partial (\mathcal{H}^*_{qg}, \hat{\Lambda}^* \mathcal{H}_{qg} + \beta y)}{\partial (x,y)} = 0,$$

with

(32,18)
$$\hat{\Lambda}^* \mathcal{H}^*_{qg} = \vec{D}^2 \mathcal{H}^*_{qg} - S \mathcal{H}^*_{qg}.$$

A method of deriving the *barotropic main equation of the quasi-geostrophic model* (32,17): it consists of taking the limit $K_0(p) \to 0$, in which limit the derivatives $\partial \mathcal{H}_{qg}/\partial s$ no longer depend on p; s=(x,y).

Therefore, for the barotropic instability, when we have $U_B \equiv U_B^*(y)$, we obtain for $\tilde{h}^*(y)$ the following equation, in place of (32,15),

(32,19)
$$\left[ U_B^*(y) - c \right] \left[ \frac{d^2 \tilde{h}^*}{dy^2} - (k^2 + S)\tilde{h}^* \right] + \left[ \beta - \frac{d^2 U_B^*}{dy^2} \right] \tilde{h}^* = 0.$$

## 33 . THE CLASSICAL
## EADY PROBLEM

According to Drazin (1978) we consider an inviscid, non-conducting, *Boussinesq fluid* (see the section 8 at the Chapter II) in an infinitely long rigid rectangular channel whose cross-section is given by

$$0 \leq y \leq L_0, \quad 0 \leq z \leq H_0.$$

The channel rotates with angular velocity $\Omega_0 \vec{k}$ and there is downwards gravitational acceleration $-g\vec{k}$.

It follows that the governing Boussinesq equations are

(33,1)
$$\frac{\partial \vec{u}}{\partial t} + (\vec{u}.\vec{\nabla})\vec{u} + 2\Omega_0 \vec{k} \wedge \vec{u} = -\frac{1}{\rho_0} \vec{\nabla} p + \alpha g(\theta - \theta_0)\vec{k},$$

(33,2)             $\dfrac{\partial \theta}{\partial t} + (\vec{u}.\vec{\nabla})\theta = 0,$

(33,3)             $\vec{\nabla}.\vec{u} = 0,$

where $\vec{u}$ is the fluid velocity relative to the roteting frame, $\rho = \rho_0[1-\alpha(\theta-\theta_0)]$ the density, p the relative pressure and $\theta$ the temperature of the fluid. We note also that $\rho_0$ a density scale, $\theta_0$ a temperature scale, $\alpha$ the constant coefficient of cubical expansion and $\vec{k}$ the unit vector in the direction of the upward vertical.

For inviscid fluid the boundary conditions are (with $\vec{u}=u\vec{i}+v\vec{j}+w\vec{k}$):

(33,4)             $v=0$ at $y=0, L_0$,

(33,5)             $w=0$ at $z=0, H_0$.

Consider now a basic flow $\vec{U}_B$ which will be perturbed to test its stability. We take the zonal flow increasing linearly with height:

(33,6)             $\vec{U}_B = \dfrac{U_0}{H_0} z\vec{i}$

where $U_0$ is a velocity scale. This is balanced geostrophically and hydrostatically by the basic temperature

(33,7)             $\theta_B = \theta_0 + \dfrac{\Delta T_0}{H_0}z - \dfrac{2\Omega_0 U_0}{\alpha g H_0}y$

and basic pressure

(33,8)             $p_B = \rho_0\left\{\dfrac{1}{2}\dfrac{\alpha g \Delta T_0}{H_0}z^2 - \dfrac{2\Omega_0 U_0}{H_0}yz\right\}$

where $\Delta T_0$ is a constant scale of basic vertical temperature difference across the channel. Finally the basic density is

(33,9)             $\rho_B = \rho_0\left\{1 - \dfrac{\alpha \Delta T_0}{H_0}\left[z - \dfrac{2\Omega_0 U_0}{\alpha g \Delta T_0}y\right]\right\}.$

We next scale the variables, denoting dimensionless ones by tildes:

$$(33,10) \quad \begin{cases} \tilde{x} = \dfrac{x}{L_0}, \ \tilde{y} = \dfrac{y}{L_0}, \ \tilde{z} = \dfrac{z}{H_0}, \ \tilde{t} = \dfrac{t}{L_0/U_0}, \\[2mm] \tilde{u} = \dfrac{u}{U_0}, \ \tilde{v} = \dfrac{v}{U_0}, \ \tilde{w} = \dfrac{w}{\varepsilon_0 U_0 Ro}, \\[2mm] \tilde{\theta} = \dfrac{\alpha g H_0}{2\Omega_0 U_0 L_0} \left\{ \theta - \theta_0 - \dfrac{\Delta T_0}{H_0} z \right\}, \\[2mm] \tilde{p} = \dfrac{1}{2\Omega_0 U_0 L_0 \rho_0} \left\{ p - \dfrac{1}{2} \dfrac{\alpha g \rho_0 \Delta T_0}{H_0} z^2 \right\}, \end{cases}$$

where $\varepsilon_0 = \dfrac{H_0}{L_0}$ and the Rossby number is defined by

$$(33,11) \qquad Ro = \frac{U_0}{2\Omega_0 L_0}.$$

The reason for scaling the vertical velocity with the Rossby number is that baroclinic instability occurs with $\tilde{w}$ of order one when the Rossby number is small, as will be seen later (see, also, the section 9).

It is also convenient at this stage to definie the *Burger number* by

$$(33,12) \qquad \mathcal{B} = \frac{\alpha g H_0 \Delta T_0}{4\Omega_0^2 L_0^2}.$$

It is an important number for large-scale meteorological problems, representing the square of the ratio of the buoyancy frequency to the Coriolis parameter.

Note also that

$$(33,13) \qquad \mathcal{B} = Ri.Ro^2$$

where the *Richardson number* is defined by

$$(33,14) \qquad Ri = \frac{\alpha g H_0 \Delta T_0}{U_0^2} \cong - \frac{g H_0^2}{U_0^2} \frac{1}{\rho} \frac{d\rho}{dz},$$

and $\rho$ is evaluated with $\theta = \theta_B$.

We wish to take $\Delta T_0 > 0$ so that hot fluid is above the cold and there is static stability. For the Westerlies in the atmosphere it is found that

$$(33,15) \qquad Ro \ll 1, \ \varepsilon_0 \ll 1 \ \text{and} \ \mathcal{B} \ll 1,$$

and therefore that the slope of the surfaces of constant density $\beta = \dfrac{2\Omega_0 U_0}{\alpha g \Delta T o} \ll 1$.

From now on we drop the tildes oven the dimensionless variables, and then the basic flow becomes:

(33,16)
$$\begin{cases} \vec{U}_B = z\vec{1}, \quad \theta_B = -y, \quad p_B = -yz, \\ \text{for } -\infty < x < +\infty, \quad 0 \le y, z \le 1. \end{cases}$$

We perturb this dimensionless basic flow (33,16), writing

(33,17) $\qquad \vec{u} = \vec{U}_B + \vec{u}', \quad \theta = \theta_B + \theta', \quad p = p_B + p',$

and substitute these expansions in the dimensionless equations corresponding to (33,1)-(33,3) to obtain for the perturbations the following governing equations:

(33,18a) $\qquad \text{Ro}\left\{\dfrac{\partial u'}{\partial t} + z\dfrac{\partial u'}{\partial x} + u'\dfrac{\partial u'}{\partial x} + v'\dfrac{\partial u'}{\partial y} + \text{Row}'\dfrac{\partial u'}{\partial z} + \text{Row}'\right\} - v' = -\dfrac{\partial p'}{\partial x};$

(33,18b) $\qquad \text{Ro}\left\{\dfrac{\partial v'}{\partial t} + z\dfrac{\partial v'}{\partial x} + u'\dfrac{\partial v'}{\partial x} + v'\dfrac{\partial v'}{\partial y} + \text{Row}'\dfrac{\partial v'}{\partial z}\right\} + u' = -\dfrac{\partial p'}{\partial y};$

(33,18c) $\qquad \varepsilon_0^2\text{Ro}\left\{\dfrac{\partial w'}{\partial t} + z\dfrac{\partial w'}{\partial x} + u'\dfrac{\partial w'}{\partial x} + v'\dfrac{\partial w'}{\partial y} + \text{Row}'\dfrac{\partial w'}{\partial z}\right\} - \theta' = -\dfrac{\partial p'}{\partial z};$

(33,18d) $\qquad \dfrac{\partial \theta'}{\partial t} + z\dfrac{\partial \theta'}{\partial x} + u'\dfrac{\partial \theta'}{\partial x} + v'\dfrac{\partial \theta'}{\partial y} + \text{Row}'\dfrac{\partial \theta'}{\partial z} - v' + \mathcal{B}w' = 0;$

(33,18e) $\qquad \dfrac{\partial u'}{\partial x} + \dfrac{\partial v'}{\partial y} + \text{Ro}\dfrac{\partial w'}{\partial z} = 0.$

The boundary conditions (33,4) and (33,5) give

(33,18f) $\qquad v' = 0$ at $y = 0, 1$;
and
(33,18g) $\qquad w' = 0$ at $z = 0, 1$.

Linearizing the perturbation equations (33,18a)-(33,18e), we find

(33,19a) $\qquad \text{Ro}\left[\dfrac{\partial u'}{\partial t} + z\dfrac{\partial u'}{\partial x} + \text{Row}'\right] - v' + \dfrac{\partial p'}{\partial x} = 0,$

(33,19b) $\qquad \text{Ro}\left[\dfrac{\partial v'}{\partial t} + z\dfrac{\partial v'}{\partial x}\right] + u' + \dfrac{\partial p'}{\partial y} = 0,$

(33,19c) $\qquad \varepsilon_0^2\text{Ro}\left[\dfrac{\partial w'}{\partial t} + z\dfrac{\partial w'}{\partial x}\right] - \theta' + \dfrac{\partial p'}{\partial z} = 0,$

(33,19d) $\qquad \dfrac{\partial \theta'}{\partial t} + z\dfrac{\partial \theta'}{\partial x} - v' + \mathcal{B}w' = 0.$

Elimination of p' from (33,19a) and (33,19b) gives the vorticity equation

(33,20)
$$\left[\frac{\partial}{\partial t} + z\frac{\partial}{\partial x}\right]\left(\frac{\partial v'}{\partial x} - \frac{\partial u'}{\partial y}\right) - \frac{\partial w'}{\partial z} = Ro\frac{\partial w'}{\partial y},$$

if we use of (33,18e).

But in the geostrophic limit as Ro$\to$0 (with t,x,y, and z fixed), equations (33,19a)-(33,19c) become

(33,21)
$$v' = \frac{\partial p'}{\partial x}, \quad u' = -\frac{\partial p'}{\partial y}, \quad \theta' = \frac{\partial p'}{\partial z}$$

and therefore equation (33,20) becomes, for Ro $\to$ 0,

(33,22)
$$\left[\frac{\partial}{\partial t} + z\frac{\partial}{\partial x}\right]\left(\frac{\partial^2 p'}{\partial x^2} + \frac{\partial^2 p'}{\partial y^2}\right) - \frac{\partial w'}{\partial z} = 0.$$

Similarly equation (33,19d) becomes

(33,23)
$$\left[\frac{\partial}{\partial t} + z\frac{\partial}{\partial x}\right]\frac{\partial p'}{\partial z} - \frac{\partial p'}{\partial x} + \mathcal{B}w' = 0.$$

Elimination of w' from equations (33,22) and (33,23) finally gives, the following equation for the perturbation of the pression p':

(33,24)
$$\left[\frac{\partial}{\partial t} + z\frac{\partial}{\partial x}\right]\left\{\frac{\partial^2 p'}{\partial z^2} + \mathcal{B}\left(\frac{\partial^2 p'}{\partial x^2} + \frac{\partial^2 p'}{\partial y^2}\right)\right\} = 0.$$

If we use (33,21) and (33,23), then the boundary conditions (33,18f) and (33,18g) give

(33,25a)
$$\frac{\partial p'}{\partial x} = 0 \quad \text{at } y=0,1;$$

(33,25b)
$$\left[\frac{\partial}{\partial t} + z\frac{\partial}{\partial x}\right]\frac{\partial p'}{\partial z} - \frac{\partial p'}{\partial x} = 0 \text{ at } z=0,1.$$

In summary, we must solve (33,24) with boundary conditions (33,25a,b). In this stability problem we use the method of normal modes, taking:

(33,26)
$$p' = \hat{p}(z)\exp[ik(x-ct)]\sin(\gamma y).$$

Now $\gamma = n\pi$ for some positive integer n in order to satisfy the boundary condition (33,25a) on the vertical walls.
Equation (33,24) gives

(33,27)
$$\frac{d^2 \hat{p}}{dz^2} - \mathcal{B}(k^2 + \gamma^2)\hat{p} = 0.$$

Therefore

(33,28)
$$\hat{p}(z) = A_0 \cosh(2qz) + B_0 \sinh(2qz),$$

where

(33,29)
$$q = \frac{1}{2}\left[\mathcal{B}(k^2 + n^2\pi^2)\right]^{1/2} \text{ for } n=1,2,\ldots \ ,$$

and $A_0$ and $B_0$ are some constants.
Boundary condition (33,25b) at z=0 gives
$$A_0 = -2cqB_0,$$
so

(33,30)
$$\hat{p} = B_0\left[\sinh(2qz) - 2cq \cosh(2qz)\right],$$

for an arbitrary constant $B_0$ of normalization.
Finally condition (33,25b) at z=1 gives the following eigenvalue relation

$$[4q^2\sinh(2q)]c^2 - [4q^2\sinh(2q)]c - \sinh(2q) + 2q\cosh(2q) = 0,$$

i.e.,

(33,31)
$$c = \frac{1}{2} \pm \frac{[(q\coth(q) -1)(q\tanh(q) -1)]^{1/2}}{2q}.$$

It follows that the mode is stable ($kc_i \leq 0$) if and only if $q \geq q_c \cong 1.2$, where $q_c$ is defined as the positive root of the transcendental equation

$$q\tanh q = 1.$$

Therefore the modes are stable if and only if

(33,32)
$$\mathcal{B} \geq \mathcal{B}^*(k,n) = \frac{4q_c^2}{k^2 + n^2\pi^2}$$

for given real k and positive integer n. Therefore the flow is stable if and only if:

(33,33)
$$\mathcal{B} \geq \max \mathcal{B}^*(k,n) \cong 0.58$$

the maximum occuring for k=0 and n=1.

The curves of marginal stability for the modes with n=1 and n=2 are shown in Figure 17 below.

Note that the longest waves are unstable for the largest values of $B$, but that their growth rates are not the largest because $kc_i \to 0$ as $k \to 0$.

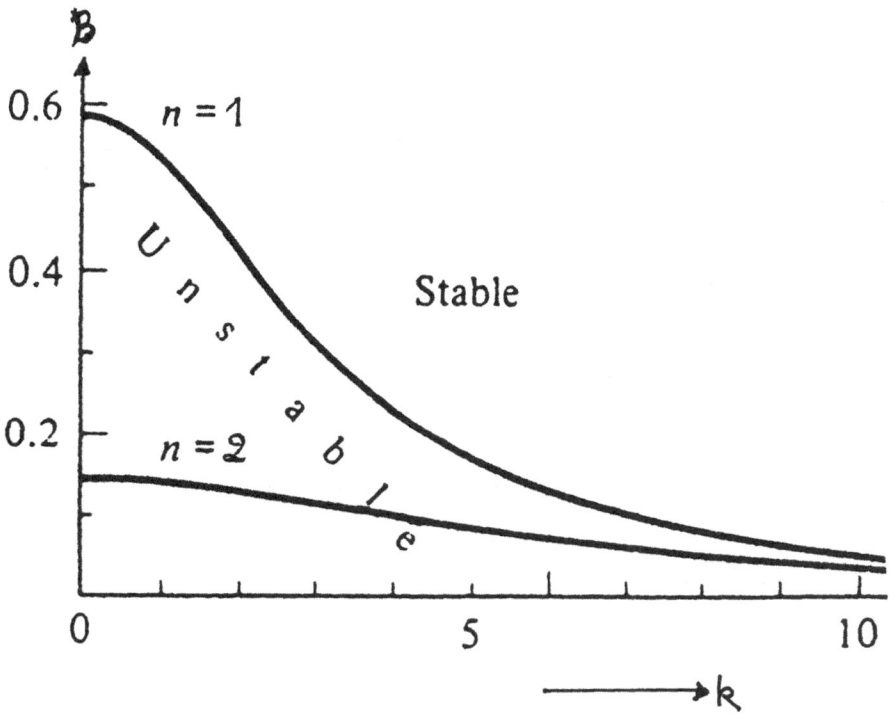

Fig. 17: The curves of marginal stability, in the $(k, B)$ plane with $B=4q_c^2/(k^2+ n^2\pi^2)$ for the first two modes (n=1 and n=2) of the Eady problem of baroclinic instability

## 34 . THE EADY PROBLEM FOR A SLIGHTLY
##    VISCOUS ATMOSPHERE

The effect of a little viscosity on baroclinic instability may be represented
approximately by taking thin viscous layers[†] on the horizontal walls.
We see that in a rapidly rotating slightly viscous fluid, a discontinuity of
the velocity of the inviscid (i.e. the "outer") solution at a rigid wall
generates a thin Ekman layer. The vorticity of the inviscid solution generates
a normal flux at the edge of the Ekman layer, so that the inviscid boundary
conditions: w=0 are replaced by

$$(34,1) \qquad w = \pm\frac{1}{2}\left(\frac{\nu_0}{\Omega_0}\right)^{1/2}\left(\frac{\partial v}{\partial x} - \frac{\partial u}{\partial y}\right), \text{ at } z = \begin{cases} 0, \\ H_0, \end{cases}$$

respectively, where $\nu_0$ is the kinematic viscosity (for the atmosphere we have
$\nu_0 \cong 5 \text{ m}^2/\text{s}$).
It may be shown (as at the section 28) that this leaves the Eady problem
unchanged except for the replacement of boundary conditions (33,18g) by

$$(34,2) \qquad Row' = \pm\left(\frac{1}{2} Ek\right)^{1/2}\left(\frac{\partial v'}{\partial x} - \frac{\partial u'}{\partial y}\right), \text{ at } z = 0,1,$$

where the Ekman number is defined as

$$(34,3) \qquad Ek = \frac{\nu_0}{2\Omega_0 H_0^2} .$$

To justify this we in fact need:

$$Ro^2 << Ek^{1/2} << 1, \quad \text{but } \frac{Ro}{Ek^{1/2}} \equiv \varepsilon_0 \frac{U_0}{\sqrt{2\Omega_0 L_0}} = O(1).$$

With (33,21) and (33,23), conditions (34,2) give respectively

$$(34,4) \qquad \left(\frac{\partial}{\partial t} + z\frac{\partial}{\partial x}\right)\frac{\partial p'}{\partial z} - \frac{\partial p'}{\partial x} = \pm\frac{\mathcal{B}}{Ro}\left(\frac{1}{2} Ek\right)^{1/2}\left[\frac{\partial^2 p'}{\partial x^2} + \frac{\partial^2 p'}{\partial y^2}\right], \text{ at } z=0,1.$$

Now we see that we can re-use almost all our calculations for the Eady
classical, inviscid, problem of the section 33. As before we find (see(33,28))

$$(34,5) \qquad \hat{p} = A_0\cosh(2qz) + B_0\sinh(2qz)$$

---

[†]  So-called, Ekman layers and see, for instance the section 28 (Chapter VII).

etc... Substituting (34,5) into (34,4), instead of (33,25b), we get

$$A_0 + 2cqB_0 = i\lambda A_0,$$

and

$$A_0 - 2q(1-c)B_0 + [B_0 - 2q(1-c)A_0]\tanh(2q) = -i\lambda[A_0 + B_0\tanh(2q)]$$

respectively, where

(34,6)
$$\lambda = \frac{\mathcal{B}}{Ro}\left[\frac{1}{2}Ek\right]^{1/2}\frac{(k^2 + n^2\pi^2)}{k}.$$

Eliminating $A_0$ and $B_0$ we deduce the eigenvalue relation

$$4q^2c(c-1)\tanh(2q) + 4i\lambda qc + 2q(1-i\lambda) - (1+\lambda^2)\tanh(2q) = 0.$$

Therefore we obtain for c the following relation:

$$c = \frac{1}{2} - \frac{i\lambda}{2q\tanh(2q)} \pm \frac{1}{2q\tanh(2q)}\left\{\left[(q^2+1)\tanh(2q) - 2q\right]\tanh(2q)\right.$$
$$\left. - \lambda^2[1-\tanh^2(2q)]\right\}^{1/2},$$

or else

(34,7)
$$c = \frac{1}{2} - i\frac{\lambda(1+T^2)}{4qT} \pm \frac{1}{4qT}\left\{4T(q-T)(qT-1) - \lambda^2(1-T^2)\right\}^{1/2},$$

where $T = \tanh q$ and $\tanh(2q) = \dfrac{2T}{(1+T)^2}$.

It can be seen that marginal stability occurs where $c=\frac{1}{2}$, there being instability of the mode if and only if $\lambda$ is less than the positive root of

$$\lambda^2 = \frac{(q-T)(1-qT)}{T}.$$

It is easier to plot the marginal curve by regarding $\lambda$ as a function of q than vice versa. Finally we note that

$$\frac{Ek\mathcal{B}^2}{Ro^2} \sim \lambda^2.$$

For further details the reader may consult the original work of Barcilon (1964).

## 35 . MORE ON BAROCLINIC
   INSTABILITY

We consider here the stability problem equation (32,15), and we will assume that $U_B$ depends only on p:

(35,1)          $U_B \equiv U_B(p)$.

We seek wave solutions $\tilde{h}(y,p)$ of equation (32,15) in the form

(35,2)          $\tilde{h}(y,p) = \tilde{\tilde{h}}(p)\exp(ily)$.

Then, setting $K^2 = k^2 + l^2$, we obtain for the complex amplitude $\tilde{\tilde{h}}(p)$ the following equation

(35,3)          $\left[U_B(p) - c\right]\left\{S\dfrac{d}{dp}\left(\dfrac{p^2}{K_0(p)}\dfrac{d\tilde{\tilde{h}}}{dp}\right) - K^2\tilde{\tilde{h}}\right\} + \mathcal{L}_B(p)\tilde{\tilde{h}} = 0,$

where

(35,4)          $\mathcal{L}_B(p) \equiv \beta - S\dfrac{d}{dp}\left(\dfrac{p^2}{K_0(p)}\dfrac{dU_B(p)}{dp}\right).$

For the equation (35,3) the boundary condition for p=1 (see (32,16)) reduces to the form:

(35,5)          $\left[U_B(1) - c\right]\left\{\dfrac{d\tilde{\tilde{h}}}{dp}\bigg|_{p=1} + \dfrac{K_0(1)}{T_0(1)}\tilde{\tilde{h}}\bigg|_{p=1}\right\} - N_B(1)\tilde{\tilde{h}}\bigg|_{p=1} = 0,$

where

(35,6)          $N_B(1) = \dfrac{dU_B(p)}{dp}\bigg|_{p=1} + \dfrac{K_0(1)}{T_0(1)}U_B(1).$

We note that the boundary condition (35,6), for p=1, is complex: it contains the eigenvalue c. Replacing it by the condition $(N_B(1)\equiv 0)$:

(35,5a)          $\dfrac{d\tilde{\tilde{h}}}{dp}\bigg|_{p=1} + \dfrac{K_0(1)}{T_0(1)}\tilde{\tilde{h}}(1) = 0,$

we can prove stability for $\mathcal{L}_B(p)>0$, or:

   if $\mathcal{L}_B(p)$ changes sign once, then the flow is stable for $(U_B(p)-K_0)\mathcal{L}_B(p)<0,$

where $K_0$ is the value of $U_B(p)$ at the point where $\mathcal{L}_B(p)$ changes sign[+].

The condition (35,5), however, must give rise to instability, but only of a type in which there cannot be more than one growing wave solution for each K. We can prove this by a finite - difference approximation of equations (35,3), with (35,5), in which the segment $0 \leq p \leq 1$ is broken up by the points $p_1, \ldots, p_{N-1}$ into N equal parts of length $\delta = \frac{1}{N}$, and the equation and condition are written in the equivalent form:

$$(35,7) \qquad \left[\left(U_B\right)_n - c\right]\left\{ r^2_{n-1/2} \frac{\tilde{\tilde{h}}_{n-1} - \tilde{\tilde{h}}_n}{\delta^2} - K^2\tilde{\tilde{h}}_n - r^2_{n+1/2} \frac{\tilde{\tilde{h}}_n - \tilde{\tilde{h}}_{n+1}}{\delta^2} \right\} + \left(\mathcal{L}_B\right)_n \tilde{\tilde{h}}_n = 0;$$

$$(35,8) \qquad \left[\left(U_B\right)_0 - c\right]\left\{ \frac{\tilde{\tilde{h}}_0 - \tilde{\tilde{h}}_1}{\delta} + s^2_0 \tilde{\tilde{h}}_0 \right\} = \left(N_B\right)_0 \tilde{\tilde{h}}_0,$$

where $r^2_{n-1/2}$ is some average value of $(S\, p^2 / K_0(p))$ between the point $p_n$ and $p_{n-1}$, while $s^2_0 = K_0(1) / T_0(1)$.

Then the following theorem holds (Dikii (1973)):

If all $(\mathcal{L}_B)_n > 0$ or all $(\mathcal{L}_B)_n < 0$ (Rayleigh condition), or if the sequence $(\mathcal{L}_B)_n$ changes sign once, and if there exists a constant $K_0$ for which $[(U_B)_n - K_0](\mathcal{L}_B)_n < 0$ (the Fjortoft condition), then equation (35,7) with the boundary conditions

$$(35,9) \qquad \tilde{\tilde{h}}_N = \tilde{\tilde{h}}_{N-1},$$

$$(35,10) \qquad \tilde{\tilde{h}}_1 - \tilde{\tilde{h}}_0 = K(c)\tilde{\tilde{h}}_0, \quad K(c) = a + \frac{K}{b-c}, \quad a \geq 0,$$

has no more than one pair of non-real complex-conjugate eingenvalues c.

The idea of the proof involves the fact that the eigenvalues are obtained graphically as intersections of the function $M(c)=(\tilde{\tilde{h}}_1 - \tilde{\tilde{h}}_0)/\tilde{\tilde{h}}_0$ and the hyperbola K(c), while M(c) is a rational fraction whose denominator is a polynomial of degree N-1 without any roots that are not real.

---

† An analog of the Fjortoft (1950) theorem; see for instance the book of Drazin and Reid (1981; §22).

*CONDITIONS FOR INSTABILITY*

The necessary conditions for instability can be derived directly from (35,3). If $c = c_r + ic_i$ and $c_i \neq 0$, then (35,3) can always be devided by $[U_B - c]$, since $[U_B - c]$ is never zero. If the resulting equation is multiplied by $\tilde{\tilde{h}}^{\circledast}$ (the complex conjugate of $\tilde{\tilde{h}}$) and integrated of p=1 at p=0, we obtain after integration by part:

$$S\int_1^0 \frac{p^2}{K_0(p)}\left|\frac{d\tilde{\tilde{h}}}{dp}\right|^2 dp - \int_1^0 \frac{\mathcal{L}_B(p)}{U_B(p)-c}|\tilde{\tilde{h}}|^2 dp + K^2\int_1^0 |\tilde{\tilde{h}}|^2 dp$$

$$+ S\lim_{p\to 0}\left[\frac{p^2}{K_0(p)}\tilde{\tilde{h}}^{\circledast}\frac{d\tilde{\tilde{h}}}{dp}\right] - \frac{S}{K_0(1)}\tilde{\tilde{h}}^{\circledast}(1)\frac{d\tilde{\tilde{h}}}{dp}\bigg|_{p=1} = 0.$$

But if we take into account the boundary condition (35,5), with (35,6), it immediately follows that:

(35,11)
$$S\int_1^0 \frac{p^2}{K_0(p)}\left|\frac{d\tilde{\tilde{h}}}{dp}\right|^2 dp + K^2\int_1^0 |\tilde{\tilde{h}}|^2 dp = \int_1^0 \frac{\mathcal{L}_B(p)}{U_B(p)-c}|\tilde{\tilde{h}}|^2 dp$$

$$+ S\lim_{p\to 0}\left[\frac{p^2}{K_0(p)}\tilde{\tilde{h}}^{\circledast}\frac{d\tilde{\tilde{h}}}{dp}\right] + S\left[\frac{1}{T_0(1)} - \frac{N_B(1)}{K_0(1)(U_B(1)-c)}\right]|\tilde{\tilde{h}}|^2_{p=1}.$$

Since

$$\frac{1}{U_B - c} = \frac{1}{|U_B - c|^2}\left\{U_B - c_r + ic_i\right\},$$

the imaginary part of (35,11) may be written

(35,12)
$$c_i\left\{\int_1^0 \frac{|\tilde{\tilde{h}}|^2}{|U_B - c|^2}\mathcal{L}_B(p)dp - \frac{N_B(1)}{K_0(1)|U_B(1)-c|^2}|\tilde{\tilde{h}}|^2_{p=1}\right\} = 0,$$

*if we suppose* that

(35,13)
$$\lim_{p\to 0}\left[\frac{p^2}{K_0(p)}\tilde{\tilde{h}}^{\circledast}\frac{d\tilde{\tilde{h}}}{dp}\right] = 0.$$

Hence, if $c_i$ is not to equal to zero, i.e., if the mode is to be unstable, then the bracket { }, in (35,12), multiplied by $c_i$ must vanish. Therefore the vanishing of the bracket { }, in (35,12), is a necessary condition for instability:

$$(35,14) \qquad \int_1^0 \frac{|\tilde{\tilde{h}}|^2}{|U_B - c|^2} \mathcal{L}_B(p)dp = \frac{N_B(1)}{K_0(1)|U_B(1)-c|^2} |\tilde{\tilde{h}}|^2_{p=1}.$$

The real part of (35,11) yields, with (35,14) for $c_i \neq 0$, and $c_0$ any constant,

$$(35,15) \qquad \int_1^0 \frac{|\tilde{\tilde{h}}|^2}{|U_B - c|^2}(U_B-c_0)\mathcal{L}_B(p)dp + S\left[\frac{1}{T_0(1)} - \frac{N_B(1)(U_B - c_0)}{K_0(1)|U_B(1)-c|^2}\right]|\tilde{\tilde{h}}|^2_{p=1}$$

$$= S\int_1^0 \left\{ \frac{p^2}{K_0(p)}\left|\frac{d\tilde{\tilde{h}}}{dp}\right|^2 + \frac{K^2}{S}|\tilde{\tilde{h}}|^2 \right\}dp > 0.$$

Therefore certain stringent conditions must be satisfied by the basic state in order that (35,14) and (35,15) may be satisfied.

## 36 . BAROTROPIC INSTABILITY

We consider the equation (32,19):

$$(36,1) \qquad [U_B^*(y) - c]\left[\frac{d^2\tilde{h}^*}{dy^2} - \mu^2\tilde{h}^*\right] + \left[\beta - \frac{d^2U_B^*}{dy^2}\right]\tilde{h}^* = 0,$$

where $\mu^2 = k^2 + S$.

We assume that at $y=\pm1$ rigid walls exist containing the region of the atmospheric flow and the perturbations. Although clearly an artifice, this effectively isolates the region from its surroundings and assures that should an instability arise, its source must lie within the region under consideration. Under these conditions it follows that[†]:

$$(36,2) \qquad \tilde{h}^* = 0, \quad y = \pm1.$$

In this case the necessary condition for instability, is easily obtained if the equation (36,1) is multiplied by the complex conjugate of $\tilde{h}^*$, after the division by $[U_B^*(y)-c]$, and integrated of $y=-1$ at $y=+1$. Hence we obtain after integration by part:

$$(36,3) \qquad \int_{-1}^{+1}\left|\frac{d\tilde{h}^*}{dy}\right|^2 dy + \mu^2\int_{-1}^{+1}|\tilde{h}^*|^2 dy - \int_{-1}^{+1}\left(\beta - \frac{d^2U_B^*}{dy^2}\right)\frac{|\tilde{h}^*|^2}{[U_B^*(y) - c]}dy = 0,$$

---

[†] If we take into account (9,7), (32,5) and (32,13) for the barotropic case.

if we take into account (36,2). The imaginary part of (36,3) may be written in the following form:

$$(36,4) \qquad c_i \int_{-1}^{+1} \left[ \beta - \frac{d^2 U_B^*(y)}{dy^2} \right] \frac{|\tilde{h}^*|^2}{|U_B^*(y) - c|^2} \, dy = 0,$$

to yield Kuo's (1949) theorem.

The real part of (36,3) yields, with (36,4) for $c_i \neq 0$ and $c_0$ any constant,

$$(36,5) \qquad \int_{-1}^{+1} \left\{ \mu^2 |\tilde{h}^*|^2 + \left| \frac{d\tilde{h}^*}{dy} \right|^2 \right\} dy = \int_{-1}^{+1} \left[ \beta - \frac{d^2 U_B^*(y)}{dy^2} \right] \frac{(U_B^*(y) - c_0)}{|U_B^*(y) - c|^2} |\tilde{h}^*|^2 dy.$$

$\frac{d\Pi^*}{dy} \equiv \beta - \frac{d^2 U_B^*(y)}{dy^2}$ is the gradient of the basic state absolute vorticity. The

relation (36,4) shows that for unstable disturbances $(c_i \neq 0)$ to exist in the barotropic system, the gradient of the absolute vorticity, $\frac{d\Pi^*}{dy}$, must change sign within the range $(-1,+1)$. That is to say, for inviscid unstable barotropic disturbances to exist , the basic velocity disturbtion must be such that $d^2 U_B^*(y)/dy^2$ is able to over-balance $\beta$ to make $\beta - d^2 U_B^*(y)/dy^2$ change its sign. This is Kuo's (1949) extension of Rayleigh's theorem of instability to include the influence of $\beta$ on the stability of the zonal flow. Thus, a positive (negative) $\beta$ has a stabilizing (destabilizing) influence on the wave motions in the westerlies and a destabilizing (stabilizing) influence on wave motions in the easterlies.

Equation (36,5) also shows that $(U_B^*(y) - c_0)$ and $d\Pi^*/dy$ must be positively correlated within $(-1,+1)$ for free disturbance to exist, whereas (36,4) together with (36,5) show that, for unstable disturbance to exist, the product $U_B^*(y) \frac{d\Pi^*}{dy}$ must be positive at least in part of $(-1,+1)$ in addition to the change of sign of $d\Pi^*/dy$. Further, this relation requires $|\tilde{h}^*|^2$ to be large in the region of positive $U_B^*(y) \frac{d\Pi^*}{dy}$ and small in the region of negative $U_B^*(y) \frac{d\Pi^*}{dy}$. This condition has often been stated incorrectly to imply that the existence of unstable disturbances requires $U_B^*(y) \frac{d\Pi^*}{dy}$ to be positive at every point.

## 37 . THE TAYLOR-GOLDSTEIN EQUATION
## AND STABILITY OF STRATIFIED
## SHEAR ISOCHORIC FLOW

We consider here two-dimensional motions ($v \equiv 0$ and $\frac{\partial}{\partial y} \equiv 0$) of stably stratified, nonrotating inviscid isochoric flow in which there is a steady mean shear flow $U_\infty(z)$ in the x-direction. If $p_\infty(z)$ and $\rho_\infty(z)$ denote the hydrostatic pressure and density as usual, the basic state $p_\infty$, $\rho_\infty$ and $\vec{u}_\infty = [U_\infty(z),0,0]$ exactly satisfies the two-dimensional (nonlinear) adiabatic, isochoric equations (see the eqs.(14,3)):

$$
(37,1) \qquad
\begin{cases}
\rho\dfrac{D\vec{u}}{Dt} + \vec{\nabla}p + \rho g\vec{k}; \\[2mm]
\vec{\nabla}.\vec{u} = 0; \\[2mm]
\dfrac{D\rho}{Dt} = 0,
\end{cases}
$$

with $\dfrac{D}{Dt} = \dfrac{\partial}{\partial t} + u\dfrac{\partial}{\partial x} + w\dfrac{\partial}{\partial z}$ and $\vec{\nabla} = \left[\dfrac{\partial}{\partial x} , \dfrac{\partial}{\partial z}\right]$.

We now suppose there are small perturbations of this basic state. Thus for the total field we let:

$$
(37,2) \qquad \vec{u} = (U_\infty + u',0,w'), \quad p = p_\infty + p', \quad \rho = \rho_\infty + \rho',
$$

where the perturbation (primed) quantities are functions of x, z and t and are assumed to be small compared to their basic-state counterparts. Substituting (37,2) into (37,1), using the hydrostatic relation for $p_\infty$ and $\rho_\infty$ and then neglecting the products of all perturbations quantities, we obtain, upon dropping the primes, the following linearized equations:

$$
(37,3) \qquad \frac{D_0 u}{Dt} + \frac{dU_\infty}{dz} w + \frac{1}{\rho_\infty} \frac{\partial p}{\partial x} = 0;
$$

$$
(37,4) \qquad \frac{D_0 w}{Dt} + \frac{1}{\rho_\infty} \frac{\partial p}{\partial z} = -\frac{\rho}{\rho_\infty}g;
$$

$$
(37,5) \qquad \frac{D_0 \rho}{Dt} + \frac{d\rho_\infty}{dz} w = 0;
$$

$$
(37,6) \qquad \frac{\partial u}{\partial x} + \frac{\partial w}{\partial z} = 0,
$$

where $\dfrac{D_o}{Dt} = \dfrac{\partial}{\partial t} + U_\infty \dfrac{\partial}{\partial x}$ .

We now assume that each dependent function is of the plane-wave form

$$(37,7) \qquad \phi(x,z,t) \equiv \begin{bmatrix} u \\ w \\ p \\ \rho \end{bmatrix} = \phi(z)e^{ik(x-ct)}, \quad k>0.$$

under this assumption (37,3)-(37,6) reduce to:

$$(37,8) \qquad \rho_\infty\left[ik(U_\infty - c)u + \dfrac{dU_\infty}{dz} w\right] + ikp = 0;$$

$$(37,9) \qquad \rho_\infty ik(U_\infty - c)w + \dfrac{dp}{dz} = -\rho g;$$

$$(37,10) \qquad ik(U_\infty - c)\rho + \dfrac{d\rho_\infty}{dz} w = 0;$$

$$(37,11) \qquad iku + \dfrac{dw}{dz} = 0,$$

where $u$, $w$, $p$, $\rho$, $U_\infty$ and $\rho_\infty$ are functions of $z$ alone.

We now use (37,11) to eliminate $u$ in (37,8) and substitute the resulting expression for $p$ into (37,9). Then, upon eliminating $\rho$ by means of (37,10), (37,9) yields the following equation for $w(z)$:

$$(37,12) \qquad \dfrac{d}{dz}\left[\rho_\infty(U_\infty - c)\dfrac{dw}{dz}\right] - \dfrac{d}{dz}\left(\rho_\infty \dfrac{dU_\infty}{dz} w\right)$$

$$- \left[\dfrac{d\rho_\infty}{dz} \dfrac{g}{(U_\infty - c)} + \rho_\infty k^2(U_\infty - c)\right]w = 0.$$

This is known as the Taylor-Goldstein equation; it was first derived by Taylor (1931) and Goldstein (1931) in their studies of the stability of stratified shear flow. When $\rho_\infty$ is held constant in the first two terms, we readily retrieve the Boussinesq form

$$(37,13) \qquad (U_\infty - c)^2 \dfrac{d^2w}{dz^2} + \left\{N_\infty^2 - (U_\infty - c)\dfrac{d^2U_\infty}{dz^2} - (U_\infty - c)^2 k^2\right\}w = 0,$$

with $\quad N_\infty^2 = -\dfrac{g}{\rho_\infty^0} \dfrac{d\rho_\infty}{dz}$, for static stability, and $\rho_\infty^0$=constant.

We note here two fundamental restrictions associated with (37,12). The first concerns the two-dimensional nature of the disturbance, i.e., waves travelling only parallel to the mean flow were introduced. This simplification has its origin in *Squire's theorem* (Squire (1933)), which for a homogeneous fluid states that:

"For each unstable three-dimensional wave there is always a more unstable two-dimensional one travelling parallel to the flow".

For a simple proof of this, see Drazin and Howard (1966). Yih (1955) extended this theorem to the case of stratified fluids, which provides the motivation for our consideration of two-dimensional disturbances only since these are always the fastest growing waves. On the other hand, waves travelling strictly normal to the flow are unaffected by the current (Yih (1965)).

The second restriction concerns the plane-wave decomposition (37,7); this approach a priori eliminates the *transient* solution (a continuous spectrum) that would arise in any initial value calculation. However, for large time the solution will be dominated by the growing unstable plane-wave modes and the latter thus deserve first consideration.

For flows with a smoothly varying shear, $\dfrac{d^2 U_\infty}{dz^2} \neq 0$, (37,12) no longer has elementary functions as solutions, independently of whether $\rho_\infty$ is continuous or discontinuous. While some analytical sollutions of (37,12) are possible in terms of special functions (e.g., see Drazin and Howard (1966); Thorpe (1969)), much effort has been expended on seeking numerical solutions (see Turner (1973) for a discussion of some of these).

## SYNGE'S GENERALIZED RAYLEIGH CRITERION

More a century ago Rayleigh (1880) showed that for flow of a homogeneous fluid with rigid boundaries (or boundaries at infinity), a necessary condition for instability is that the profil $U_\infty(z)$ should have *at least one point of inflection*. An analogous but more complicated necessary condition for instability was obtained by Singe (1933) for the case of a stratified fluid. However, his paper was overlooked for several decades and the same result was proved independently by Yih (1957) and Drazin (1958).

To prove[†] Synge's *necessary* condition, we start with the Taylor-Goldstein equation (37,12) written in the following form:

(37,14)
$$\frac{d}{dz}\left[\rho_\infty \omega \frac{dw}{dz}\right] - \frac{d}{dz}\left[\rho_\infty \frac{dU_\infty}{dz} w\right] + \rho_\infty \left[\frac{N_\infty^2}{\omega} - k^2\omega\right] w = 0, \quad (z_1 \le z \le z_2),$$

where $\omega \equiv U_\infty - c$. Since $\frac{d\omega}{dz} \equiv \frac{dU_\infty}{dz}$ the first two terms, in the equation (37,14) can be written as

$$\omega \frac{d}{dz}\left[\rho_\infty \frac{dw}{dz}\right] - \frac{d}{dz}\left[\rho_\infty \frac{dU_\infty}{dz}\right]w;$$

hence (37,14) becomes

(37,15)
$$\frac{d}{dz}\left[\rho_\infty \frac{dw}{dz}\right] - \left\{\frac{d}{dz}\left[\rho_\infty \frac{dU_\infty}{dz}\right]\frac{1}{\omega} - \rho_\infty \frac{N_\infty^2}{\omega^2} + \rho_\infty k^2\right\} w = 0.$$

Next we multiply (37,15) by $w^*$ (the complex conjugate of $w$), and its complex conjugate by $w$ substract and then integrate from $z_1$ to $z_2$:

(37,16)
$$\int_{z_1}^{z_2}\left[w^* \frac{d}{dz}\left[\rho_\infty \frac{dw}{dz}\right] - w \frac{d}{dz}\left[\rho_\infty \frac{dw^*}{dz}\right]\right]dz$$

$$- \int_{z_1}^{z_2}|w|^2\left\{\frac{d}{dz}\left[\rho_\infty \frac{dU_\infty}{dz}\right]\left[\frac{1}{\omega} - \frac{1}{\omega^*}\right] - \rho_\infty N_\infty^2\left[\frac{1}{\omega^2} - \frac{1}{\omega^{*2}}\right]\right\}dz = 0,$$

where $|w|^2 \equiv ww^*$.

If the boundaries are rigid or at infinity, $w$ and $w^*$ vanish there and the first integral in (37,16) drops out after an integration by parts. Putting $c = c_r + ic_i$ in the second integral and simplifying yields

(37,17)
$$c_i\int_{z_1}^{z_2}\frac{|w|^2}{[(U_\infty - c_r)^2 + c_i^2]^2}\left\{\frac{d}{dz}\left[\rho_\infty \frac{dU_\infty}{dz}\right][(U_\infty - c_r)^2 + c_i^2]\right.$$

$$\left. - 2\rho_\infty N_\infty^2(U_\infty - c_r)\right\}dz = 0.$$

Hence if $c_i > 0$ (unstable waves), the expression in the curly brackets must change sign and hence must vanish for some $z \in [z_1, z_2]$. Thus formally, we obtain *Synge's theorem*:

---

[†] According to Leblond and Mysak (1978; §43).

"A necessary condition for a stratified shear flow to be *unstable* is that

(37,18)    $\dfrac{d}{dz}\left[\rho_\infty \dfrac{dU_\infty}{dz}\right][(U_\infty - c_r)^2 + c_i^2] = 2\rho_\infty N_\infty^2(U_\infty - c_r)$

for at least one value of $z \in [z_1, z_2]$".

When $\rho_\infty$ = constant (homogeneous fluid, $N_\infty^2 \equiv 0$), (37,18) reduces to

$$\dfrac{d^2 U_\infty}{dz^2} = 0,$$

which is Rayleigh's well-known necessary condition for instability. The condition (37,18) has no simple interpretation, however, since it involves the unknown eigenvalue $c = c_r + ic_i$.

*MILES' SUFFICIENCY CONDITION*
*FOR STABILITY*

Of the many stability properies Miles (1961) established for stratified shear flow, the most celebrated is undoubtedly the stability criteria involving the Richardson number (see, also, the formula (33,14)):

(37,19)    $Ri = \dfrac{N_\infty^2}{\left[\dfrac{dU_\infty}{dz}\right]^2}$ .

On the assumption of analyticity of $\dfrac{dU_\infty}{dz}$ and $\rho_\infty$, Miles showed that a *sufficient condition* for stability is that:

" $Ri \geq \dfrac{1}{4}$ *everywhere in the flow.* "

Here we shall present Howard's (1961) proof of this theorem, which is simpler and does not require the analyticity assumption.
We make, first, two transformations in the equation (37,12). First let $w=\chi\omega$, where $\chi=\chi(z)$ represents a new dependent variable. Then (37,12) becomes

(37,20)    $\dfrac{d}{dz}\left[\rho_\infty \omega\left[\omega \dfrac{d\chi}{dz} + \dfrac{d\omega}{dz}\chi\right]\right] - \dfrac{d}{dz}\left[\rho_\infty \omega \dfrac{d\omega}{dz}\chi\right] + \rho_\infty(N_\infty^2 - k^2\omega^2)\chi = 0, \quad z_1 \leq z \leq z_2.$

The rigid-wall boundary conditions imply that $\chi(z_1) = \chi(z_2) = 0$ provided $\omega \neq 0$ at $z = z_1$ and $z_2$, which we will assume to be the case. Carrying out the differentiation in (37,20) gives

(37,21)
$$\frac{d}{dz}\left[\rho_\infty \omega^2 \frac{d\chi}{dz}\right] + \rho_\infty(N_\infty^2 - k^2\omega^2)\chi = 0.$$

Suppose now that $\chi(z)$ is an *unstable* solution. Then $c = c_r + ic_i$ is complex and $\omega = U_\infty - c \neq 0$, for any $z$, and we can choose one branch of $\sqrt{w}$ for all $(z_1, z_2)$ which be as differentiable as is $U_\infty$. Now set

$$\chi = \psi/\sqrt{w}$$

in (37,21). After a little algebra it follows that

(37,22)
$$\frac{d}{dz}\left[\rho_\infty \omega \frac{d\psi}{dz}\right] - \left[\frac{1}{2}\frac{d}{dz}\left(\rho_\infty \frac{dU_\infty}{dz}\right) + \rho_\infty k^2 \omega + \rho_\infty \frac{1}{\omega}\left(\frac{1}{4U_\infty} - N_\infty^2\right)\right]\psi = 0, \quad z_1 \le z \le z_2,$$

with $\psi(z_1) = \psi(z_2) = 0$. Multiplying by $\psi^*$ (the complex conjugate of $\psi$) and integrating over $(z_1, z_2)$, (37,22) yields after integrating by parts

(37,33)
$$\int_{z_1}^{z_2} \rho_\infty \omega \left[\left|\frac{d\psi}{dz}\right|^2 + k^2|\psi|^2\right]dz + \frac{1}{2}\int_{z_1}^{z_2}\frac{d}{dz}\left(\rho_\infty \frac{dU_\infty}{dz}\right)|\psi|^2 dz$$

$$+ \int_{z_1}^{z_2} \rho_\infty \omega^* \left|\frac{\psi}{\omega}\right|^2 \left[\frac{1}{4}\left(\frac{dU_\infty}{dz}\right)^2 - N_\infty^2\right]dz = 0.$$

Equating the imaginary part of (37,23) to zero, we obtain

(37,24)
$$c_i\left\{\int_{z_1}^{z_2}\rho_\infty\left[\left|\frac{d\psi}{dz}\right|^2 + k^2|\psi|^2\right]dz + \int_{z_1}^{z_2}\rho_\infty\left|\frac{\psi}{\omega}\right|^2\left[-\frac{1}{4}\left(\frac{dU_\infty}{dz}\right)^2 + N_\infty^2\right]dz\right\} = 0.$$

Hence if $c_i > 0$ (unstable waves), (37,24) implies that $-\frac{1}{4}\left(\frac{dU_\infty}{dz}\right)^2 + N_\infty^2 < 0$ for some range of $z$. Thus, as Howard put it, a necessary condition for instability is that $-\frac{1}{4}\left(\frac{dU_\infty}{dz}\right)^2 + N_\infty^2$ be somewhere negative.

On the other hand if $-\frac{1}{4}\left(\frac{dU_\infty}{dz}\right)^2 + N_\infty^2 \ge 0$ everywhere, then (37,24) implies that $c_i = 0$. Thus we obtain (for $\frac{dU_\infty}{dz} \neq 0$) Miles' theorem. Unlike Singe's necessary condition for instability, Miles' sufficient condition for stability has a simple physical interpretation. Since the Richardson number represents the ratio of buoyancy to inertia, Miles' theorem effectively state that:

*if the stabilizing influence of stratification dominates the destabilizing influence of the nonlinear terms, then the flow is stable.*

*HOWARD'S SEMICIRCLE THEOREM*

This theorem defines a semicircular region in the complex c-plane in which the eigenvalue c for an unstable wave is located. This result was first established by Howard (1961).

Our starting point is (37,21), the equation for $\chi$, where $w=\omega\chi=(U_\infty - c)\chi$. If this equation is multiplied by $\chi^*$ and the result integrated over $(z_1,z_2)$, we find (again, for rigid boundaries or boundaries at infinity)

$$(37,25) \qquad \int_{z_1}^{z_2} \rho_\infty \omega^2 \left[\left|\frac{d\chi}{dz}\right|^2 + k^2|\chi|^2\right]dz - \int_{z_1}^{z_2} \rho_\infty N_\infty^2 |\chi|^2 dz = 0.$$

Setting the real and imaginary parts of (37,25) equal to zero gives

$$(37,26a) \qquad \int_{z_1}^{z_2} \rho_\infty[(U_\infty - c_r)^2 - c_i^2]\left\{\left|\frac{d\chi}{dz}\right|^2 + k^2|\chi|^2\right\}dz - \int_{z_1}^{z_2} \rho_\infty N_\infty^2 |\chi|^2 dz = 0,$$

$$(37,26b) \qquad c_i\int_{z_1}^{z_2} \rho_\infty(U_\infty - c_r)\left[\left|\frac{d\chi}{dz}\right|^2 + k^2|\chi|^2\right]dz = 0.$$

The second relation immediately gives another of Synge's results (Synge (1933)): "*If $c_i > 0$, then there exists some point $z_r$ such that $c_r = U_\infty(z_r)$*". This is sometimes rephrased to read:

"*Instability implies that $c_r$ must lie in the range of $U_\infty$*".

Now let

$$(37,27) \qquad Q = \rho_\infty\left[\left|\frac{d\chi}{dz}\right|^2 + k^2|\chi|^2\right] > 0.$$

Then (37,26b) becomes, for $c_i > 0$,

$$(37,28) \qquad \int_{z_1}^{z_2} U_\infty Q dz = c_r\int_{z_1}^{z_2} Q dz,$$

and by virtue of (37,28), (37,26a) reduces to

$$(37,29) \qquad \int_{z_1}^{z_2} U_\infty^2 Q dz = (c_r^2 + c_i^2)\int_{z_1}^{z_2} Q dz + \int_{z_1}^{z_2} \rho_\infty N_\infty^2|\chi|^2 dz.$$

We now suppose that $(U_\infty)_{min} \equiv a \le U_\infty(z) \le b \equiv (U_\infty)_{max}$.
Then

(37,30)
$$0 \geq \int_{z_1}^{z_2} Q(U_\infty - a)(U_\infty - b)dz = \int_{z_1}^{z_2} U_\infty^2 Q dz - (a+b)\int_{z_1}^{z_2} U_\infty Q dz$$

$$+ ab\int_{z_1}^{z_2} Q dz = \left[ c_r^2 + c_i^2 - (a+b)c_r \right.$$

$$\left. + ab\right]\int_{z_1}^{z_2} Q dz + \int_{z_1}^{z_2} \rho_\infty N_\infty^2 |\chi|^2 dz$$

using (37,28) and (37,29).

But $\int_{z_1}^{z_2} Q dz > 0$ and $\int_{z_1}^{z_2} \rho_\infty N_\infty^2 |\chi|^2 dz > 0$; hence the inequality (37,30) implies

that $c_r^2 + c_i^2 - (a+b)c_r + ab \leq 0$ or equivalently

(37,31)
$$\left[ c_r - \frac{1}{2}(a+b) \right]^2 + c_i^2 \leq \left[ \frac{1}{2}(a-b) \right]^2.$$

Thus we have *Howard's semicircle theorem*:

"*The complex wave velocity c for any unstable mode ($c_i > 0$) must lie on or inside the semicircle in the upper half c-plane with center at $[(a+b)/2, 0]$ and diameter equal to the range of $U_\infty$*".


# 38 . THE CONVECTIVE INSTABILITY PROBLEM[†]

Here in nondimensional variables we consider the following equations, for the atmospheric motion:

(38,1)
$$\begin{cases} \dfrac{\partial \rho}{\partial t} + \vec{\nabla}.(\rho\vec{u}) = 0; \\[2mm] \rho\dfrac{D\vec{u}}{Dt} + \dfrac{1}{\gamma M^2}\vec{\nabla}p + \dfrac{1}{Fr^2}\rho\vec{k} = \dfrac{1}{Re}\left[\Delta\vec{u} + \dfrac{1}{3}\vec{\nabla}(\vec{\nabla}.\vec{u})\right]; \\[2mm] \dfrac{1}{\Xi}\dfrac{D\Xi}{Dt} = \dfrac{1}{\rho}\left\{\dfrac{1}{RePr}\Delta T + \dfrac{(\gamma-1)}{Re}M^2\phi + Q_0 Q(z)\right\}; \\[2mm] p = \rho T. \end{cases}$$

[†] According to Bois (1984).

Note that these equations (38,1) are normalized with the same length ($L_0$) for horizontal and vertical scales ($\varepsilon_0 \equiv 1$).

The quantity $\Xi$, which is the *potential* temperature of the atmospheric medium, is related to p, $\rho$ and T by the formulae:

$$(38,2) \qquad \Xi = \frac{T}{P^{(\gamma-1)/\gamma}} = \frac{p^{1/\gamma}}{\rho}.$$

The quantity $\phi$ appearing in the energy equation is the viscous dissipation and the terme Q(z) is a heat source term. We shall see later on that the presence of this term is necessary.

First let us write the equations of equilibrium ($\vec{u} \equiv 0$):

$$(38,3) \qquad \begin{cases} \dfrac{dp_\infty}{dz} + \dfrac{M^2}{Fr^2}\, \gamma \rho_\infty = 0; \\[2mm] p_\infty = \rho_\infty T_\infty; \\[2mm] \dfrac{1}{PrRe}\, \dfrac{d^2 T_\infty}{dz^2} + Q_0 Q(z) = 0. \end{cases}$$

From the (38,3) the function $T_\infty(z)$ is related to the intensity Q(z) of the heat source. The first equation of the system (38,3) show that the hypothesis that $T_\infty$ depends on z by the intermediate of $\alpha z$, where

$$(38,4) \qquad \alpha \equiv \frac{M^2}{Fr^2} = \frac{U_0^2/\gamma RT_\infty(0)}{U_0^2/gL_0} = \frac{gL_0}{\gamma RT_\infty(0)} \equiv \frac{Bo}{\gamma \varepsilon_0}$$

is analogous at the Boussinesq number Bo (see, the formula (3,5)), is no longer natural. It appears as necessary in order that the Boussinesq approximation be satisfied (see, the section 8).

Thus we assume: Q(z) is a function which depends on z by the intermediate of

$$(38,5) \qquad \zeta = \alpha z;$$

moreover

$$(38,6) \qquad Q_0 = \alpha^2 \overline{Q}_0,$$

and with this hypothesis we have

(38,7)
$$\frac{d^2 T_\infty(\zeta)}{d\zeta^2} + Pr Re \bar{Q}_0 Q(\zeta) = 0.$$

The equation (38,7) provides $T_\infty(\zeta)$ for a given $Q(\zeta)$.

This equation (38,7) also shows why the existence of $Q(\zeta)$ is necessary: if $Q(\zeta) = 0$, then

(38,8)
$$T_\infty(\zeta) = a\zeta + b.$$

For $T_\infty$ defined $\forall$ $\zeta > 0$, and decreasing if $\zeta$ increases, then there exists a point $\zeta_0$ where $T_\infty(\zeta_0) = 0$.

For a realistic distribution of $T_\infty(\zeta)$ (see the Figure 18 below) we can assume that $Q(\zeta) = 0$ in the troposphere. But the Boussinesq approximation is then only locally valid. For a tropospheric flow the uniformly valid Boussinesq approximation is useless and, in particular, it is useless to simultaneously introduce the two scales z and $\zeta$ (see, the section 12, for instance).

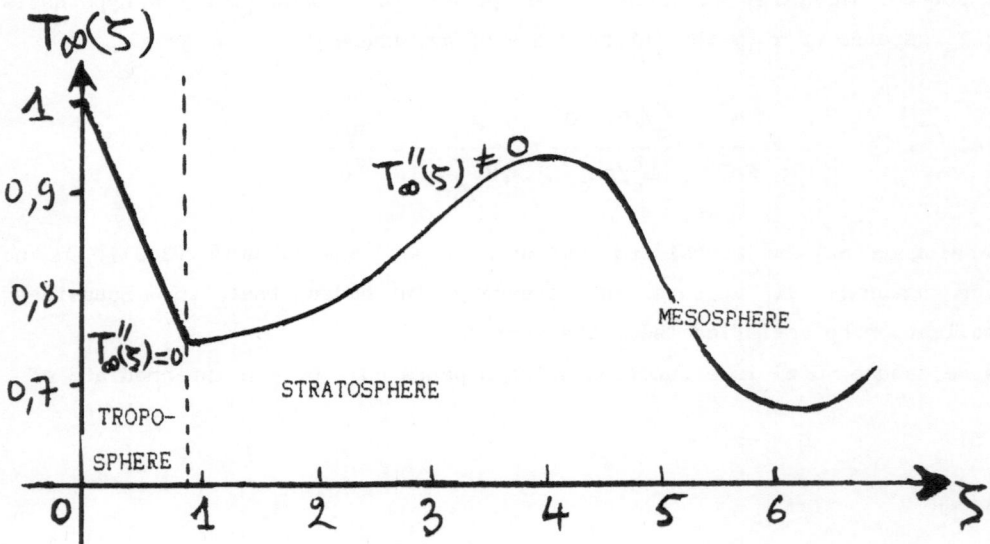

Fig. 18: For $T_\infty(\zeta) = 1$ the real temperature is 288°K.
For $\zeta = 1$ the real altitude is 11,8 km.

Now, we assume that

(38,9) $\qquad\qquad \alpha=M \implies Fr^2=M, \ \zeta=Mz,$

and we set in (38,1)

(38,10) $\qquad \begin{cases} \vec{u} \equiv \vec{u} \ , \ \vec{u}.\vec{k} \equiv \bar{w}; \\[2mm] p = p_\infty(\zeta) + M^2\bar{p}; \\[2mm] \rho = \rho_\infty(\zeta) + M^2\bar{\rho}; \\[2mm] T = T_\infty(\zeta) + M\bar{T}. \end{cases}$

Then (38,1) takes the forms:

(38,11) $\quad \begin{cases} \gamma\rho_\infty(\zeta)\dfrac{D\vec{u}}{Dt} + \vec{\nabla}\bar{p} + \gamma\bar{\rho} \ \vec{k} = -\gamma \ M \ \dfrac{D\vec{u}}{Dt} + \dfrac{\gamma}{Re}\left\{\Delta\vec{u} + \dfrac{1}{3}\vec{\nabla}(\vec{\nabla}.\vec{u})\right\}; \\[4mm] \rho_\infty(\zeta)\vec{\nabla}.\vec{u} = -M\left\{\dfrac{\partial\bar{\rho}}{\partial t} + \vec{\nabla}.(\bar{\rho} \ \vec{u}) + \dfrac{d\rho_\infty}{d\zeta} \ \bar{w}\right\}; \\[4mm] \rho_\infty(\zeta)\dfrac{D\bar{T}}{Dt} + p_\infty(\zeta)\mathcal{H}_\infty(\zeta)\bar{w} - \dfrac{1}{PrRe}\Delta\bar{T} = M\left\{\dfrac{\gamma-1}{\gamma} \ \dfrac{D\bar{p}}{Dt} - \dfrac{dT_\infty}{d\zeta} \ \bar{\rho} \ \bar{w}\right. \\[6mm] \qquad\qquad\qquad\qquad\qquad\qquad\qquad\qquad\qquad\qquad \left. -\bar{\rho} \ \dfrac{D\bar{T}}{Dt} + \dfrac{\gamma-1}{Re} \ \phi\right\} + O(M^2); \\[4mm] \rho_\infty\bar{T} + T_\infty\bar{\rho} = M(\bar{p} - \bar{\rho} \ \bar{T}), \end{cases}$

where

(38,12) $\qquad\qquad \mathcal{H}_\infty(\zeta) = \dfrac{1}{\Xi_\infty} \ \dfrac{d\Xi_\infty}{d\zeta} \ , \quad \text{with } \Xi_\infty = \dfrac{p_\infty^{1/\gamma}}{\rho_\infty} \ .$

The system (38,11) is the uniformly valid Boussinesq system for a dissipative flow.

We examine now the *natural* instability of a atmospheric medium, which is related to "*inverse*" temperature profiles, namely, temperature profiles for which the function $\mathcal{H}_\infty(\zeta)$ is *negative*. Usual measurements show that this situation can effectively arise in the troposphere, where $\mathcal{H}_\infty(\zeta)$ can considerably vary from a day to another because of the radiation from the ground.

When $\mathcal{R}_\infty(\zeta)$ is negative, the propagation of periodic waves is impossible (see, the section 12) that there appear instability effects due to the wave propagation for which the velocity is of the form

$$(38,13) \qquad \vec{\bar{u}} = \vec{U}(z)\exp\{\sigma t + i\vec{k}.\vec{x}\},$$

with *real* $\sigma$.

The correponding flows are unstable flows of the *Rayleigh-Bénard* type, and the question of their existence is a Bénard problem. This existence has been experimentally placed in evidence (see, for example, Warner and Telford (1963)). The theoritical justifications of this existence (Manton (1974)) were proposed, in general, by assuming that the atmospheric medium is confined between two levels $\zeta_0$ and $\zeta_1$ ($\zeta_0$ being eventually the ground) considered either as walls, or free surfaces. If the interval $[\zeta_0,\zeta_1]$ is small, the atmospheric medium can be considered as incompressible with a constant temperature gradient. A Rayleigh number can be defined as in the case of the classical Bénard problem. Cellular flows appear from a critical value of this Rayleigh number.

The aim of the present section 38 is, by a more detailed analysis, to place in evidence a variable Rayleigh number, positive in the zone of convective flow, negative in the stable region, and which vanishes at the boundary.

Since the problem deals with the *linear* stability of the atmospheric medium, we seek the critical Rayleigh number.

*THE EIGENVALUE PROBLEM*

We seek $\vec{\bar{u}}$ in the form (38,13) and for the system (38,11) we search a solution

$$(38,14) \qquad \begin{cases} \bar{\rho} = R(z)e^{\sigma t + i\vec{k}.\vec{x}} \ , \quad \bar{p} = P(z)e^{\sigma t + i\vec{k}.\vec{x}} , \\ \bar{T} = T_\infty(\zeta)\theta(z)e^{\sigma t + i\vec{k}.\vec{x}} \ . \end{cases}$$

We note that in the relation (38,13) we have

$$(38,15) \qquad \vec{U}(z) = \begin{bmatrix} U(z) \\ V(z) \\ W(z) \end{bmatrix},$$

and after eliminating the other quantities we obtain the following system for the vertical velocity W(z) and the pseudo-temperature $\theta(z)$:

(38,16)
$$\sigma \rho_\infty D^2 W + K^2 \rho_\infty \theta = \frac{1}{Re} D^4 W + M \left\{ K^2 \frac{P}{T_\infty} - 2\sigma \frac{d\rho_\infty}{d\zeta} \frac{dW}{dz} + \sigma^2 \frac{d\theta}{dz} \right.$$
$$\left. + \frac{1}{Re}\left[ -\sigma D^2 \frac{d\theta}{dz} + \frac{1}{\rho_\infty} \frac{d\rho_\infty}{d\zeta} D^2 \frac{dW}{dz} \right] \right\} + O(M^2);$$

(38,17)
$$\frac{1}{PrRe} D^2 \theta - \sigma \rho_\infty \theta = \rho_\infty \mathcal{H}_\infty(\zeta) W - M \frac{\sigma}{T_\infty} \frac{\gamma-1}{\gamma} P + M \frac{1}{PrRe} \left\{ \frac{1}{P_\infty} D^2 P \right.$$
$$\left. - \left[ a(\zeta) - c(\zeta) \right] \frac{d\theta}{dz} \right\} + O(M^2),$$

with $D^2 = \dfrac{d^2}{dz^2} - K^2$, $a(\zeta) = \dfrac{1}{T_\infty}\left[ 3\dfrac{dT_\infty}{d\zeta} + 2\gamma \right]$, $c(\zeta) = -\dfrac{1}{\rho_\infty}\dfrac{d\rho_\infty}{d\zeta}$ .

In (38,16) and (38,17) the pressure P(z) has the value:

(38,18)
$$P(z) = \frac{\gamma}{K^2}\left[ \frac{1}{Re} D^2 \frac{dW}{dz} - \sigma \rho_\infty \frac{dW}{dz} \right] + O(M).$$

The system $\{(38,16), (38,17)\}$ can be reduced to one equation for W(z), which is:

(38,19)
$$\frac{1}{PrRe^2 \rho_\infty} D^6 W - \frac{\sigma}{Re}\left(1 + \frac{1}{Pr}\right) D^4 W + \sigma^2 \rho_\infty D^2 W - K^2 \rho_\infty \mathcal{H}_\infty W$$
$$= -\frac{M}{PrRe^2 \rho_\infty} a(\zeta) D^4 \left(\frac{dW}{dz}\right) - \frac{M\sigma}{Re} b(\zeta) D^2 \left(\frac{dW}{dz}\right)$$
$$+ M\sigma^2 c(\zeta) \rho_\infty \frac{dW}{dz} + O(M^2),$$

where

$$b(\zeta) = \frac{1}{T_\infty}\left[\frac{\gamma}{Pr} + 2\right] \text{ and } \mathcal{H}_\infty(\zeta) = \frac{d}{d\zeta}\left[ Log\left( \frac{P_\infty^{1/\gamma}}{\rho_\infty} \right) \right].$$

In fact, it will be here as much convenient to consider the system $\{(38,16),(38,17)\}$ as equation (38,19) because of the associated boundary conditions. These conditions are those of equilibrium, namely:

(38,20)
$$W = \frac{dW}{dz} = 0 \quad \text{at } z=0 \text{ and } z \rightarrow +\infty;$$

condition of zero velocity at the wall (bottom) and vanishing of the perturbation at infinity,

(38,21) $\qquad\qquad \theta = 0$ at z=0 and z$\longrightarrow$+∞;

prescribed temperature at the wall and vanishing of the perturbation at infinity.

We finally have six homogeneous boundary conditions. The problem then is to look for couple $(K^2, \sigma)$ for which the solution of (38,19) is *not* identically zero. These couple can be defined by relations of the form

$$K = \mathcal{K}(\sigma, M).$$

Hence we assume that K can be written in the form

(38,22) $\qquad\qquad K = K_0(\sigma) + MK_1(\sigma) + \ldots,$

and in the following we compute only $K_0$.

*PRINCIPLE OF EXCHANGE OF STABILITIES*

The aim of the present section is to eastablish the assertion that:

"*If* $Real(\sigma) > 0$, *then* $Imag(\sigma) = 0$" (*principle of exchange of stabilities*).

However, in order to establish this property, it is necessary to study more precisely the behavior of unstable solutions for large altitudes z's. In particular, the following theorem will be used.

We assume that $T_\infty(\zeta)$ is a positive function, which is n times continuously differentiable, and having for large $\zeta$'s the behavior of $\zeta^\delta$ with $\delta \geq 0$. We suppose that its m-th derivative has, for large $\zeta$, following the behavior:

$$\delta (\delta - 1) \ldots (\delta - m+1)\zeta^{\delta-m}, \quad \forall m \leq n.$$

Let us set $\sigma = p + iq$.

Then:

"*For* p>0, *there exist two real constants* B *and* $\tilde{K}(\sigma) > 0$ *which do not depend of* M, *such that any solution of equation* (38,19) *which vanishes at infinity satisfies the inequalities*

$$(38,23) \quad \begin{cases} |W| < Be^{-\tilde{K}(\sigma)z}, \\[2ex] \left|\dfrac{dW}{dz}\right| < Be^{-\tilde{K}(\sigma)z}, \\[1ex] \cdots\cdots\cdots\cdots \\[1ex] \left|\dfrac{d^{n}W}{dz^{n}}\right| < Be^{-\tilde{K}(\sigma)z}, \end{cases}$$

*when* $z \rightarrow +\infty$. "

The proof of this theorem is very technical because of the particular cases[†] which must be considered in detail. We refer the reader to Bois (1979) for a detailed proof. The fundamental inequalities (38,23) set in evidence by the above theorem, allow one to establish the principle of exchange of stabilities. In this view we form an energy equation deduced from equation (38,17). By multiplying (38,17) by the complex conjugate $\theta^{*}$ of $\theta$ and by integrating from 0 to $+\infty$, we obtain:

$$(38,24) \quad \int_{0}^{+\infty} \theta^{*}\left\{\frac{1}{PrRe}D^{2}\theta - \sigma\rho_{\infty}\theta\right\}dz = \int_{0}^{+\infty}\theta^{*}\rho_{\infty}\mathcal{H}_{\infty}Wdz + M\int_{0}^{+\infty}\mathcal{L}_{1}(z,M)dz,$$

where $\mathcal{L}_{1}$ is a quadratic form of $(\theta,W)$ and their successive derivatives.
Now estimate the integrals in the relation (38,24). By integrating by parts the term $\theta^{*}\dfrac{d^{2}\theta}{dz^{2}}$ and by taking the boundary conditions (38,21) into account, we have:

$$(38,25) \quad \int_{0}^{+\infty}\theta^{*}\left\{\frac{1}{PrRe}D^{2}\theta - \sigma\rho_{\infty}\theta\right\}dz = \int_{0}^{+\infty}\left\{\frac{1}{PrRe}\left[\left|\frac{d\theta}{dz}\right|^{2} + K^{2}|\theta|^{2}\right]\right.$$

$$\left. + \sigma\rho_{\infty}|\theta|^{2}\right\}dz.$$

In order to estimate the integral $\mathcal{I} = \displaystyle\int_{0}^{+\infty}\theta^{*}\rho_{\infty}\mathcal{H}_{\infty}(\zeta)Wdz$ we replace $\rho_{\infty}\theta^{*}$ with the

---

[†] Indeed, it is necessary to examine the particular case where $\sigma^{3}$ is real.

aid of (38,16). Hence by setting $D^2W = G$:

$$\mathcal{F} = \int_0^{+\infty} \frac{\mathcal{H}_\infty(\zeta)}{K^2} W\left\{\frac{1}{Re}D^2 G^* - \sigma^* \rho_\infty G^*\right\}dz + M\int_0^{+\infty} \mathcal{L}_2(z,M)dz,$$

where $\mathcal{L}_2$ is also a quadratic form of $(W,\theta)$ and their derivatives. By integrating twice by parts and taking into account that $\mathcal{H}_\infty(\zeta)$ (depends on z only through $\zeta$), we obtain:

(38,26)
$$\mathcal{F} = \int_0^{+\infty}\left\{\frac{\mathcal{H}_\infty(\zeta)}{K^2 Re}\left[G^*\frac{d^2W}{dz^2} - K^2 G^* W\right] - \frac{1}{K^2}\sigma^* \rho_\infty G^* \mathcal{H}_\infty(\zeta)W\right\}dz$$
$$+ M\int_0^{+\infty} \mathcal{L}_3(z,M)dz,$$

where, in (38,26), we have set

$$\mathcal{L}_3(z,M) \equiv \mathcal{L}_2(z,M) + \frac{d\mathcal{H}_\infty}{d\zeta}\left[G^*\frac{dW}{dz} - \frac{dG^*}{dz}W\right],$$

and so that $\mathcal{L}_3$ is also a quadratic form of $(\theta,W)$ and their derivatives.
Now by exhibiting G in the first term of (38,26) and by integrating the term $\sigma^* G^* W$ by parts, we finally obtain:

$$\mathcal{F} = \int_0^{+\infty}\left\{\frac{1}{ReK^2}|G|^2 \mathcal{H}_\infty + \frac{\sigma^* \rho_\infty \mathcal{H}_\infty}{K^2}\left[\left|\frac{dW}{dz}\right|^2 + K^2|W|^2\right]\right\}dz$$
$$+ M\int_0^{+\infty} \mathcal{L}_4(z,M)dz.$$

Relation (38,24) takes then form:

(38,27)
$$\int_0^{+\infty}\left\{\frac{1}{PrRe}\left[\left|\frac{d\theta}{dz}\right|^2 + K^2|\theta|^2\right] + \sigma\rho_\infty|\theta|^2 + \frac{1}{ReK^2}\mathcal{H}_\infty|G|^2\right.$$
$$\left. + \frac{\sigma^* \rho_\infty \mathcal{H}_\infty}{K^2}\left[\left|\frac{dW}{dz}\right|^2 + K^2|W|^2\right]\right\}dz = -M\int_0^{+\infty} \mathcal{L}(z,M)dz.$$

Let us set $\mathcal{L}(z,M) = \mathcal{P}(z,M) + iQ(z,M)$ in (38,27) and let us separate the real and the imaginary part of (38,27). Then, setting $\sigma = p + iq$ we obtain:

(38,28)
$$q\int_0^{+\infty}\left\{\rho_\infty|\theta|^2 - \frac{\rho_\infty \mathcal{H}_\infty}{K^2}\left[\left|\frac{dW}{dz}\right|^2 + K^2|W|^2\right]\right\}dz = M\int_0^{+\infty} Q(z,M)dz;$$

$$(38,29) \qquad \int_0^{+\infty} \left\{ \frac{1}{\Pr \mathrm{Re}} \left[ \left| \frac{d\theta}{dz} \right|^2 + K^2 |\theta|^2 \right] + \frac{1}{\mathrm{Re} K^2} \mathcal{H}_\infty |G|^2 \right\} dz$$

$$= -p \int_0^{+\infty} \left\{ \rho_\infty |\theta|^2 + \rho_\infty \frac{\mathcal{H}_\infty}{K^2} \left[ \left| \frac{dW}{dz} \right|^2 + K^2 |W|^2 \right] \right\} dz$$

$$+ M \int_0^{+\infty} \mathcal{P}(z, M) dz.$$

From the two equalities (38,28) and (38,29) we can deduce two important assertions:

first suppose that $\mathcal{H}_\infty(\zeta) > 0$ $\forall \zeta$ and $p > 0$. We can apply the theorem (by taking $n=6$). From this theorem the integral:

$$M \int_0^{+\infty} \mathcal{P}(z, M) dz$$

is *uniformly* bounded by a term $O(M)$. In agreement with (38,22) we replace K by $K_0$, and thus:

$$(38,30) \qquad p = - \frac{\displaystyle\int_0^{+\infty} \left\{ \frac{1}{\Pr \mathrm{Re}} \left[ \left| \frac{d\theta}{dz} \right|^2 + K_0^2 |\theta|^2 \right] + \frac{\mathcal{H}_\infty(\zeta)}{K_0^2 \mathrm{Re}} |G|^2 \right\} dz}{\displaystyle\int_0^{+\infty} \left\{ \rho_\infty |\theta|^2 + \frac{\rho_\infty \mathcal{H}_\infty(\zeta)}{K_0^2} \left[ \left| \frac{dW}{dz} \right|^2 + K_0^2 |W|^2 \right] \right\} dz} < 0.$$

This assertion is in contradiction with the assumption. Hence

"If $\mathcal{H}_\infty(\zeta) \geq 0$ $\forall \zeta$, then at zeroth order in M, p is either negative or zero". This theorem, in fact, establishes the well known property that the *atmosphere is stable when $\mathcal{H}_\infty(\zeta)$ is always positive.*
Now consider the case where $\mathcal{H}_\infty(\zeta)$ changes its sign. In fact we assume that $\mathcal{H}_\infty(\zeta)$ is continuous $\leq 0$ if $\zeta \leq \zeta_0$, and $\mathcal{H}_\infty(\zeta) \geq 0$ if $\zeta \geq \zeta_0$; $\zeta_0$ being the *only* root of the equation $\mathcal{H}_\infty(\zeta) = 0$.
In this case we have the following result:

"*If $\mathcal{H}_\infty(\zeta) \leq 0$, for $\zeta \leq \zeta_0$, and $\mathcal{H}_\infty(\zeta) \geq 0$, for $\zeta \geq \zeta_0$, $\mathcal{H}_\infty(\zeta_0) = 0$, only and being continuous if $\zeta < \zeta_0$, then at zeroth order in M, if p > 0, q is zero".*

This theorem proves the principle of exchange of stabilities. In particular is establishes that the threshold between a stable state and an unstable state is a stationary state $(\sigma = 0)$.

## THE BÉNARD PROBLEM

The Bénard problem is the practical research of the thresholds of stability, namely the search of the relation $K=K(\sigma)$ when $\text{Real}(\sigma)=0$.

We now introduce the local Rayleigh number of the convective flow:

$$(38,31) \qquad Ra(\zeta) = -PrRe^2\rho_\infty^2(\zeta)\mathcal{H}_\infty(\zeta).$$

Equation (38,19) takes the simplified form (*when* $\sigma=0$):

$$(38,32) \qquad D^6W + K^2Ra(\zeta)W = -Ma(\zeta)D^4\left[\frac{dW}{dz}\right] + O(M^2),$$

and we look for K when $\sigma=0$.

We assume that there exists $\zeta_0$ such that:

$$(38,33) \qquad \zeta<\zeta_0 \implies Ra(\zeta)>0, \quad \zeta\geq\zeta_0 \implies Ra(\zeta)\leq 0.$$

The solution of the equation (38,32) can be computed and we give here only the results of the computation. One obtains:

1) if $\zeta>\zeta_0$: $Ra(\zeta)$ is negative and the solution which is damped at infinity is the sum of three monochromatic waves:

$$(38,34) \qquad W(z,M) = \sum_{j=1}^{3} A_j(\zeta)\exp\left[ -\frac{\varphi_j(\zeta)}{M} \right],$$

with

$$(38,35) \qquad \begin{cases} \dfrac{d\varphi_j}{d\zeta} = \left[K^2 + N(\zeta)\exp\left[\dfrac{2}{3}i\pi(j+2)\right]\right]^{1/2}; \\[2em] N(\zeta) = \left|K^2 Ra(\zeta)\right|^{1/3}, \end{cases}$$

and

$$(38,36) \qquad A_j(\zeta) = \frac{B_j^0\,\exp[-g(\zeta)]}{\sqrt{6\dfrac{d\varphi_j}{d\zeta}}\,\left[\left[\dfrac{d\varphi}{d\zeta}\right]^2 - K^2\right]}, \qquad g(\zeta) = \int_0^\zeta \frac{a(\bar{\zeta})}{6}d\bar{\zeta}.$$

The square roots in (38,35) and (38,36) are those of which real part is positive. The constants $B_j^0$ are undetermined.

If $\zeta \leq \zeta_0$, because of the change of sign of $Ra(\zeta)$, the $\dfrac{d\varphi_j}{d\zeta}$'s in (38,35) now must read:

$$(38,37) \qquad \frac{d\varphi_j}{d\zeta} = \left[ K^2 - N(\zeta)\exp\left[\frac{2}{3}i\pi(j+2)\right] \right]^{1/2}.$$

Hence there exists one root $\dfrac{d\varphi_j}{d\zeta}$ (corresponding to j=1) for which $\dfrac{d\varphi_j}{d\zeta}$ can vanish. From (38,37) $\dfrac{d\varphi_j}{d\zeta}$ vanishes at the point $\zeta_1$ defined by the equation:

$$(38,38) \qquad K^2 = N(\zeta_1)$$

and $\zeta_1$ is a turning point for the equation (38,32)[†]. In what follows we assume that $N(\zeta)$ is a monotonous function. Two cases must be considered, according to whether $\zeta$ is smaller or greater than the *unique* root of the equation (38,38):

    2) if $\zeta_1 < \zeta \leq \zeta_0$: the solutions have the form (38,34) where $\dfrac{d\varphi_j}{d\zeta}$ is given by (38,37). The relation (38,36) remains valid,

    3) if $\zeta < \zeta_1$: the waves of phases $\varphi_2$ and $\varphi_3$ are unchanged.
Now W reads

$$(38,39) \qquad W = \sum_{j=1}^{2} A_{1j}(\zeta)\exp\left[-\frac{\varphi_{1j}(\zeta)}{M}\right] + \sum_{j=2}^{3} A_j(\zeta)\exp\left[-\frac{\varphi_j(\zeta)}{M}\right].$$

The $d\varphi_{1j}/d\zeta$'s and $A_{1j}$'s read:

---

[†] Concerning the turning point see, for instance, the section 12 (equations (12,19) to (12,30)).

$$(38,40) \quad \begin{cases} \dfrac{d\varphi_{11}}{d\zeta} = -\dfrac{d\varphi_{12}}{d\zeta} = i\left[N(\zeta) - K^2\right]^{1/2}, \\[4mm] A_{1j}(\zeta) = \dfrac{B_{1j}\exp[-g(\zeta)]}{\sqrt{6\dfrac{d\varphi_{1j}}{d\zeta}}\left[\left(\dfrac{d\varphi_{1j}}{d\zeta}\right) + K^2\right]}. \end{cases}$$

Note that only the waves of phases $\varphi_{11}$ and $\varphi_{12}$ are really oscillatory; the other phases ($\varphi_1$, $\varphi_2$ or $\varphi_3$) contain a term of fast damping. Hence these waves are inexistant except in the immediate neighborhood of singular points (the bottom for $\varphi_2$ and $\varphi_3$, and the turning point $\zeta_1$ for $\varphi_1$). The waves of phases $\varphi_2$ and $\varphi_3$ represent classical boundary layers, while the wave of phase $\varphi_1$ represents a free boundary layer.

4) In the neighborhood of $\zeta_1$, the waves of phases $\varphi_{11}$ and $\varphi_{12}$ must be matched to the wave of phase $\varphi_1$ which exists only for $\zeta>\zeta_1$. The matching provides two relations between $B_{11}$, $B_{12}$ and $B_1$. By eliminating $B_1$ between these relations, we obtain between $B_{11}$ and $B_{12}$ the reflection relation

$$(38,41) \quad B_{12} = B_{11}\exp\left[2i\left(\frac{\psi_1}{M} - \frac{\pi}{4}\right)\right] = -B_{11}\exp\left[2i\left(\frac{\psi_1}{M} + \frac{\pi}{4}\right)\right],$$

with

$$(38,42) \quad \psi_1 = \int_0^{\zeta_1}\sqrt{N(\zeta) - K^2}\,d\zeta = \varphi_{11}(\zeta_1).$$

This reflection formula (38,41) can be compared to the formula (12,42) for periodic flows (see the section 12).
Finally W depends linearly on three constants. These constants are determined by writing the boundary conditions at z=0. From (38,20) and (38,21) these conditions now are:

$$(38,43) \quad W = \frac{dW}{dz} = D^2W = 0 \quad \text{at} \quad z=0.$$

## SOLUTION OF THE BÉNARD PROBLEM

Writing the condition (38,43) we obtain for the constants $B_j$ and $B_{1j}$ an algebraic linear system, of which the righthand side is zero. This system is a system for $B_1$, $B_2$, $B_3$ if $\zeta_1$ does not exist, and for $B_{11}$, $B_{12}$, $B_2$, $B_3$ if $\zeta_1$ exists. In this last case the additional equation is the equation (38,41). The differentiation of each function $A_j(\zeta)\exp[-\varphi_j(\zeta)/M]$ with respect to $z$ are at the zeroth order in M, the same functions, but multiplied by: $-d\varphi_j/d\zeta$. This allows one to determine completely the equations. Note again that $d\varphi_3/d\zeta$ is the complexe conjugate $d\varphi_2^*/d\zeta$ of $d\varphi_2/d\zeta$, and that $d\varphi_{12}/d\zeta = -d\varphi_{11}/d\zeta$. By cancelling the determinants we finally obtain the following equations:

1) if $\zeta_1$ does not exist

(38,44)    $\Delta_1 = 0,$

where

$$
\Delta_1 = \begin{vmatrix}
1 & 1 & 1 \\[2mm]
\sqrt{\dfrac{d\varphi_2}{d\zeta}}\left[\left(\dfrac{d\varphi_2}{d\zeta}\right)^2 - K^2\right] & \sqrt{\dfrac{d\varphi_2^*}{d\zeta}}\left[\left(\dfrac{d\varphi_2^*}{d\zeta}\right)^2 - K^2\right] & \sqrt{\dfrac{d\varphi_1}{d\zeta}}\left[\left(\dfrac{d\varphi_1}{d\zeta}\right)^2 - K^2\right] \\[4mm]
\dfrac{-\sqrt{d\varphi_2/d\zeta}}{\left(\dfrac{d\varphi_2}{d\zeta}\right)^2 - K^2} & \dfrac{-\sqrt{d\varphi_2^*/d\zeta}}{\left(\dfrac{d\varphi_2^*}{d\zeta}\right)^2 - K^2} & \dfrac{-\sqrt{d\varphi_1/d\zeta}}{\left(\dfrac{d\varphi_1}{d\zeta}\right)^2 - K^2} \\[4mm]
\dfrac{\left(\dfrac{d\varphi_2}{d\zeta}\right)^2 - K^2}{\sqrt{d\varphi_2/d\zeta}} & \dfrac{\left(\dfrac{d\varphi_2^*}{d\zeta}\right)^2 - K^2}{\sqrt{d\varphi_2^*/d\zeta}} & \dfrac{\left(\dfrac{d\varphi_1}{d\zeta}\right)^2 - K^2}{\sqrt{d\varphi_1/d\zeta}}
\end{vmatrix} \quad ;
$$

2) if $\zeta_1$ exists

(38,45)    $\Delta_2 = 0,$

where

$$\Delta_2 = \begin{vmatrix} \Delta_1^{11} & \Delta_1^{21} & \dfrac{1}{\sqrt{\dfrac{d\varphi_{11}}{d\zeta}\left[\left(\dfrac{d\varphi_{11}}{d\zeta}\right)^2 + K^2\right]}} & \dfrac{1}{\sqrt{\dfrac{d\varphi_{11}}{d\zeta}\left[\left(\dfrac{d\varphi_{11}}{d\zeta}\right)^2 + K^2\right]}} \\[40pt] \Delta_1^{12} & \Delta_1^{22} & \dfrac{i\sqrt{d\varphi_{11}/d\zeta}}{\left(\dfrac{d\varphi_{11}}{d\zeta}\right)^2 + K^2} & \dfrac{-i\sqrt{d\varphi_{11}/d\zeta}}{\left(\dfrac{d\varphi_{11}}{d\zeta}\right)^2 + K^2} \\[40pt] \Delta_1^{13} & \Delta_1^{23} & \dfrac{\left(\dfrac{d\varphi_{11}}{d\zeta}\right)^2 + K^2}{\sqrt{d\varphi_{11}/d\zeta}} & \dfrac{\left(\dfrac{d\varphi_{11}}{d\zeta}\right)^2 + K^2}{\sqrt{d\varphi_{11}/d\zeta}} \\[30pt] 0 & 0 & \exp\left[i\left(\dfrac{\psi_1}{M} + \dfrac{\pi}{4}\right)\right] & \exp\left[-i\left(\dfrac{\psi_1}{M} + \dfrac{\pi}{4}\right)\right] \end{vmatrix}$$

and we note that $\Delta_1 \equiv \left[\Delta_1^{ij}\right]$, $i,j = 1,2,3$.

Let us show that equation (38,44) has no solution. To that purpose we introduce the quantities:

$$\lambda = \sqrt{\frac{d\varphi_2}{d\zeta}} \;,\quad \mu = \left(\frac{d\varphi_2}{d\zeta}\right)^2 - K^2, \quad \alpha = \sqrt{\frac{d\varphi_1}{d\zeta}} \;,\quad \beta = \left(\frac{d\varphi_1}{d\zeta}\right)^2 - K^2.$$

Note that $\alpha$ and $\beta$ are real. By remarking that

$\dfrac{\mu}{\mu^*} = e^{-i2\pi/3}$, the determinant $\Delta_1$, after some computations, reads:

$$(38,46) \qquad \Delta_1 = 2i\frac{\beta\sin(\theta)}{|\mu|^2\alpha} + \frac{1}{|\lambda|^2\alpha\beta}\left\{(\lambda^{*2} - \alpha^2)e^{-2i\pi/3} - (\lambda^2 - \alpha^2)e^{2i\pi/3}\right\},$$

where we have set: $\theta = \mathrm{Arg}\left(\left[\dfrac{d\varphi_2}{d\zeta}\right]^2\right)$.

From the relations

$$\left(\frac{d\varphi_2}{d\zeta}\right)^2 = K^2 - N(\zeta)e^{i\pi/3}, \quad \left(\frac{d\varphi_1}{d\zeta}\right)^2 = K^2 - N(\zeta)$$

it can then be deduced that

$$\left|\left(\frac{d\varphi_2}{d\zeta}\right)^2\right| > \left|\left(\frac{d\varphi_1}{d\zeta}\right)^2\right| \implies \left|\frac{d\varphi_2}{d\zeta}\right| > \left|\frac{d\varphi_1}{d\zeta}\right|.$$

By remarking that $\mathrm{Arg}\left(\left[\frac{d\varphi_2}{d\zeta}\right]^2\right) < 0$ and that the roots chosen are such that

$\mathrm{Real}\left(\frac{d\varphi_j}{d\zeta}\right)^2 > 0$ we deduce that

$$\frac{d\varphi_2}{d\zeta} - \frac{d\varphi_1}{d\zeta} = n + i\ell, \quad n>0, \ell>0,$$

so that (38,46) also reads:

$$i\Delta_1 = -2\frac{\beta\sin(\theta)}{|\mu|^2\alpha} + \frac{1}{|\lambda|^2\alpha\beta}(\ell - n\sqrt{3}).$$

We deduce from this expression that $i\Delta_1$ is a real quantity of which the sign is constant (this sign is that of $\alpha\beta$).

In particular, this shows that $\Delta_1$ *never* vanishes. A consequence of this property is that in the $(K, Ra(0))$ plane, the curve of equation $\zeta_1=0$ bounds a *stable* region. From (38,38) this curve has the equation:

(38,47)    $Ra(0) = K^4$,

and the only solutions of the problem are the roots of the equation (38,45). As for the classical Bénard problem (see, for instance, Chandrasekhar (1961)) these roots can be studied only by trials and errors: denote by $K_0$ the roots of (38,45) for a given $Ra(\zeta)$. We can draw the curve $\Delta_2(K)$. This curve is discontinuous at the points 0 and $K_0$ (see figure 19 below) and for given $Ra(\zeta)$ we can numerically determine its smallest zero $K_1$ and its largest zero $K_2$.

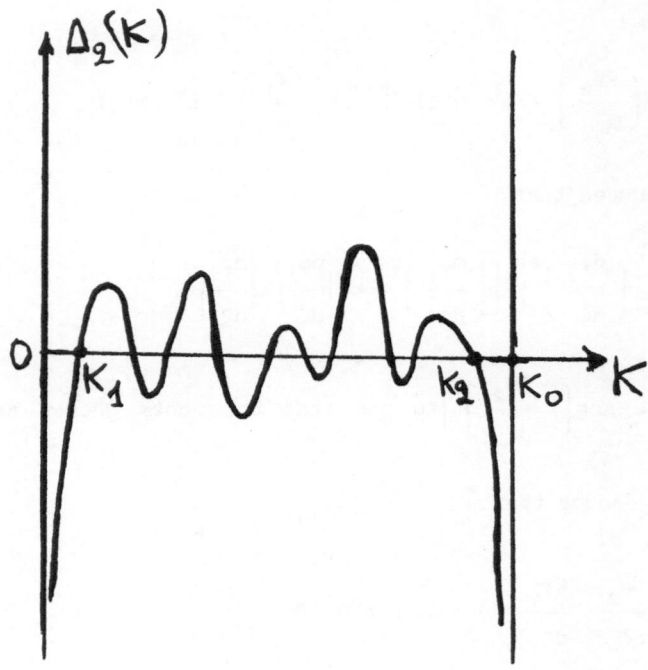

Fig. 19: Zeros of the function $\Delta_2(K)$ for given $Ra(\zeta)$

*RELATION WITH THE CLASSICAL*
*BÉNARD PROBLEM*

In the classical Bénard problem the stability of the flow can be determined by drawing a curve in the $(K, \bar{R}a)$ plane, where $\bar{R}a$ is the Rayleigh number of the flow. This curve separates two regions: one where the flow is stable, the other where the flow is unstable. The absolute stability is obtained when the flow is stable for any $K$, so that we obtain a stability threshold $\bar{R}a_c$ such that,

$$\forall \; \bar{R}a < \bar{R}a_c ,$$

the flow is stable for any K. The number $\bar{\mathrm{Ra}}$ is defined by:

(38,48)
$$\bar{\mathrm{Ra}} = \alpha_0 \beta_0 L_0 \mathrm{Pr} \left(\frac{\mathrm{Re}}{\mathrm{Fr}}\right)^2,$$

where Fr, Re and Pr are the Froude, Reynolds and Prandtl numbers of the flow and $L_0$ is the characteristic length. The coefficient $\alpha_0$ is related to the law of the fluid medium and $\beta_0$ is related to the temperature gradient. We have for the variation of density $\rho$ with respect to the altitude:

(38,49)
$$\frac{\rho}{\rho_0} = 1 + \alpha_0 \beta_0 (z - z_0),$$

(see, for instance, Drazin and Reid (1981)).

The drawing of a stability curve is possible here only if the number $\mathrm{Ra}(\zeta)$ is defined with the help of a unique parameter. For example, if $\mathcal{H}_\infty(\zeta)$ is negative in an interval of small length (with respect to the atmospheric one) then denoting by $z_0^*$ the altitude (with dimension) where $\mathcal{H}_\infty(\zeta)$ vanishes, we have

(38,50)
$$\varepsilon = z_0^*/H_0,$$

since $\varepsilon$ is small the Boussinesq approximation corresponds to the value $M \equiv \varepsilon$, hence

(38,51)
$$M = z_0^*/H_0,$$

and $\mathcal{H}_\infty(\zeta)$ vanishes at the point $\zeta = M$. On the other hand, approximating $\mathcal{H}_\infty(\zeta)$ by its derivative in this interval, for $z^* < z_0^*$ we can set:

(38,52)
$$\mathcal{H}_\infty(\zeta) = a_0 M(z-1) = a_0(\zeta - M).$$

The Rayleigh number, in this case, is defined with the help of one parameter only, which is $a_0$, stratification parameter of the atmospheric medium.
From (38,50) we have:

(38,53)
$$\mathrm{Ra}(\zeta) = M \mathrm{Pr} \mathrm{Re}^2 a_0 \rho_\infty^2(\zeta)(1-z).$$

For such a $\mathrm{Ra}(\zeta)$ it is possible to draw a stability curve in the $(K, \mathrm{Ra}(0))$ plane. Such a curve has been drawn in Bois (1979) for numerical values which correspond to those of air at the ground level.

The results (see, the figures 20 and 21 below) show that there exists a critical Rayleigh number $\mathrm{Ra}_c = 740$, from which the cellular flows appear. The cellular flows are really cellular if $\zeta < \zeta_1$ and exponentially small if $\zeta > \zeta_1$.

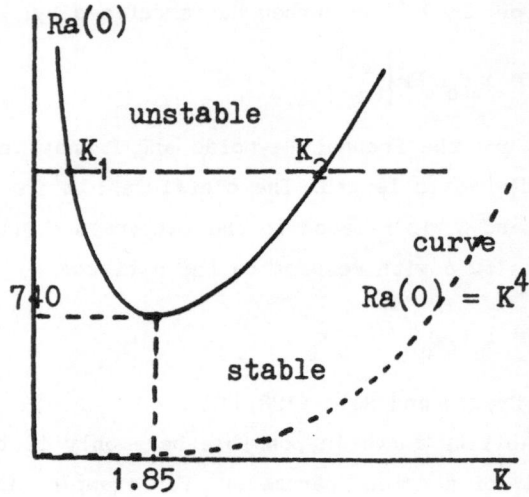

Fig. 20: Stability curve in the (K, Ra(0)) plane.

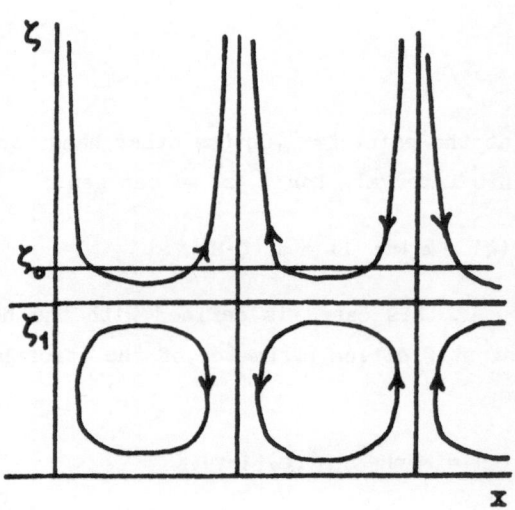

Fig. 21: Cellular flow in the (x, z) plane.

When the cellular flows appear, the only part of W which is not damped is the term including $\varphi_{11}$ and $\varphi_{12}$. Hence the only solution which is distinguishable, except in the immediate neighborhood of the ground, is the solution:

$$(38,54) \qquad W = \sum_{j=1}^{2} B_{1j} \exp(-\varphi_{1j}(\zeta)/M).$$

Let us introduce the relation (38,41), then:

$$(38,55) \qquad W = C\ \mathscr{G}(\zeta)\sin(\xi - \xi_1 - \pi/4), \quad \left(\xi \equiv \frac{\varphi_{11}}{M}\right),$$

so that $W = D^2 W = 0$ at the altitude $\zeta$ such that

$$(38,56) \qquad \xi = \xi_1 + \pi/4, \quad \left(\xi_1 \equiv \frac{\psi_1}{M}\right).$$

The flow hence as it were reflected by a free surface, located at the altitude $\xi_1 + \pi/4$. It can be numerically observed that this altitude is very near $\zeta_0$. Observe that the important difference which appears numerically in the determination of the stability threshold (740 instead of 1100 for a classical Bénard problem with corresponding boundary conditions) proceeds from the manner which as been used to define the Rayleigh number: the definition is not the same in both problems. Moreover the Rayleigh number of the present problem is not constant in the whole flow.

At least, note that the Bénard problem, treated here for convenience in a fluid at the rest, would be more realistic if it were treated for a basic stationary shear flow. Such problems have been treated by Tveitereid (1974) for incompressible confined flows, and have shown that the first perturbations, which occur in the flow, are longitudinal rolls.

*BACKGROUND READING*

For a extensive treatment of the concept of the stability in fluids the reader is referred to:

CHANDRASEKHAR, S. (1961) _ *Hydrodynamic and Hydromagnetic stability.*
                                Clarendon Press, Oxford, England.
and

DRAZIN, P. and REID, W.H. (1981) _ *Hydrodynamic stability.* Cambridge
                                       University Press.

Concerning the baroclinic and barotropic instability, see the Chapter 7 of the book of:

PEDLOSKY, J. (1982) _ *Geophysical Fluid Dynamics.* Springer-Verlag, New York Inc

*REFERENCES TO WORKS CITED IN THE TEXT*

BARCILON, V. (1964) _ J. Atmos. Sci., 21, 291-299.

BOIS, P.A. (1979) _ Journal de Mécanique, 18, 633-660.

BOIS, P.A. (1984) _ "*Asymptotic theory of Boussinesq waves in the atmosphere*".
                     Publ. IRMA, Lille, vol.VI, Fasc.4, n°2, pages II,1-II,89.

DIKII, L.A. (1973) _ Izv. Acad. Sci. SSSR. Physics of the atmosphere and ocean,
                     9, n°12, 1312-1315, (in Russian).

DRAZIN, P.G. (1958) _ J. Fluid Mech. 4, 214-224.

DRAZIN, P.G. (1978) _ in "*Rotating Fluids in Geophysics*".
                     Academic Press Inc., London (see, pages 139-169).

DRAZIN, P.G. and HOWARD, L.N. (1966) _ In "*Advances in Applied Mechanics*";
                     Accademic Press, New York, 9, 1-89.

EADY, E.A. (1949) _ Tellus, 1, 33-52.

FJORTOFT, R. (1950) _ Geophys. Publ., 17, n°6, Oslo, 1-52.

GOLDSTEIN, S. (1931) _ Proc. Roy. Soc. London, A, 132, 524-548.

HOWARD, L.N. (1961) _ J. Fluid Mech.,10, 509-512.

KUO, H.L. (1949) _ J. Meteorol., 6; 105-122.

LEBLOND, P.H. and MYSAK, L.A. (1978) _ *Waves in the Ocean.*

                                       Elsevier Scientific Publ. Company, Amsterdam.

MANTON, M.J. (1974) _ Austr. J. Phys., 27, 495-509.

MILES, J.W. (1961) _ J. Fluid Mech., 10, 496-508.

RAYLEIGH, J.W.S. (1880) _ Proc. London Math. Soc., 9, 57-70.

SQUIRE, H.B. (1933) _ Proc. Roy. Soc. London, A, 142; 621-628.

SYNGE, J.L. (1933) _ Trans. Roy. Soc. Can., 27 (III); 1-18.

TAYLOR, G.I. (1931) _ Proc. Roy. Soc. London, A, 132; 499-523.

THORPE, S.A. (1969) _ J. Fluid Mech., 36; 673-683.

TURNER, J.S. (1973) _ *Buoyancy Effects in Fluids.* Cambridge University Press.

TVEITEREID, M. (1974) _ Z.A.M.M., 54, 533-540.

WARNER, J. and TELFORD, J.W. (1967) _ J. Atm. Sci., 24; 374-382.

YIH, C-S. (1955) _ Q. Appl. Math., 12; 434-435.

YIH, C-S. (1957) _ Tellus, 9; 220-227.

YIH, C-S. (1965) _ *Dynamics of Nonhomogeneous Fluids.* Mac Millan, New York.

# CHAPTER IX

## DETERMINISTIC CHAOTIC
## BEHAVIOUR OF ATMOSPHERIC
## MOTIONS

### 39. ATMOSPHERIC EQUATIONS
### AS A FINITE-DIMENSIONAL
### DYNAMICAL SYSTEM

In a non-adiabatic, viscous, atmosphere, a finite number of lower (large-scale) modes of motion determines all the remaining modes, since the higher (small-scale) modes are strongly damped due to frictional force and dissipation function and only replicate with decreased amplitude the fundamental modes of oscillation (in particular, they have the same kind of spectrum).

In connection with these issues Hopf(1948)has advanced the hypothesis that every set of phase trajectories of the Navier-Stokes equations is attracted for t $\longrightarrow$ +∞ to *finite-dimensional* set. Thus the equations of a non-adiabatic, viscous, atmosphere can be written in the following form:

$$(39,1) \qquad \frac{d\mathcal{U}}{dt} = \mathcal{F}(\mathcal{U};\mu),$$

with $\mathcal{U} = \{\mathcal{U}^1(t),\dots,\mathcal{U}^N(t)\}$ and where $\mathcal{U}^k(t)$, e.g., are the coefficients in the Galerkin approximation[†] and $\mu$ is the bifurcation parameter.

Non adiabaticity and viscosity of the atmospheric motions gives rise not only to finite-dimensionality of the phase space but also to dissipation of phase flow, i.e., an *average compression of the phase volume* "downstream" as t→+∞.

At different phase point $\mathcal{U}_0$ the quantity div$\mathcal{F}(\mathcal{U}_0)$ can be either positive (expansion) or negative(compression).

The phase flow is called *dissipative* if for every $\mathcal{U}_0$ we have:

$$(39,2) \qquad \hat{\Lambda}(\mathcal{U}_0) \equiv \text{div } \mathcal{F}(\mathcal{U}_0) < 0.$$

---

[†] In the section 41 we consider the classical Bénard problem for the internal free convection and as example of finite-dimensional dynamical system we obtain the Lorenz system (section 42).

Because of dissipation, attractors have *zero* phase volume and dimensionality smaller than N.

More recently interest has been aroused in attractors which are neither critical points nor closed curves (limit cycles),called *strange attractors*. They belong essentially to third-order and higher-order systems,and include *horseshoe* and the *Lorenz attractor* (see the section 42). They are aptly called strange attractors because their properties are much more subtle than those of critical points and limits cycles (closed curves in the phase plane which represent periodic solutions which are approached by neighbouring solutions as t→±∞ respectively). The solutions are not periodic, althouth   entirely deterministic, they share some properties of random systems.

For example, several solutions which are initially arbitrarily close together may develop in substantially different way as time increases .The trajectories originating in the appropriate *domain of attraction*[†] tend to a subset   of phase space in which they wander for ever in a way that may appear to be random.

The correlation functions of quantities describing strange attractors, e.g. the second-order correlation

$$(39,3) \qquad \mathcal{R}_{ij}(t) = \lim_{T \to \infty} \frac{1}{2T} \int_{-T}^{+T} \mathcal{U}^i(s)\mathcal{U}^j(s+t)ds,$$

decay rapidly as t increase from zero, as to similar correlations of quantities describing turbulence itself.

Ruelle and Takens (1971) used strange attractors to model transition, in marked contrast to Landau's model of spectral evolution (see the section 40).

In Landau's model there is a succession of bifurcations as the Reynolds number increases, so that at each stage of transition the solution has components of differents period: $2\pi/\omega_1$, $2\pi/\omega_2$,etc..., or, precisely the solution is a *quasi-periodic* function of time.

Spectral analysis of the solution gives resonance peaks at the frequencies $\omega_1, \omega_2, \ldots$ etc ;yet spectral analysis of the *aperiodic* solution for a strange attractor gives a broad band of "noise" rather than high peaks.

So a strange attractor is suggestive of turbulence itself,and Landau's spectral evolution is suggestive of the more orderly transition seen in the cellular motion of Benard convection.

---

[†] The domain of attraction of the solution $\mathcal{U} \equiv 0$ is the set of the point $\mathcal{U}^0$ of $\mathbb{R}^N$ such that $\mathcal{U}(0) = \mathcal{U}^0$ then $\mathcal{U}(t) \to 0$ as $t \to \infty$. So if the solution $\mathcal{U} \equiv 0$ is asymptotically stable then its domaine of attraction includes some neighbourhood of 0 in $\mathbb{R}^N$.

In all dynamical systems where the general solution is aperiodic and the attractor has been determined, it has proven to be a strange attractor. That is, it is not topologically the product of several one-dimensional continua, in the sense that a smooth surface in three-dimensional space is the product of two continua. Instead it is the product of several continua and one or more *Cantor* sets, and an arbitrary intersecting curve, such as a line parallel to a coordinate axis, intersects it in a Cantor set.

This is an uncountable nowhere-dense set; an example is the set of all numbers between 0 and 1 whose decimal expansions contain only 0's and 1's. For N=3 a strange attractor would be an *infinite* complex of surfaces.

*STOCHASTICITY*

Of special interest to us here are the strange attractors, on which phase trajectories display the following properties of stochasticity.

(A)- An extremely *sensitive dependence on initial conditions*, due to exponential divergence of trajectories which are initially close together and leading to their unpredictability or non reproductibility for initial conditions which are given with arbitrarily high (but finite) precision.

(B)-The *every-where denseness at the attractor* of almost all *trajectories*, i.e, their arbitrarily close approach to any of the attractor's points-which implies that their return infinitely often to the attractor-and the property that any initial nonequilibrium probality distribution (measure) over the phase space (or, more precisely, over the region of attraction of the strange attractor) reduces to some limiting equilibrium distribution at the attractor (an invariant measure).

(C)-The *mixing property*: for any (measurable) subset A and B of the attractor, the probability after emerging from A of arrival at B is proportional after a long time to the measure of B. A consequence of the mixing property is the fact that the time-averaged value:

$<\phi[U(t)]>$ of any function $\phi[U(t)]$ defined on the strange attractor

is independent of the initial conditions $U^0$ (for almost all $U^0$), and that this average value coincides with the average:

$\overline{\phi[U(t)]}$ over the invariant measure (*ergodicity*):

$$(39,4) \qquad <\phi> \equiv \lim_{T \to \infty} \frac{1}{T} \int_0^T \phi[U(t)]dt$$

$$= \int \phi(U) P(dU) \equiv \overline{\phi}.$$

One mark of the mixing property is a rather *rapid decay* of the correlation function as $\tau \longrightarrow +\infty$:

(39,5) $\qquad\qquad B^{j1}(\tau) = <|\mathcal{U}^j(t) - <\mathcal{U}^j>||\mathcal{U}^1(t+\tau) - <\mathcal{U}^1>|>$

which is to say continuity of their Fourier transforms with respect to $\tau$, i.e., their spectral functions.

According to Monin (1986) it appears expedient to have the term TURBULENCE refer to the stochastic(random) evolution (in the sense of (A)-(C), above) of the flow of a non-adiabatic, viscous, atmosphere which possesses VORTICITY.

We note also that the existence of a broad band spectrum implies that then an auto-correlation function which tends to zero when the time increases. Because the auto-correlation function measures the time similarity, then the presence of turbulence implies a loss of the similarity, a loss of memory of the initial states. In other words , the knowledge of the time behaviour in the past does not permit to predict the behaviour in the future: TURBULENCE CORRESPONDS TO UNPREDICTABILITY.

A very important consequence of this loss of similarity (or memory) of the initial states is the divergence of the trajectories in phase space : two very neighbouring (then very similar) trajectories diverge (and then loose their similitude) if the regime is turbulent.

This fundamental property (in the sense of (A), above) of any strange attractor(turbulence)related to the divergence of very neighbouring trajectories is called "sensitivity to initial conditions" or *S.I.C.* It corresponds to the fact that if *in principle* the system is predictible (the *same* iteration performed twice gives indeed the same final state) *in practice* it is *not* due to even infinitesimal errors which always affect the initial conditions and are exponentially amplified when the time runs.

Lorenz (1963), who invented the first and most famous example of deterministic chaos through the appearance of a strange attractor in the solution of his three equations (see the section 42), gave this striking illustration of the S.I.C.

Referring to the general unpredictability for (long-term) weather forecasting, Lorenz said that even a very small change of initial conditions such as that produced by the motion of butterfly wings would have (unpredictible) large consequence for the next future behaviour of the atmosphere (considered as a dynamical system).

*BIFURCATIONS AND INSTABILITIES*

In the framework of the atmosphere-dynamical equations, the origin of irregularity of turbulence is attributed to the instability, or the S.I.C, of the solution of the equation (39,1) at supercritical of $\mu$.

The instability of such a dynamical system is characterized by the positive Lyapunov number (see the definition in the section 40) at a point of the phase space and the irregularity of the solution is interpreted as a nature of the strange attractor.

Thus the irregular nature of the atmospheric turbulence (chaotic behaviour) is now fully understood in terms of the theory of dynamical systems. It is not obvious that a turbulent state of the atmosphere can be described by a dynamical system of finite dimensions, but it is guaranted by the center manifold theorem[†] at the stage of instability where only a finite number of unstable modes are involved.

Consider a dynamical system (39,1), assume that there is a periodic self oscillating solution; this happens, for instance, in Rayleigh-Bénard convention problem (see the section 41).

Now consider cases when this periodic flow loses its stability for $\mu$ crossing a critical value $\mu^*$.

Elementary possible bifurcations which might take place are well known: it can appear a new periodic motion with a frequency half of the fundamental one("*flip* bifurcation"), or a quasi periodic motion with two natural frequencies ("*Hopf bifurcation*" into an invariant torus), or the periodic motion disappears eventually leaving place to a type of intermittency ("*sadlenode* bifurcation"). Eventual symmetries and group invariances of the atmospheric problem could lead to richer possibilities (see, for example, Iooss (1984)).

To study the stability of a periodic orbit, it is necessary to look at the *Floquet multipliers*, i.e., the eigenvalues of linearized *Poincaré map* (after elimination of eigen-direction tangent at the origin of the orbit).

The total multiplicity of the eigenvalues crossing the unit circle determines the relevant dimension where asymptotic dynamical phenomena lie, for the full system (dimension of the center manifold). More precisely, in the typical situation (for example, in the Rayleigh-Bénard instability in the *small boxes*), at a definite threshold $\mu^*$ the steady state is replaced by an

---

[†] For a complete and rigorous description of this method and obtention of companion amplitude equations see the paper by Coullet and Spiegel (1983).

oscillating regime at a well defined frequency $f_1$ (including eventually higher harmonics $nf_1$). This means that the velocity in any point of the box is modulated with a constant period (periodic regime).On the other hand the Fourier spectrum of the velocity contains only sharp peaks at $f_1$ ($2f_1, 3f_1, \ldots$); finally a phase space diagram obtained consists in a closed loop and such a closed loop is called a limit cycle and we say that,at $\mu^*$, the system bifurcates from the steady state(a fixed point)to a stable limit cycle. The second step breaking the (mono)-periodic regime corresponds to the appearance at a new threshold $\mu^{**} > \mu^*$ of a new frequency $f_2$ superimposed to the former $f_1$ and more generally the two frequencies $f_1$ and $f_2$ are *unrelated*. But for simplificity let us assume that only $f_1$ and $f_2$ are present and *incommensurate*[†].

This new regime is called biperiodic or more precisely *quasi-periodic* to recall that its properties are intermediate between a periodic regime and chaotic one (aperiodic regime)[††].

---

[†]    Incommensurate means that one cannot find integers n and m, for two frquencies $f_1$ and $f_2$, as
$$mf_1 = nf_2$$
and $f_1$ and $f_2$ are unrelated.

[††]    The equation (39,1) possesses a unique solution
$$\mathcal{U} = \varphi(t), \qquad \mathcal{U}^0 \equiv \varphi(t^0),$$
satisfying the initial condition. Here $\varphi$ is assumed to be continuously differentiable with respect to t for $t \geq 0$.

If for any $\varepsilon > 0$ and any $\tau_0 > 0$, there exists a time interval $\tau(\varepsilon, \tau_0) > \tau_0$ and a time $t_1(\varepsilon, t_0)$ such that $t_2 > t_1$ implies
$$|\varphi(t_2 + \tau) - \varphi(t_2)| < \varepsilon,$$
then $\mathcal{U} = \varphi(t)$ is called quasi-periodic. This simply means that a trajectory $\varphi(t)$ is called quasi-periodic if for some arbitrarily large time interval $\tau$, $\varphi(t+\tau)$ ultimately (i.e. large t or $t_2 > t_1$) remains arbitrarily close to $\varphi(t)$. If $\varphi(t)$ is periodic, it is obviously also quasi-periodic. Quasi-periodic trajectories include multiple periodic trajectories with incommensurable periods. Thus if $\varphi(t) = \varphi_1(t) + \varphi_2(t)$ and $\varphi(t+\tau_1) = \varphi_1(t)$, $\varphi(t+\tau_2) = \varphi_2(t)$, $\tau_1$ and $\tau_2$ incommensurable, then $\varphi(t)$ is quasi-periodic. A trajectory which is *not* periodic or quasi-periodic is non periodic (aperiodic); *an aperiodic trajectory is unstable: " if the system being considered has an aperiodic behaviour, its futur is essentially unpredictible even if the system is deterministic"*.

In order to draw the corresponding phase space trajectories a two-dimensional representation can no longer be used. The limit cycle representing $f_1$ has to wind with the frequency $f_2$ and thus the representation, in a three-dimensional phase space, takes the form of a torus. At $\mu^{**}$ we say that the stable limit cycle bifurcates to an invariant torus $T^2$.

The fact that the phase space is no more representable in two-dimensions leads us to use a transformation which allows to represent the section (*Poincaré section*) of a three-dimensional phase space. The interest of the Poincaré section is not only to obtain a good two-dimensional representation of a complex three-dimensional phase diagram, but also to make useful comparaison with very simple mathematical models labelled generically as "iterated map". Very simply the Poincaré section is a cut, a section, of all the trajectories (here in 3 diemensions) by a given plane $\pi$. Then, an ensemble of points are obtained which corresponds to the Poincaré section of phase diagram. Obviously this simple transformation lowers by one the dimensionality of the representation. A limit cycle (phase space in 2 dimensions) is transformed into a single point (analogous to a fixed point). Much more interesting is the following:

The Poincaré section for the trajectories in a quasi-periodic regime (torus $T^2$) is a simple closed loop (analogous to a limit cycle) . The presence of harmonics of the fundamental frequencies may give rise to some more complicated section than a simple ellipse, but as far as the regime remains quasi-periodic (torus $T^2$; phase space at 3 dimensions) the Poincaré section is a closed loop.

Reciprocally if the Poincaré section is a closed loop one can affirm that the dimensionality of the corresponding phase space is 3.

The existence of attractors is closely related to the existence of dissipation. In the phase space, the trajectory is attracted, converges toward the fixed point (corresponding to the equilibrium state) or to an attractive limit cycle (corresponding to the oscillating regime).

Note that dissipation *contracts* phase space, more precisely dissipation lowers the measure of the phase space. More generally if we consider the quasi-periodic regime in a dissipative system (as the atmosphere) we deal with a quasi-periodic attractor, or an attractiv torus.

What happens by a further increase of the $\mu$ parameter? At $\mu^{***} > \mu^{**}$ turbulence or chaos appears; the spectrum is no more composed of sharp lines but also broad band noise begin to appear especially near zero frequency. In the phase space, the trajectories are attracted on a complicated structure of "strange" aspect (and strange properties) which is called *strange attractor*. This kind of route to turbulence is in very good agreement with the ideas of Ruelle and Takens (see the section 40), through it is not yet proved that at $\mu^{***}$ it is appearance of a third frequency which produces chaos!

We note also that the appearance of an invariant torus $T^2$ at a bifurcation from a periodic orbit does not imply the appearance of orbits *dense* on that torus. Instead, the appearance of finitely many periodic orbits and fixed points is generic. The appearance of an invariant $T^3$ torus at the next bifurcation depends on existence of an orbit dense on the $T^2$ torus and hence, the bifurcation to an invariant $T^3$ torus seems unlikely.

Finally, if a periodic orbit on the $T^2$ torus goes round the long way "n" times before closing, then the bifurcation is *subharmonic* with a sudden n-folding of the period at the bifurcation.

## 40 . SCENARIOS

A first bifurcation may be followed by further bifurcations, and we may ask what happens when a certain sequence of bifurcations has been encountered.

In principle there is an infinity of further possibilities, but, in some sense to be specified, not all of them are equally probable.

The more likely ones will be called *scenarii*, and below we shall examine three prominent scenarii which have had theoretical and experimental success.

In general, a scenario deals with the description of a few attractors. On the other hand a given dynamical system may have many attractors (see, for example the Lorenz system at the section 41). Therefore, several scenarii may evolve concurrently in different regions of phase space. Finally, a scenario does not describe its domain of applicability.

*THE LANDAU-HOPF "INADEQUATE" SCENARIO*

After the first bifurcation the motion is generally periodic; after the second it is generally qusi-periodic with two priod, and so on. It was shown that if the first bifurcation leads to closed orbits, the second can lead to an attracting invariant torus in the phase space. If, furthermore, the motion is such that its orbits covers the torus densely, then a resulting function of time, such as one of the coordinates in the phase space, is quasi-periodic with two periods. Specifically, one can define two intrinsic angle -coordinate $\theta$ and $\varphi$ on the torus such that

$$\theta = \omega_1 t + \text{Const} \ , \quad \varphi = \omega_2 t + \text{Const},$$

and the orbits is dense on the torus if and only if $\omega_1$ and $\omega_2$ are *incommensurable*.

After the next bifurcation there may be motion on a $T^3$ torus, and so on. The idea behind the Landau-Hopf scenario was that as soon as there are many *independent* frequencies, the motion is so irregular in appearence that it must be regarded for pratical purposes as chaotic.

Obviously the appearence of turbulence is related to a system with a large number of degrees of freedom (N).

There are various way in which this scenario can be *inappropriate*:

(a) One of the bifurcation may be *subcritical*; then, as soon as the corresponding critical value of the $\mu$ is exceeded, there is no nearby stable motion for the system to follow, and there is a so-called *explosive transition* to a motion involving more or less remote parts of the phase space,

(b) Although an invariant torus generally appears at the second bifurcation, the orbit need *not be dense* on it; it may return to its starting point after winding finitely many times around _ then the orbit is closed and the motion is periodic. In fact it is now believed, on the basis of Peixoto's theorem that closed orbits on the torus are more likely than dense ones; this may lead to the Feigenbaum (1978) scenario,

(c) A possibility discussed by Ruelle and Takens is that, after a few bifurcations, there appears an invariant point set in the phase space, which is not a torus but a so-called *strange attractor*; then, as explained below, the motion is not quasi-periodic, but *aperiodic*.

*THE RUELLE-TAKENS "STRANGE ATTRACTOR" SCENARIO*

In the scenario of the early onset of turbulence proposed by Ruelle and Takens in 1971, the first four bifurcations are assumed, as in the Landau-Hopf scenario, to be supercritical and to lead to invariant tori $T^k$, k=1,2,3 and 4, each of which is attracting between its appearance and be next bifurcation. Concerning the existence of these tori, see the discussion of the Feigenbaum scenario in the next subsection.

Ruelle and Takens prove that, on $T^4$, motion on a particular kind of strange attractor contained in $T^4$ is rather likely. The attractor is locally the Cartesian product of a two-dimensional *Cantor set* and a two-dimensional

surface. The vector field that yield the strange attractor cannot be dismissed
as unlikely, however their particular choice of strange attractor is somewhat
arbitrary; one can imagine many variations of it, each having the property
stated. Apparently, no one has found a specific vector field on a specific
manifold that leads to a strange attractor precisely according to the Ruelle
and Takens scenario!

The important idea their paper is that motions on strange attractors are in
some sense likely, or at least not unlikely and are possibly even generic in
certain circumstances. Their theorem does not say that the existence of a
strange attractor is a generic property of vector field on $T^4$: *it say simply
that motion on a strange attractor is more likely, once an invariant $T^4$ has
been established than the quasi-periodic motion on $T^4$.*

A *strange attractor* is one on which the motions are *unstable* in the sense of
Lyapunov and hence are characterized by a continuous power spectrum. The
strangeness of the attractor is *stable* under small perturbations of the
dynamical system; in other words, it is no exceptional.

While it is true that the set of vector field with strange attractor is open
near the constant vector fields, this does not mean that this set is large in
the measure theoretic sense. In order to describe how the appearance of the
scenario manifest itself in measurements and to show the measurable
consequences of the presence of strange attractor, let us reformulate the
scenario (see, Eckmann (1981)):

*if a system undergoes three Hopf bifurcations, starting from a stationary
motion, as a parameter is varied, then it is likely that the system possesses
a strange attractor with S.I.C. after the third bifurcation.*

The power spectrum of such system will exhibit one, then two, and possibly
three independent basic frequencies. When the third frequency is about to
appear, simultaneously some broad-band noise will appear if there is a strange
attractor. This we interpret as chaotic, turbulent evolution of the system. The
RT scenario *is not* destroyed by the addition of small external noise to the
evolution equations. The nature of chaotic systems may be totally insensitive
to small external noise. The systems most sensitive to noise seem to be
deterministic systems near transition (bifurcation) points. This insensitivity
to noise is surprising and at first sight counter intuitive and it has been
discovered by Kifer (1974). In effect, the chaos of the scenario is so strong
that order cannot be accidentally established by small noise terms, much like
a very attracting fixed point is locally not much altered by noise, and
globally there is at most a small probability to change stochastically from on

bassin (domain) of attraction to another. Although the flow (for the dissipative systems) contracts volume, it need not contract lengths. If we take snapshots of the flow at t=0, 1 and 2, say, we may have (see, the Figure 22) the picture shown in (a) but could also get that of (b) or even that of (c).

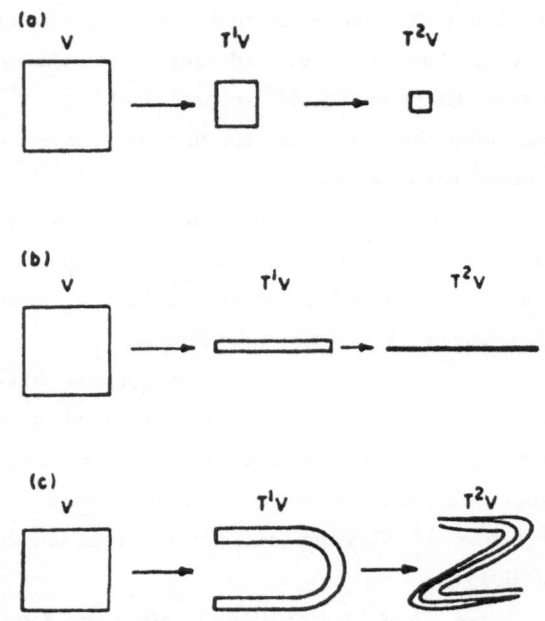

$$\textit{Fig. 22 : Contraction of volume in phase space}$$
$$\textit{(a) "normal case",}$$
$$\textit{(b) with stretching of length,}$$
$$\textit{(c) with stretching of length, and folding.}$$

In particular, even if all points in $V$ (a finite volume in state space $\mathbb{R}^N$) converge to a single attractor $\mathcal{A}$, one still may find that points which are arbitrarily close initially may get macroscopically separated on the attractor after sufficiently large time interval. This property is called "sensitive dependence on initials conditions".

It is *not* excluded for area-contracting flows, i.e., it can, and will, occur in dissipative dynamical systems. An attractor exhibiting this property will be called a strange attractor.

The solution of dissipative dynamical systems separate exponentially with time, having a positive Lyapunov characteristic exponent and thus the motion is characterized as chaotic with the appearance of a strange attractor.

The *Lyapunov* characteristic $\sigma$ *exponent* of the *flow* is defined as:

$$(40,1) \qquad \sigma = \lim_{\substack{t \to \infty \\ \mathcal{D}(0) \to 0}} \left\{ \frac{1}{t} \ln \frac{\mathcal{D}(t)}{\mathcal{D}(0)} \right\},$$

where the $\mathcal{D}(t)$ are values of distance between initially neighboring solutions. This gives us a measure of the mean exponential rate of divergence of two initially neighboring solutions, or of the *chaoticity* of the turbulence. The important is that $\sigma$ *is positive*, indicating that the movement is chaotic. It would be expected that as Reynolds number increases the movement would become more chaotic, with a consequence increase in $\sigma$.

The next property, after the Lyapunov exponent, which might be calculated to characterize the turbulent motion is the *dimension* of the strange attractor on which the resides. That dimension should be a measure of the number of *active degrees of freedom* or actives modes of the turbulence. The problem of chosing which modes are active may prove to be formidable.

The early discussion of atmosphere-dynamic chaos was largely based on a geometrical reconstruction of strange attractors, which is possible only in *low* dimension (i.e., at the *onset* of chaos). For moderately excited atmospheric systems, other tools are available from the *ergodic* theory of differential dynamical systems. To an ergodic measure $\rho$, various parameters are associated:

(a) characteristic exponents $\lambda_1 \geq \lambda_2 \geq \ldots$ (also called *Lyapunov exponents*) from the multiplicative ergodic theorem (Oseledec). The $\lambda_1$ give the rate of exponential divergence of nearby orbits of the dynamical system;

(b) *entropy*, h(p); this is mean rate of creation of information by the system, or *Kolmogorov-Sinaï* invariant;

(c) *information dimension*, $\dim_H \rho$ _ smallest Hausdorff dimension of a set $\mathcal{E}$ such that $\rho(\mathcal{E})=1$ (see, Eckmann and Ruelle (1985)).

We consider, now, dynamical systems such as map (discrete time, n):

$$(40,2) \qquad \hat{X}_{n+1} = \hat{F}(\hat{X}_n),$$

where $\hat{X}$ is a p-dimensional vector. To define the Lyapunov numbers, let

$$J_n = [J(\hat{X}_n)J(\hat{X}_{n-1})\ldots J(\hat{X}_1)],$$

where $J(\hat{X}) \equiv (\partial \hat{F}/\partial \hat{X})$ is the Jacobian matrix of the map, and let

$$j_1(n) \geq j_2(n) \geq \ldots \geq j_p(n)$$

be the magnitudes of the eingenvalues of $J_n$. The Lyapunov numbers are:

(40,3) $\qquad \ell_i = \lim_{n \to \infty} [j_i(n)]^{1/n}, \quad i=1,2,\ldots,p,$

where the positive real nth root is taken, and the Lyapunov exponents $\lambda_i$ are simply the logarithms of the $\ell_i$.

If the system is *ergodic* the Kolmogorov entropy

(40,4) $\qquad h = \sum_{i=1}^{N} \lambda_i$

where the $\sum_{n=1}^{N}$ is extended over all *positive* $\lambda_i$.

Dimension is perhaps the most basic property of an attractor. The relevant definitions of dimensions are of two general types, thoses that depend only on metric properties, and those that depend on the frequency with which a typical trajectory visists different region of the attractor.

We define here the dimension of chaotic attractor, d, by:

(40,5) $\qquad d = N - \sum_{i=1}^{N} \dfrac{\lambda_i}{\lambda_{N+1}},$

where $\qquad \sum_{i=1}^{N+1} \lambda_i < 0 \quad$ and $\quad \sum_{i=1}^{N} \lambda_i \equiv h \geq 0.$

we have that:

(40,6) $\qquad 0 \geq -\sum_{i=1}^{N} \dfrac{\lambda_i}{\lambda_{N+1}} < 1.$

For the Lorenz chaotic, strange attractor (see the section 42), we have: d=2,06. The dimension of an attractor provides a way of quantifying the number of relevant degrees of freedom present in dynamical motion. The dimension d of an attractor, if it is small and non integral, confirms that the dynamics admits a low-dimensional deterministic mathematical description characterized by a strange attractor.

The key to understanding chaotic behaviour lies in understanding a simple *stretching* and *folding* operation, which takes place in the state space.

Exponential divergence is a local feature: because attractors have finite size, two orbits on a chaotic attractor cannot diverge exponentially forever. Consequently the attractor must fold over onto itself. Although orbits diverge and follow increasingly different paths, they eventually must pass close to one another again.

The orbits on a chaotic attractor are *shuffled* by this process, much as a deck of cards is shuffled by a dealer. The *randomness* of the chaotic orbits is the result of the shuffling process. The process of the stretching and folding happens repeatedly, creating *folds within folds and infinitum*. A chaotic attractor is, on other words, a *fractal*[†]: an object that reveals more detail as it is increasingly magnified.

The stretching and folding operation of a chaotic attractor systematically removes the initial information and replaces it with new information: the stretch makes small-scale uncertainties larger, the fold brings widely separated trajectories together an erases large-scale information. Thus chaotic attractors act as a kind of pump bringing microscopic fluctuations up to a macroscopic expression. In this light it is clear that no exact solution, no short cut to tell the future, can exist. After a brief time interval the uncertainty specified by the initial measurement covers the entire attractor and all predictive power is lost: there is simply *no causal connection between past an future*. Finally, we note that initially, the two conditions of attraction and divergence of the trajectories might appear uncompatible! Indeed it is impossible to get a strange attractor if the phase space trajectories are attracted on an finite object at two dimensions: in a bounded two-dimensional area the divergence of trajectories is not possible.

On the contrary if we want to keep the possibility to form a strange attractor we have to consider an attractive object with 3 dimensions: for example in the case of the attractor described by Rossler (1976) the trajectories can diverge on a two-dimensional spiral, escape by emerging into space and return toward the centrum diverges again etc...

For the same raisons, we can expect that, from a torus $T^3$ (3 frequencies), the attracting region being in 3 dimensions an instability may lead to trajectories which can be attracted along a direction but diverge along the perpendicular other one (*hyperbolicity*). Then, on the contrary to what happens on a torus $T^2$ $(f_1, f_2)$ whose instability lead to synchronisation (limit cycle), an instability on a torus $T^3(f_1, f_2, f_3)$ may lead to a strange attractor.

It is through topological consideration of this kind that one can understand the *highly non intuitive* idea that a deterministic system with 3-independent frequencies (3 degrees of freedom) may lead to turbulent behaviour.

---

[†] See, for example, Sreenivasan and Meneveau (1986) and also Lovejoy and Schertzer (1986). In the Miscellanea ( section 47 in the chapter X) we consider briefly the fractals in atmospheric turbulence.

We do not want to leave the reader on the (false) idea that deterministic chaos in atmospheric motions always occurs through strange attractors an according to the Ruelle-Takens mechanism (or _ at least through biperiodism). Another kind of route-through intermittencies-has been proposed by Pomeau and Manneville (1980). In this later case, there is not a strange attractor, though the chaos is deterministic (phase space at 3 dimensions). A third kind of route has been proposed by Feigenbaum (1978) and corresponds to cascade of subharmonics bifurcations.

## THE FEIGENBAUM "CASCADE OF PERIOD DOUBLINGS" SCENARIO[†]

While the Lorenz (strange) attractor appears in connection with a subcritical Hopf bifurcation, the Landau-Hopf scenario and the Ruelle and Takens scenario both require a sequence of supercritical bifurcations leading to invariant tori of successively higher dimension, arbitrarily high in the former scenario and of dimension at least 4 in the latter. However, such a sequence is unlikely according to Peixoto's theorem.

Feigenbaum (1978,1980) has developped a scenario based on a sequence of subharmonic bifurcations with period doubling. It turns out that such doubling occur in many examples of iterated mappings and simple dynamical systems. Furthermore, as the number n of doublings increases, the behavior of the system is governed by certain asymptotic laws that involve universal constants and functions, independent of the system under study. In addition, the asymptotic laws appear to hold quite accurately for rather small values of n. In particular, the values $\mu_n$ of the dimensionless parameter $\mu$, in (39,1), at which the bifurcations (doublings) take place converge to a value $\mu_\infty$ geometrically, with

$$(40,7) \qquad \frac{\mu_{n+1} - \mu_n}{\mu_n - \mu_{n-1}} \sim 0,21416938\ldots$$

for large n. An n→∞, at least in the cases studied, the power spectrum of the motion approaches a continuous spectrum with certain universal features. At $\mu \cong \mu_\infty$, the motion is presumably aperiodic on a strange attractor. There is evidence for an example of this behavior in the Lorenz system (42,9) at considirably higher values of the dimensionless parameter r than values

---

[†] See also the work of Coullet and Tresser (1984) for an analysis of the cascade of period doubling bifurcations.

studied by Lorenz. Namely, the strange attractor that appears at r=24,74 persist up to a value r=r$^*$(~250). For r considirably greater than r$^*$, there is a periodic orbit, and as r is *decreased* toward r$^*$, there is a sequence of doubling at values $r_n$ of r that converge to r$^*$ from above, with:

$$\frac{r_{n+1} - r_n}{r_n - r_{n-1}} \sim 0,214\ldots .$$

After the cascade of period doublings, one expects beyond the accumulation point $\mu_\infty$ an *inverse cascade* of noisy periods.

In an experiment, if one observes subharmonic bifurcations at $\mu_1$ and $\mu_2$, then, according to the scenario, it is very probable for a further bifurcation to occur near

$$\mu_3 = \mu_2 - (\mu_1 - \mu_2)\frac{1}{\delta} ,$$

where $\delta \cong 4,66920\ldots$ . In addition, if one has seen three bifurcations, a fourth bifurcation become more probable than a third after only two, etc. . We note that $\delta$ is a universal number such that

$$(40,8) \qquad \lim_{j \to \infty} \frac{1}{j} \text{Log}|\mu_j - \mu_\infty| = -\text{Log}\delta$$

and one even has

$$(40,9) \qquad |\mu_j - \mu_\infty| \sim \text{Constant } \delta^{-j}, \text{ as } j \to \infty.$$

At the accumulation point, one will observe aperiodic behavior, but not broad-band spectrum. This Feigenbaum scenario is extremely well tested on numerical and physical grounds. The periods doublings have by now been observed in most current *low* dimensional dynamical systems (Henon map, Lorenz equations, forced oscillator with friction, Rayleigh-Bénard convection, etc..).

Now we recall the main steps of the Renormalization Group analysis of the cascade of period-doubling bifurcations according to Argoul and Arneodo (1984, p.274).

*RENORMALIZATION GROUP*
*ANALYSIS*

Dynamical systems that exhibit such a cascade of period-doubling bifurcations are in practice well modelled by one-dimensional maps with a single smooth maximum such as:

$$(40,10) \qquad f_R(x) = Rx(1-x).$$

As we increase the parameter R which determines the height of the maximum of $f_R(x)$, we observe successive steps of the cascade and a continuous transition to chaos which presents a strong analogy with second-order phase transition (see, Ma (1976)).

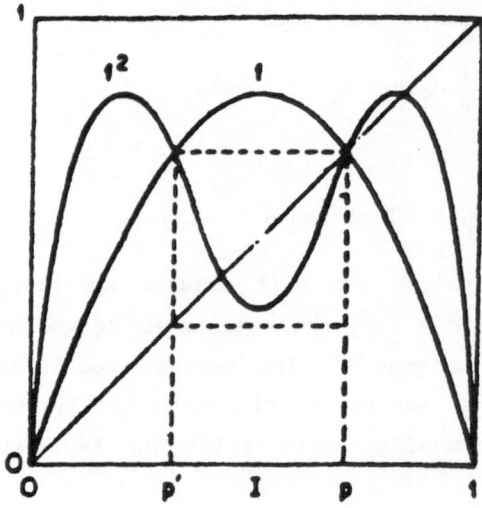

*Fig. 23: The renormalization group operation for maps of the interval $[0,1]$ of the form (40,10). Under a coordinate change, this one-parameter map family coincides with maps of $[-1,+1]$ into itself of the form $f_R(x)=1-Rx^2$, with a maximum equal to one as implicitly assumed in the text.*

In the neighborhood of the trasnition (R=R$_c$) one can define a *correlation time* which diverges according to the scaling law:

(40,11)          $T(R) \cong (R_c - R)^{-\nu}$,

where T is the period of the bifurcation cycles, and an "order" parameter which displays the universal behavior:

(40,12)          $L(R) \cong (R - R_c)^{\nu}$

where L is the envelope of the Lyapunov characteristic exponent. In (45,16) and (45,17) we have

(40,13)          $\nu = \dfrac{\ln 2}{\ln \delta}$,   $\delta \cong 4,66920...$,

and $\nu$ can be assimilated to a critical exponent in the sense that it *does not* depend on the explicit form of $f_R(x)$ but only on the quadratic nature of its maximum. Like in critical phenomena, one can use Renormalization Group technique in order to understand these universal properties. As sketched in Figure 23 above the renormalization operation

(40,14)          $\Lambda(f(x)) = \alpha f\left[f\left[\dfrac{x}{\alpha}\right]\right]$

with $\alpha = \dfrac{1}{f(1)}$, results from the similarity in the shape that characterizes f(x) and its *second iterate* $f^2(x)$ in neighborhood of its extremum.

The asymptotic scale invariance displayed by the dynamics at R=R$_c$, suggests that one look for a fixed point g(x) of renormalization operation, which must satisfy the equation:

(40,15)          $\alpha^{-1}g(\alpha x) = g(g(x))$;   $\alpha = 1/g(1)$.

The coefficient in the Taylor series of g(x) have been found numerically. In the quadratic case, one gets

$$g(x) = 1 - 1,5276 \ x^2 + 0,10481 \ x^4 + \ ... \ ,$$

with $\alpha = -2,5029...$

To handle the approach to such a critical situation, we need to study the spectrum of the renormalization operator linearized around g(x). Only one eigenvalue lies outside of the unit circle. This single relevant eigenvalue is associated with an eigenvector $e_\lambda(x)$ which is an even function of x.

As illustrated in Figure 24 the unstable manifold $W_u$ of the fixed point $g(x)$ is of dimension 1 while the stable manifold $W_s$ is of codimension 1.

Therefore any one parameter path obtained by varying R in $f_R(x)$ will intersect transversally $W_s$ for $R=R_c$.

On the way to criticality, near $W_s$, one thus feels essentially the unstable direction $e_\lambda$, which explains why the critical exponents in (40,11) and (40,12) depend only one universal constant, namely the unstable eigenvalue $\delta=4,66920...$ An experimental estimate of $\delta$ consists in measuring the parameter $R_n$ where a $2^n$-periodic cycle bifurcates into a $2^{n+1}$-periodic cycle; these values are predicted to scale according to $(R_c - R_n) \sim \delta^{-n}$.

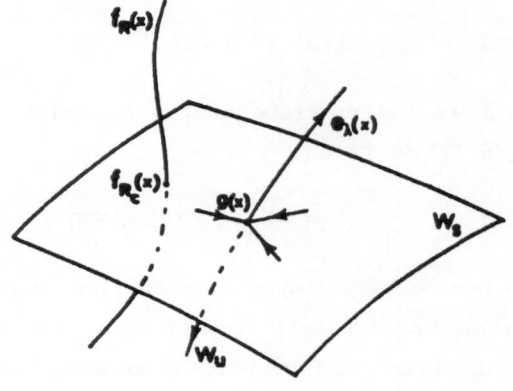

Fig. 24: Three-dimensional sketch of a codimension 1 critical surface $W_s$ corresponding to the stable manifold of $g(x)$. A generic path obtained by varying R in $f_R(x)$ cuts transversally this surface for $R=R_c$.

At the accumulation point of the period-doubling cascade, $R=R_c$ and we are on $W_s$. Wherever this intersection point lies on $W_s$, the successive iterations of the renormalization operation will converge to $g(x)$ which illuminate the universal scale invariance of the dynamics at $R=R_c$ where the adherence of the asymptotic orbit displays a Cantor set structure (see the figure 25 below). Rigourous results have been obtained by several authors (Eckmann(1983), Lanford III(1982), Epstein and Lascoux (1981)).

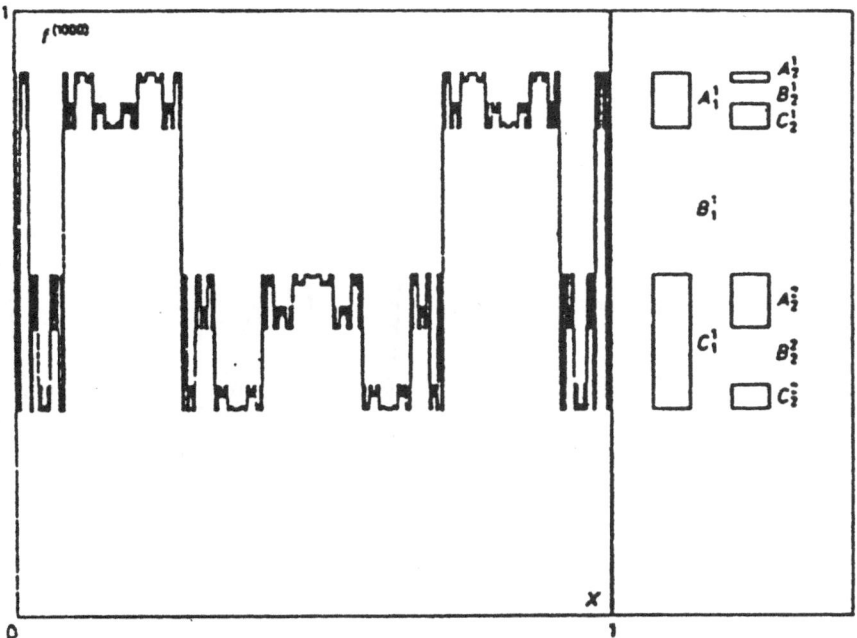

*Fig.25: Graph of* $f_{R_c}^{1000}$ *and corresponding Cantor set.*

## THE POMEAU-MANNEVILLE
## "INTERMITTENTLY TURBULENT"
## SCENARIO

While the two other scenarii have been associated with Hopf bifurcations
(Ruelle-Takens) and pitchfork[†] bifurcations (Feigenbaum), this one is
associated with a "*saddle node bifurcation*", i.e., the collision of a stable
and unstable fixed point which then both disappear (into complex fixed
points). The general idea is best explained for the simple example of a
one-parameter family of iterated maps on the unit interval, $x_{n+1} = f_{\mu}(x_n)$ and
we take $f_{\mu}(x) = 1-\mu x^2$, which for $\mu \in [0,2]$ maps $[-1,+1]$ into itself.

---

[†] In this case a stable point loses its stability and gives rise to a stable
periodic orbit as the parameter is changed.

The function $f_\mu^3 = f_\mu \circ f_\mu \circ f_\mu$ can be shown to have a saddle node for $\mu = 7/4$. For $\mu > 1,75$, $f_\mu^3$ has a stable periodic orbit of period three, and an unstable one nearby. The two collid at $\mu=1,75$, and both have then eigenvalue 1 (see Figure 26) opposite.

For $\mu$ slightly below 1,75, the local picture near x=0 is shown in Figure 27.

It can be shown that if $\mu-1,75=O(\varepsilon)$ then a typical orbit will need $O(\varepsilon^{-1/2})$ iterations to cross a fixed small x interval around x~0. As long as the orbit is in this small interval , an observer will have the impression of seeing a periodic orbit of period three.

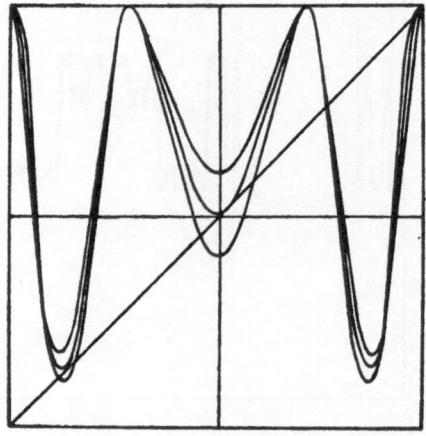

Fig. 26: Graph of $f_\mu^3$ for three values of $\mu$.

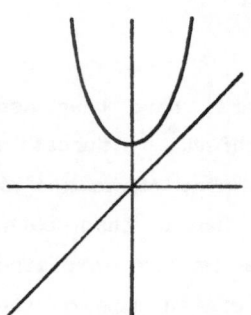

Fig. 27: Graph of $f_\mu^3$ in the vicinity of the origin.

Once one has left the small interval, the iterations of the map will look rather like those of a chaotic map. Thus this map can be called intermittently turbulent (see Figure 28 below).

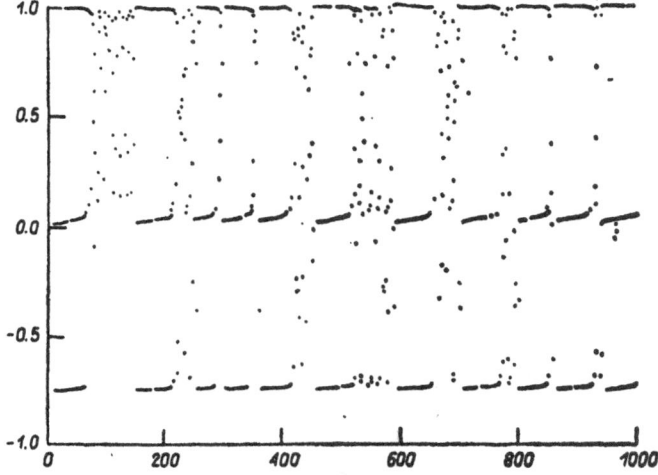

*Fig. 28: Graph describing* $f_\mu^n(0)$ *as a function of* n *in the neighborhood of* $\mu \cong 1,75$ *, and indicating the existence of an intermittent turbulence.*

The problem with this arguments comes in the splitting into two regions. It is true that the interated map may have sensitivity to initial conditions for x ∉ small intervals around contact points. But this destabilizing effect may be lost whenever one passes near the contact point. In fact, we conjecture that this will happen for an infinity of parameter values near to, and just below μ=1,75. For these parameter values, on will have (very long) stable period, but not chaos. On the hand, we also conjecture that a modification of the proof of Jakobson (1980) would show that truly aperiodic behavior with sensitivity to initial conditions (S.I.C.) occurs for a set of a parameter value of positive Lebesgue measure near 1,75 (according to Eckmann (1981; p.651)).

We can now formulate a reasonable version of this scenario for general dynamical systems:

Assume a one-parameter family of dynamical systems has Poincaré maps close to a one-parameter family of maps of the interval, and that these maps have a stable and unstable fixed point which collid as the parameter is varied. Then, as the parameter is varied further to μ from the critical parameter value $\mu_c$, one will see intermittently turbulent behavior of random duration, with laminar phases of mean duration

$$\sim \; |\mu - \mu_c|^{-1/2}$$

in between.

The difficulty with this scenario is that it does not have any clear-cut precursors, because the unstable fixed point which is going to collid with the stable fixed point (respectively periodic orbit) may not be visible. On can think of two ways out of this problem. The first would be that increasingly long transients can be observed before the two fixed points (periodic orbits) collid. The second kind of precursor is a cascade of inverse pitchfork bifurcations, and, at the "end" of this, the intermittent transition to turbulence (Collet and Eckmann (1980)). We note that Pomeau and Manneville based their work on observation for the Lorenz system. Intermittent transitions to turbulence can be seen in many physical experiments. The intermittent trasition to turbulence is now well understood for systems with a few degrees of freedom, such as that describing fluid motions in bounded geometries. Nevertheless, the best known example of intermittent transition to turbulence is in parallel flows (the so-called "transition flows"). The structure of these transition flows is usually attributed to the subcritical character of the instabilities described in their linear version by the Orr-Sommerfeld equations. However, the precise connection between intermittency at the onset and subcriticality of the instability needs to be made more precise. Intermittency in transition flows probably has little to do with linear stability analysis. In pipe flows for instance, it seems likely that localized, steady and finite-amplitude fluctuations can exist with a given relative velocity along the flow lines (say along the x-variable) when the Reynolds number reaches a well defined value, dependent on the relative velocity of the fluctuation. If one considers the fluid equations as dynamical system where the variable x plays the role of time. This sort of localized fluctuation is represented (according to Pomeau (1983)) by an exceptional trajectory starting from and returning to the undistributed parallel flow at $x \rightarrow +\infty$. The fact that such trajectories exist in a manifold of codimension one (which fixes the values of the Re for a given relative velocity) in space of all flows it still true for a dynamical system in function space (one must consider function spaces as the local state of the system, i.e., the value of all functions for a given x is an element of an infinite dimension space spanned by all admissible functions of the space variables independent of x). This helps one to understand how a well defined nonlinear stability criterion may exist and why, in transition flows, the *bursts* tend to widely separated at the threshold. This picture, if true, needs to be made more precise. It could provide, for instance, a detailed structure for critical fluctuations in concrete examples and the behaviour of the intermittency parameter at the threshold.

Finally, for a renormalization group and universality point of view of the Pomeau and Manneville intermittency, see the book of Guckenheimer and Holmes (1986; section 6.8).

Concerning these three main scenarii, the reader can consult the recent paper of Monin (1986).

## 41 . BENARD PROBLEM
## FOR THE INTERNAL FREE
## CONVECTION

Thermal instability often arises when a fluid is heated from below and the classical example of this, described in this section, is a horizontal layer of fluid with its lower side hotter than its upper. The first quantitative experiments were made by Bénard (1900). Stimulated by Bénard's experiments, Rayleigh (1916) formulated the theory of convective instability of a layer of fluid between horizontal planes.

Hence, we consider internal free convection between two horizontal planes with the separation $d'_0$ and having temperatures $T'_0$=constant, for the upper plane, and $T'_0+ \Delta T'_0$ ($\Delta T'_0$=constant>0) for the lower plane. The main body of the present section is concerned, specifically with the convection in the fairly deep layer of expansible liquid with an equation of state of the following simple form

(41,1)        $\rho'=\rho'(T')$

where $\rho'$ is the density and $T'$ is the temperature.

For our expansible liquid we have possibility to construct the Froude and Mach numbers:

$$(41,2) \qquad Fr = \frac{(\nu'_0/d'_0)^2}{g'd'_0}, \quad M = \frac{(\nu'_0/d'_0)^2}{C'_0 \Delta T'_0}$$

where $\nu'_0$ is the kinematic viscosity at the temperature $T'_0$, $g'$ is the acceleration due to gravity and $C'_0$ is the specific heat at the temperature $T'_0$. In the *exact*, Navier-Stokes, equations for the expansible liquid we have a fundamental small parameter, namely

$$(41,3) \qquad \varepsilon_0 = \beta'_0 \Delta T'_0 \ll 1$$

where

$$(41,4) \qquad \beta'_0 = -\left[\frac{1}{\rho'} \frac{d\rho'}{dT'}\right]_{T'=T'_0}.$$

The parameter

$$(41,5) \qquad \frac{\varepsilon_0}{Fr} = Gr$$

is the Grashof number and if $\sigma = \dfrac{\rho'_0 \nu'_0 C'_0}{k'_0}$ is the Prandtl number ($k'_0$ is the thermal conductivity at the temperature $T'=T'_0$ and $\rho'_0 \equiv \rho'_0(T'_0)$), then:

$$(41,6) \qquad Ra = \sigma\, Gr \equiv \frac{\sigma}{Fr}\varepsilon_0,$$

is the Rayleigh number.
Finally

$$(41,7) \qquad Bo = \frac{M}{Fr}$$

is the Boussinesq number.

## THE EXACT NAVIER-STOKES EQUATIONS

We present here the equations that govern a wide class of thermal convection problems. A primed quantity will denote a dimensional quantity, while a non-primed quantity will denote a dimensionless quantity.

Let there be established an $\{x'_i\}$, i=1,2,3, Cartesian co-ordinate system in which gravity acts in the negative $x'_3$ direction. Then, for $x'_i$ component of velocity $u'_i$, the pressure $p'$, the density $\rho'$ and the temperature $T'$, we can write the following exact equations:

(41,8)  $$\rho'\frac{Du'_i}{Dt'} + \frac{\partial p'}{\partial x'_i} + g'\rho'\delta_{13} = \frac{\partial}{\partial x'_i}\left[\lambda'\frac{\partial u'_k}{\partial x'_k}\right] + \frac{\partial}{\partial x'_j}\left[\mu'\left(\frac{\partial u'_i}{\partial x'_j} + \frac{\partial u'_j}{\partial x'_i}\right)\right],$$

(41,9)  $$\frac{D}{Dt'}(\text{Log }\rho') + \frac{\partial u'_k}{\partial x'_k} = 0,$$

(41,10)  $$\rho'C'(T')\frac{DT'}{Dt'} + p'\frac{\partial u'_k}{\partial x'_k} = \frac{\partial}{\partial x'_i}\left[k'\frac{\partial T'}{\partial x'_i}\right] + \lambda'\left(\frac{\partial u'_k}{\partial x'_k}\right)^2 + \frac{\mu'}{2}\left(\frac{\partial u'_i}{\partial x'_j} + \frac{\partial u'_j}{\partial x'_i}\right)^2.$$

We have postulated that the expansible liquid is a newtonian compressible, heat conducting, viscous fluid and as a consequence of equation of state (41,1) we have

(41,11)  $$\frac{De'}{Dt'} \equiv C'(T')\frac{DT'}{Dt'}, \quad \lambda' \equiv \lambda'(T'), \quad \mu' \equiv \mu'(T'), \quad k' \equiv k'(T'),$$

where e' is the specific internal energy, $\mu'$ and $\lambda'$ are the two dynamic viscosity coefficients and k' is the thermal conductivity. Finally, C'(T') is the specific heat for the expansible liquid, only function of T' since we have postulated the equation of state (41,1). $\frac{D}{Dt'} \equiv \frac{\partial}{\partial t'} + u'_j\frac{\partial}{\partial x'_j}$ , where t' is the time.

For the temperature we have, for the Bénard problem, the following boundary conditions:

(41,12)  $\quad T' = T'_0 + \Delta T'_0$  at  $x'_3 = 0$  and  $T' = T'_0$  at  $x'_3 = d'_0$.

Let $C'_0$, $\lambda'_0$, $\mu'_0$, $k'_0$ and $\rho'_0$ be values of C'(T'), $\lambda'(T')$, $\mu'(T')$, k'(T') and $\rho'(T')$ at $T' = T'_0$. Define

(41,13)  $$C = \frac{C'}{C'_0}, \quad \lambda = \frac{\lambda'}{\lambda'_0}, \quad \mu = \frac{\mu'}{\mu'_0}, \quad k = \frac{k'}{k'_0}, \quad \rho = \frac{\rho'}{\rho'_0}$$

$$\nu'_0 \equiv \mu'_0/\rho'_0 \ .$$

Let $d'_0$, $\nu'_0/d'_0$, $\Delta T'_0$, $d'^2_0/\nu'_0$ and $\rho'_0 g' d'_0$ be scaling units for length, velocity, temperature, time and pressure, respectively. New dimensionless variables are defined as follows:

(41,14)  $$x_i = \frac{x'_i}{d'_0}, \quad u_i = \frac{u'_i}{\nu'_0/d'_0}, \quad t = \frac{t'}{d'^2_0/\nu'_0}, \quad T = \frac{T'}{\Delta T'_0}, \quad p = \frac{p'}{\rho'_0 g' d'_0} \ .$$

Equations (41,8)-(41,10) transform to:

$$\begin{cases} \dfrac{D}{Dt}(Log\rho) \ + \ \dfrac{\partial u_k}{\partial x_k} \ = \ 0; \\[4mm] \rho \ \dfrac{Du_i}{Dt} \ + \ \dfrac{1}{Fr}\left(\dfrac{\partial p}{\partial x_i} \ + \ \rho\delta_{13}\right) \ = \ \dfrac{\partial}{\partial x_j}\left[\mu\left(\dfrac{\partial u_i}{\partial x_j} \ + \ \dfrac{\partial u_j}{\partial x_i}\right)\right] \ + \ \dfrac{\lambda_0'}{\mu_0'} \ \dfrac{\partial}{\partial x_i}\left(\lambda\dfrac{\partial u_k}{\partial x_k}\right); \\[4mm] \rho C(T)\dfrac{DT}{Dt} \ + \ Bo \ p \ \dfrac{\partial u_k}{\partial x_k} \ = \ \dfrac{1}{\sigma}\dfrac{\partial}{\partial x_i}\left(k \ \dfrac{\partial T}{\partial x_i}\right) \ + \ \dfrac{\lambda_0'}{\mu_0'} \ M\lambda\left(\dfrac{\partial u_k}{\partial x_k}\right)^2 \\[4mm] \qquad\qquad\qquad\qquad + \ M \ \dfrac{\mu}{2}\left(\dfrac{\partial u_i}{\partial x_j} \ + \ \dfrac{\partial u_j}{\partial x_i}\right)^2; \\[8mm] \rho \ = \ \rho(T), \end{cases}$$

(41,15)

where $\lambda=\lambda(T)$, $\mu=\mu(T)$ and $k=k(T)$.

In the dimensionless system (41,15) we have four non-dimensional parameters:

$$Fr \ = \ \dfrac{(\nu_0'/d_0')^2}{g'd_0'}, \quad M \ = \ \dfrac{(\nu_0'/d_0')^2}{C_0'\Delta T_0'}, \quad Bo \ = \ \dfrac{g'd_0'}{C_0'\Delta T_0'}, \quad \sigma \ = \ \dfrac{\mu_0'C_0'}{k_0'} \ .$$

The boundary conditions (41,12) lead

(41,16)
$$\begin{cases} T \ = \ T_0 + 1 \quad \text{at } x_3 = 0, \\[2mm] T \ = \ T_0 \qquad\;\; \text{at } x_3 = 1, \end{cases}$$

with $T_0 \equiv T_0'/\Delta T_0'$.

## THE DIMENSIONLESS DOMINANT EQUATIONS

When no motions are presents ($u_i \equiv 0$), the equations (41,15) require only that the pressure distribution is governed by the equation

(41,17)
$$\dfrac{\partial p}{\partial x_3} \ = \ -\rho(T)$$

and the temperature distribution is governed by the equation

(41,18)
$$\dfrac{\partial}{\partial x_i}\left(k \ \dfrac{\partial T}{\partial x_i}\right) \ = \ 0.$$

But if we suppose that the solution of equation (41,18) is simply $T \equiv T_0$, then the corresponding distribution of the density is given by $\rho \equiv 1$ and with the expression for $\rho$, equation (41,17) can be integrated to give : $p = 1 - x_3$.

Let the initial state:

(41,19)        $u_i \equiv 0$, $T \equiv T_0$, $\rho \equiv 1$ and $p \equiv 1 - x_3$

be slightly perturbed. Let $u_i$ denote the velocity in the perturbed state and let the altered temperature distribution be[†]:

(41,20)        $T = T_0 + \theta$.

Finally, let

(41,21)        $p = 1 - x_3 + A_0 Fr\, \pi$

denote the altered pressure distribution, where

(41,22)        $A_0 = \dfrac{1}{Fr} \dfrac{\Delta p'_0}{g' \rho'_0 d'_0}$ ,

with $\Delta p'_0$ the pressure fluctuation in slightly perturbed convective motion.

We shall now consider the problem of approximate expressions for the density $\rho$ and the coefficients $\lambda$, $\mu$, $k$ and $C$.

For density $\rho$ we obtain the following relation:

(41,23)        $\rho = 1 - \varepsilon_0 (\theta + \tau_0 \theta^2 + \ldots)$

where

(41,24)        $\tau_0 = \dfrac{\Delta T'_0}{2} \dfrac{d}{dt'} \left[ Log \beta' \rho' \right]_{T' = T'_0}$ .

If the expression

(41,25)        $x_\rho \equiv \dfrac{1}{2} \left[ \dfrac{1}{\beta'^2_0} \dfrac{d\beta'}{dT'} \Big|_{T' = T'_0} - 1 \right]$

is *bounded* when $\varepsilon_0 \to 0$, then:

(41,26)        $\rho = 1 - \varepsilon_0 \omega$,        $\omega = \theta + x_\rho \varepsilon_0 \theta^2 + \ldots$ .

In the same way we can obtain, for $\lambda$, $\mu$, $k$ and $C$, the following analogous approximate relations:

---

[†] With the dimensions we have

$$\theta = (T' - T'_0)/\Delta T'_0$$

and we suppose that $\theta$ is of the order of unity.

$$(41,27) \quad \begin{cases} \lambda = 1 - \varepsilon_0\varphi, \qquad \varphi = \dfrac{\Lambda'_0}{\beta'_0}\,(\theta + x_\lambda \varepsilon_0 \theta^2 + \ldots), \\[2em] \mu = 1 - \varepsilon_0 g, \qquad g = \dfrac{M'_0}{\beta'_0}\,(\theta + x_\mu \varepsilon_0 \theta^2 + \ldots), \\[2em] k = 1 - \varepsilon_0 f, \qquad f = \dfrac{K'_0}{\beta'_0}\,(\theta + x_k \varepsilon_0 \theta^2 + \ldots), \\[2em] C = 1 - \varepsilon_0 \ell, \qquad \ell = \dfrac{N'_0}{\beta'_0}\,(\theta + x_c \varepsilon_0 \theta^2 + \ldots), \end{cases}$$

where $\Lambda'_0$, $M'_0$, $K'_0$, and $N'_0$ are the quantities similar to $\beta'_0$ (see , the formula (41,4)), while $x_\lambda$, $x_\mu$, $x_k$ and $x_c$ are the quantities similar to $x_\rho$ (see, the formula (41,25)), but respectively for $\lambda$, $\mu$, $k$ and $C$; these last four quantities, $x_\lambda$, $x_\mu$, $x_k$ and $x_c$, remains bounded when $\varepsilon_0 \to 0$.

Now, if we take into account the preceding results, then, we obtain for $u_i$, $\pi$ and $\theta$ the following set of dominant dimensionless perturbations equations:

$$(41,28) \qquad \frac{\partial u_k}{\partial x_k} = \varepsilon_0\,\frac{D\theta}{Dt} + O(\varepsilon_0^2);$$

$$(41,29) \qquad (1-\varepsilon_0\theta)\frac{Du_i}{Dt} + A_0\,\frac{\partial \pi}{\partial x_i} - Gr\,\theta\,\delta_{i3} = \Delta u_i + \varepsilon_0\left\{\left[1+\frac{\lambda'_0}{\mu'_0}\right]\frac{\partial}{\partial x_i}\left[\frac{D\theta}{Dt}\right]\right.$$

$$\left. - \frac{M'_0}{\beta'_0}\,\frac{\partial}{\partial x_j}\left[\theta\left(\frac{\partial u_i}{\partial x_j} + \frac{\partial u_i}{\partial x_i}\right)\right]\right\} + O(\varepsilon_0^2).$$

$$(41,30) \qquad \left[1-\left(1+\frac{N'_0}{\beta'_0}\right)\varepsilon_0\theta\right]\frac{D\theta}{Dt} + \varepsilon_0\left[Bo(1-x_3) + A_0 M\,\pi\right]\frac{D\theta}{Dt} = \frac{1}{\sigma}\,\Delta\theta$$

$$+ \frac{1}{2}\,M\left(\frac{\partial u_i}{\partial x_j} + \frac{\partial u_j}{\partial x_i}\right)^2 - \varepsilon_0\left[\frac{K'_0}{\beta'_0}\,\frac{1}{\sigma}\,\frac{\partial}{\partial x_i}\left(\theta\frac{\partial\theta}{\partial x_i}\right)\right.$$

$$\left. + \frac{1}{2}\,M\,\frac{M'_0}{\beta'_0}\,\theta\left(\frac{\partial u_i}{\partial x_j} + \frac{\partial u_j}{\partial x_i}\right)^2\right] + O(\varepsilon_0^2).$$

The solution of these equations $(41,28)$-$(41,30)$ must be sought which satisfy the boundary conditions:

$$(41,31) \qquad \theta=1 \text{ at } x_3=0 \quad \text{ and } \theta=0 \text{ at } x_3=1.$$

*THE "DEEP" CONVECTION*

*LIMITING EQUATIONS*

If we introduce, in place of $\theta$, the new perturbation for the temperature

(41,32)     $\Xi = \sigma\, Gr(\theta + x_3 - 1)$,

in such a case we obtain for $\Xi$ the classical homogeneous boundary conditions:

(41,33)     $\Xi = 0$ at $x_3 = 0$   and $x_3 = 1$,

in place of (41,31).

For simplicity we can assume in the equation (41,29) that:

(41,34)     $A_0 \equiv 1$   and in this case $\Delta p_0' \equiv \rho_0' \left(\dfrac{v_0'}{d_0'}\right)^2$ .

Then, the corresponding expression for the pressure perturbation (according to (41,32)) is given by:

(41,35)     $\Pi \equiv \sigma\left[\pi + Gr\, x_3\left(\dfrac{x_3}{2} - 1\right)\right]$.

Finally, we change $u_i$ into $\dfrac{v_i}{\sigma}$ and t into $\sigma\,\tau$ and we suppose that $\varepsilon_0 \to 0$ (with $x_i$ and $\tau$ fixed) in new equations, for $v_i$, $\Pi$ and $\Xi$, with[†]:

(41,36)     $M\, Gr \equiv \varepsilon_0 Bo = \delta_0 = O(1)$.

Entering the formal limiting process[††], we obtain the following set of *deep convection equations*:

(41,37)
$$
\begin{cases}
\dfrac{\partial v_k}{\partial x_k} = 0; \\[2ex]
\dfrac{1}{\sigma}\dfrac{Dv_1}{D\tau} + \dfrac{\partial \Pi}{\partial x_i} - \Xi\delta_{13} = \Delta v_1, \quad (i=1,2,3); \\[2ex]
\left[1+\delta_0(1-x_3)\right]\left[\dfrac{D\Xi}{D\tau} - Rav_3\right] = \Delta\Xi + \dfrac{\delta_0}{2}\left(\dfrac{\partial v_1}{\partial x_j} + \dfrac{\partial v_j}{\partial x_1}\right)^2,
\end{cases}
$$

with the boundary conditions, for $\Xi$:

(41,38)     $\Xi = 0$ on $x_3 = 0$ and $x_3 = 1$.

We note that: $\dfrac{D}{D\tau} \equiv \dfrac{\partial}{\partial \tau} + v_j\dfrac{\partial}{\partial x_j}$ and $\Delta = \dfrac{\partial^2}{\partial x_1^2} + \dfrac{\partial^2}{\partial x_2^2} + \dfrac{\partial^2}{\partial x_3^2}$ and $\delta_{13} \equiv 0$ for $i \neq 3$, $\delta_{33} \equiv 1$.

---

[†] In this case we suppose that $Bo \gg 1$.
[††] Zeytounian (1983,1989).

In the deep convection equations (41,37) we have the following three parameters:

$$(41,39) \quad \begin{cases} \sigma = \dfrac{\mu'_0 C'_0}{k'_0}, \quad Ra = \dfrac{\rho'_0 g' \beta'_0 \Delta T'_0 C'_0 d'^3_0}{k'_0 \nu'_0}, \\[2em] \delta_0 = \dfrac{g' \beta'_0 d'_0}{C'_0}. \end{cases}$$

Hence if

$$(41,40) \quad d'_0 \simeq \frac{C'_0}{g' \beta'_0},$$

then it is necessary to consider a *Bénard problem for the deep convection* and in this problem we have a new parameter $\delta_0$.

## THE "SHALLOW" CONVECTION
## LIMITING BOUSSINESQ
## EQUATIONS

If $\delta_0 \to 0$ in the equations (41,37) we find again, instead of (41,37), the classical Boussinesq equations for Bénard shallow convection problem.

In this case, when $\varepsilon_0 \to 0$ (with $x_i$ and $\tau$ fixed), the Boussinesq number Bo remain bounded (of the order of unity)[†].

Entering the formal limiting process, $\delta_0 \to 0$, we obtain the following set of classical equations for the Rayleigh-Bénard problem:

$$(41,41) \quad \begin{cases} \dfrac{\partial v_k}{\partial x_k} = 0; \\[1.5em] \dfrac{1}{\sigma} \dfrac{D v_i}{D \tau} + \dfrac{\partial \sqcap}{\partial x_i} - \Xi \delta_{13} = \Delta v_i; \\[1.5em] \dfrac{D \Xi}{D \tau} - Ra v_3 = \Delta \Xi. \end{cases}$$

The Rayleigh-Bénard problem for the convective instability consists in the investigation of the stability of the following basic convective flow:

---

[†]In this case we have that : $d'_0 \approx C'_0 \Delta T'_0 / g'$.

$$(41,42) \quad \begin{cases} u'_i \equiv 0; \\[2mm] T' = T'_0 + \Delta T'_0 \mathrm{Ra}(1-x'_3/d'_0); \\[2mm] p' = g' \rho'_0 d'_0 (1-x'_3/d'_0) + \Delta p'_0 \mathrm{Pr}(1-x'_3/d'_0)(x'_3/2d'_0), \end{cases}$$

starting with the equations (41,41) for the dimensionless perturbations $v_i$, $\Pi$ and $\Xi$.

## 42 . THE LORENZ DYNAMICAL SYSTEM

A simplified two-dimensional model:
$v_2 \equiv 0$ and the variables are *not* function of $x_2$,
permits the introduction of a stream function $\psi$ such as

$$(42,1) \quad v_1 = \frac{\partial \psi}{\partial x_3}, \quad v_3 = -\frac{\partial \psi}{\partial x_1} .$$

the vorticity $\Omega$ is defined as

$$(42,2) \quad \Omega = \vec{\nabla}^2 \psi \quad \text{with} \quad \vec{\nabla}^2 = \frac{\partial^2}{\partial x_1^2} + \frac{\partial^2}{\partial x_3^2} .$$

In the starting equations (41,41), the pressure $\Pi$ may be eliminated and we get:

$$(42,3) \quad \begin{cases} \dfrac{1}{\sigma} \dfrac{\partial}{\partial \tau} \vec{\nabla}^2 \psi = \dfrac{1}{\sigma} \dfrac{\partial(\psi, \vec{\nabla}^2 \psi)}{\partial(x_1, x_3)} - \dfrac{\partial \Xi}{\partial x_1} + \vec{\nabla}^2(\vec{\nabla}^2 \psi); \\[4mm] \dfrac{\partial \Xi}{\partial \tau} - \dfrac{\partial(\psi, \Xi)}{\partial(x_1, x_3)} + \mathrm{Ra}\,\dfrac{\partial \psi}{\partial x_1} = \vec{\nabla}^2 \Xi. \end{cases}$$

Let us describe convective movement in a rectangular domain $0 \le x_3 \le 1$ and $0 \le x_1 \le \ell_0$, where $\ell_0$ may be the dimensionless length of the "box", or the horizontal size of one "convective cell".
In this case the adopted boundary conditions are:

$$(42,4a) \quad v_3 = 0, \quad \frac{\partial^2 v_3}{\partial x_3^2} = 0, \quad \Xi = 0 \text{ at } x_3 = 0 \text{ and } x_3 = 1;$$

$$(42,4b) \qquad v_1 = 0, \quad \frac{\partial^2 v_1}{\partial x_1^2} = 0, \quad \frac{\partial \Xi}{\partial x_1} = 0 \text{ at } x_1=0 \text{ and } x_1=\ell_0.$$

The solution of equations (42,3) satisfying all the boundary conditions (42,4a,b) may be expanded in a double Fourier series, as:

$$(42,5) \quad \begin{cases} \psi(\tau,x_1,x_3) = \sum_{i=1}^{\infty} \sum_{j=1}^{\infty} A_{ij}(\tau)\sin\left(\frac{i\pi x_1}{\ell_0}\right)\sin(j\pi x_3); \\ \Xi(\tau,x_1,x_3) = \sum_{i=0}^{\infty} \sum_{j=1}^{\infty} B_{ij}(\tau)\cos\left(\frac{i\pi x_1}{\ell}\right)\cos(j\pi x_3), \end{cases}$$

since the solution of the nonlinear problem (42,3)-(42,4a,b) will be considered as the superposition of the eigenfunctions of the linear problem. Concerning this linear convection classical problem see, for instance, the book of Drazin and Reid (1981; pp.37-62).

The next step is to substitute expansions (42,5) into equations (42,3) and to use the Galerkin technique (see, for instance, the book of Platten and Legros (1984; chapter VI, §3)), requiring the residue to be orthogonal to each function of the set (42,5). After this substitution, the first equation of (42,3) is multiplied by $\sin\left(\frac{p\pi x_1}{\ell_0}\right)\sin(q\pi x_3)$ and the second equation of (42,3) by $\cos\left(\frac{p\pi x_1}{\ell_0}\right)\sin(q\pi x_3)$. Then these new equations are integrated over $x_1$ between 0 and $\ell_0$, and over $x_3$ between 0 and 1. The orthogonality condition lead to the evolution equations for each Fourier coefficient $A_{pq}$ and $B_{pq}$, with $p=0,\dots\infty$ and $q=1,\dots\infty$.

Thus we have replaced a system of nonlinear partial differential equations by an infinite set of nonlinear ordinary differential equations (n.o.d.eqs.) for the coefficients $A_{pq}(\tau)$ and $B_{pq}(\tau)$.

For numerical use, the expansion (42,5) has to be truncated and infinite set of n.o.d.eqs. reduces to a finite set of n.o.d.eqs. to be numerically integrated. Many types of truncations may be adobted such as

$$i \leq M \text{ and } j \leq N$$
or
$$i+j \leq K,$$

which lead to different truncation errors.

Here we adopt the truncation scheme: $i+j \leq K$ with $k=2$ ("minimum" representation). Moreover, it may be shown that the Fourier coefficients with $(p+q)$ odd do not contribute (they tend to zero when $\tau\to\infty$ even it initially they were different from zero). Therefore, for $K=2$ we have:

$$(42,6) \quad \begin{cases} \psi = A(t)\sin(\pi r_0 x_1)\sin(\pi x_3) \\ \Xi = B(t)\cos(\pi r_0 x_1)\sin(\pi x_3) + C(t)\sin(2\pi x_3), \end{cases}$$

using only three Fourier coefficients, where:

$$r_0 = \frac{1}{\ell_0}, \quad t = \sigma \tau, \quad A = \frac{1}{\sigma} A_{11}, \quad B = \frac{1}{Ra} B_{11}, \quad C = \frac{1}{Ra} B_{02}.$$

The evolution equations for $(A(t), B(t)$ and $C(t))$ are:

$$(42,7) \quad \begin{cases} -\sigma \dfrac{dA}{dt} \pi^2(r_0^2+1) = Ra\, B\, \pi\, r_0 + \sigma\, A\, \pi^4(r_0^2+1)^2; \\[2mm] \sigma \dfrac{dB}{dt} = -\sigma\, A\, \pi\, r_0 - \pi^2(r_0^2+1)B - \sigma\, A\, C\, \pi^2 r_0; \\[2mm] \sigma \dfrac{dC}{dt} = \sigma \dfrac{AB}{2} \pi^2 r_0 - 4\, C\, \pi^2. \end{cases}$$

Finally, if we introduce new variables:

$$(42,8) \quad \begin{cases} \tau = \dfrac{\pi^2}{\sigma}(1+r_0^2)t \quad, \quad X = \dfrac{1}{\sigma} \dfrac{r_0}{\sqrt{2}(1+r_0^2)} A, \\[4mm] Y = -\dfrac{Ra}{\sigma^2} \dfrac{\pi r_0^2}{\sqrt{2}(1+r_0^2)} B, \quad Z = -Ra \dfrac{\pi r_0^2}{1+r_0^2} C, \end{cases}$$

we obtain the following Lorenz system[†]:

$$(42,9) \quad \begin{cases} \dfrac{dX}{d\tau} = -\sigma X + \sigma Y; \\[2mm] \dfrac{dY}{d\tau} = rX - Y - XZ; \\[2mm] \dfrac{dZ}{d\tau} = -bZ + XY, \end{cases}$$

where

$$(42,10) \quad r = \frac{r_0^2}{\pi^4(1+r_0^2)^3} Ra, \quad b = \frac{4}{1+r_0^2}.$$

It has been proved by Lorenz that a *nonperiodic* (aperiodic) trajectory of the system (42,9) is *unstable*[††] and its futur is essentially unpredictible evan if the system (42,9) is deterministic.

This simply means that if *two* initial states differ by an *imperceptible* amount, they will inevitably elvove towards two considerably different states for large t.

---

† See Lorenz (1963)

†† A trajectory $\varphi(t)$ is stable at a point $\vec{x}_0 = \varphi(t_0, \vec{x}_0)$ if any other trajectory passing sufficiently close to $\varphi(t_0, \vec{x}_0)$ at $t=t_0$ remains close to $\varphi(t)$ as $t \longrightarrow \infty$.

In atmosphere there are always errors in observing an instantaneous state, i.e., initial conditions are not known sufficiently accurately, and therefore an acceptable prediction of the distant future is impossible if the behaviour is nonperiodic, even in deterministic systems. This statement is of some importance in the study of *atmospheric turbulence*.

## LANDAU MODEL EQUATION

If we suppose that in the Lorenz system (42,9) we have for Y and Z the steady solutions:

$$Z = \frac{XY}{b} \quad \text{and} \quad Y = rX - XZ,$$

then we obtain for X the following evolution equation

(42,11)
$$\frac{dX}{d\tau} = -\sigma X + \sigma r \frac{X}{1+X^2/b}.$$

Near threshold (near the critical point when $r \approx 1$) we obtain, from the evolution equation (42,11), the Landau model equation (Landau (1944)):

(42,12)
$$\frac{dX}{d\tau} = \sigma\left(\varepsilon X - \frac{X^3}{b}\right),$$

where $\varepsilon = r-1$, if we neglect the terms $X^5, \dots$ .

This explains the perfect agreement near the critical point between analysis using a double Fourier expansion, truncated to the lowest order, and the Landau model.

Rewriting the Landau equation (42,12) as a linear in $1/|X|^2$, namely

$$\frac{d}{dt}\left(\frac{1}{|X|^2}\right) + \frac{2\varepsilon}{|X|^2} = \frac{2}{b}, \quad t=\sigma\tau,$$

we find the explicit general solution

(42,13)
$$\frac{1}{|X|^2} = \frac{1}{b\varepsilon} + \left[\frac{1}{X_0^2} - \frac{1}{b\varepsilon}\right]\exp(-2\varepsilon t)$$

if $\varepsilon \neq 0$, where $X_0$ is the initial value of $|X|$.

Therefore

(42,14)
$$|X|^2 = \frac{X_0^2}{\frac{1}{b\varepsilon}X_0^2 + (1 - \frac{1}{b\varepsilon}X_0^2)\exp(-2\varepsilon t)}.$$

We have that b>0; if $\varepsilon > 0$ then the solution (42,15) give

$$|X| \sim X_0 e^{\varepsilon t} \quad \text{as} \quad t \longrightarrow -\infty$$

and $X_0 \to 0$, just as in the linear theory, but

(42,16)  $\qquad |X| \to |X_e| \equiv \sqrt{b\varepsilon} \quad$ as $t \to +\infty$,

whatever the value of $X_0$. This is called *supercritical stability*, the basic flow being linearly unstable for $\varepsilon > 0$ but settling down as a new linear flow eventually. The new flow is, moreover, independent of the initial conditions except through the phase of the complex amplitude X of the dominant mode; it has period $2\pi/\omega_1$ if $\omega_1 \neq 0$ or is steady if $\omega_1 = 0$. The disturbance is said to equilibrate, because its amplitude tends to $X_e$ after a long time:

(42,17)  $\qquad X \sim (b\varepsilon)^{1/2} \quad$ as $r \downarrow 1$.

Thus, when $0 < \varepsilon \ll 1$, $X_e$ is small and even if higher-order terms, for example one in $|X|^6$, were included on the right-hand side of the Landau equation,

(42,18)  $\qquad \dfrac{d|X|^2}{dt} = \varepsilon |X|^2 - \dfrac{1}{b}|X|^4, \qquad t = 2\sigma \tau,$

they would remain small and the qualitative character of the solution would be unchanged. The typical developement of $|X|$ with time is sketched in Figure 29 below and the dependence of the amplitudes of the equilibrium solutions $|X| = 0$, $|X| = X_e$ upon $r = 1 + \varepsilon$ in Figure 30 below[†].

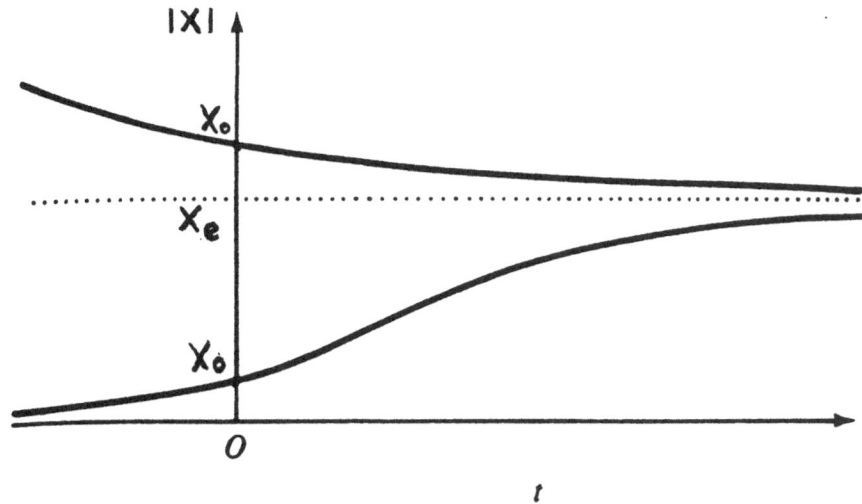

*Fig. 29: Graph of the amplitude* $|X(t)|$.

---

[†] According to Drazin and Reid (1981; p.373).

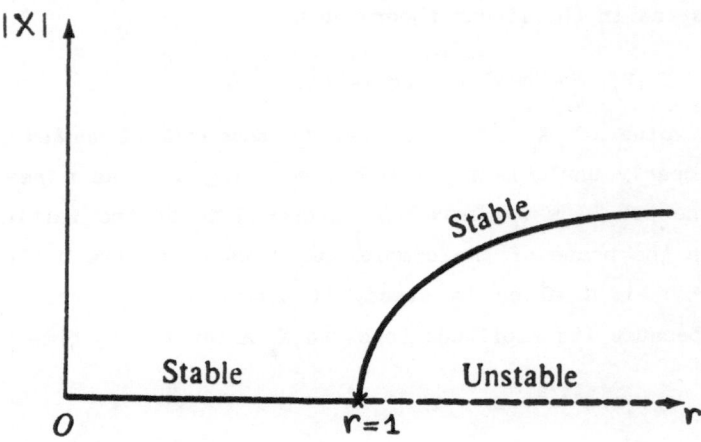

$\mathcal{F}ig.$ 30: $\mathcal{B}ifurcation$ $diagram$ $for$ $|X|=|X(r)|$.

The branching of the curve of the equilibrium solutions at r=1, $|X|$=0 is called a *bifurcation*. The Landau equation implies that the solution $|X|$=0, which represents the steady basic flow, is stable for r<1 but unstable for r>1 ($\varepsilon$>0) and that $|X|=X_e$, which represents the new laminar flow, is stable where it exists, i.e. for $\varepsilon$>0.

In more complete models of atmosphere-dynamic stability we shall see that there may be further bifurcations from the solution $|X|$=0, e.g. where the next least stable normal mode of the basic flow becomes unstable, and from the solution $|X|=X_e$.

If $\varepsilon$<0 (b>0) equation (42,12) confirms that the disturbance decays in accord with the linear theory, i.e.

$$|X| \sim X_0 e^{\varepsilon t} \quad \text{as } t \to +\infty \text{ and } X_0 \to 0.$$

In this case the term $-\frac{1}{b}|X|^4$ of equation (42,18) due to the nonlinearity remains small for all time if it is initially small.

## 43 . THE LORENZ (STRANGE)
## ATTRACTOR

Now we consider the Lorenz system (42,9). It is hard to imagine a much simpler system that is neither linear nor two-dimensional, but the solutions to these equations (42,9) nevertheless do very complicated things. Lorenz (1963) discovered numerically a striking mathematical structure which has come to be known as the Lorenz (strange) attractor and which occurs for these equations with

$$b=\frac{8}{3}, \quad \sigma=10, \quad r=28.$$

The exact parameter values are not crucial, but the behavior of typical solutions definitely does depend on the parameters and is quite different in other regions of parameter space. It should also be noted that, in spite of overwhelming numerical evidence, there is to my knowledge no complete proof that the structure about to the described actually does occur for these specific equations. It is not hard to see that it does occur for *some* equations.

The phenomenology is as follows: The equations admit three stationary solutions, one at the origin and the other two (which we will denote by $C_{\pm}$ and refer to as centers) at $X=Y=\pm\sqrt{(b(r-1))}$; $Z=r-1$. All three stationary solutions are unstable. Orbits that start near the origin escape monotonically; those that start near the centers escape through growing oscillations. If a solution is computed starting from some more or less randomly chosen initial point, what is found without exception is that the orbit will, after an initial transient regime of variable length, settle down to a motion in which, most of the time, it can be thought of as performing oscillations about one of the centers. The oscillation grows in amplitude; when it reaches a critical size, the orbit abruptly makes a transition to oscillation about the other center. This oscillation again grows and the orbit eventually makes a transition back to oscillating about the first center, and so on.

The amplitude of oscillation immediately after transition varies from transition to transition, and it in turn determines the number of oscillations before the next transition. The sequence of number of oscillations between transitions appears random, and the power spectra of the coordinates are continuous (see Figure 31 below). Thus, the motion bot appears chaotic and satisfies the standard operational test for chaotic behavior.

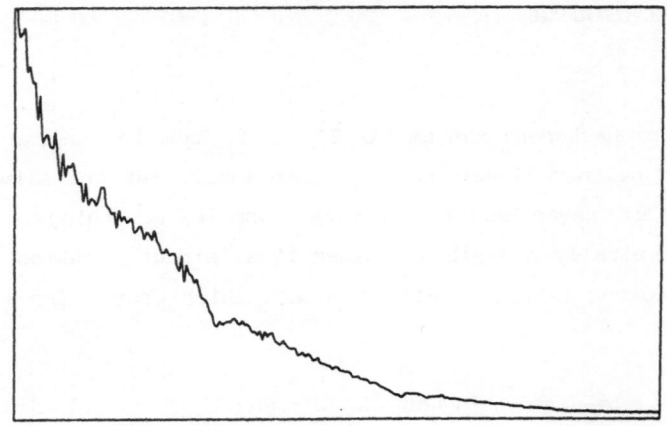

*Fig. 31: The power spectrum of the X coordinate of
the Lorenz system (42,9) (on a linear scale).
The frequency ranges from 0 to 5.*

The mathematical object responsible for this behavior is sketched
schematically in Figure 32 below.

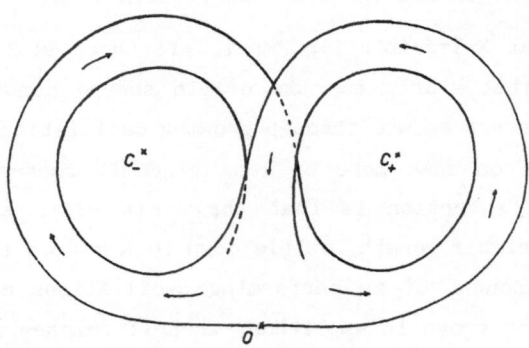

*Fig. 32: A schematic view of the Lorenz attractor.*

This sketch (see, above) represents a family of orbits in the three-dimensional state space for the Lorenz system. It is not even approximately to scale; proportions have been distorted in the hope of making the mathematical structure more transparent. To a first approximation, the structure looks like two reasonably flap loops of ribbon, one lying above the other along a central band, and the two glued together at the bottom of that band. The motion flows arounds the loops, clockwise on the left and counter-clockwise on the right. Going once around the right-hand loop constitutes a single oscillation around $C_+$. Orbits beginning either above or below the ribbon are attracted quickly down to its immediate vicinity and then follow the flow on it. The double-loop structure is strictly invariant under the solution flow; any point on it has an orbit that can be traced both forward and backward for all time without leaving it.

The central band is divided in half by orbits that flow essentially straight down to the stationary solution at the origin; these orbits are exceptions to the pattern of growing oscillations followed by transitions displayed by typical orbits. Orbits to the left of this boundary will make their next oscillation around $C_-$; those to the righ will go next around $C_+$. The fact that oscillations around the centers are growing in amplitude means that for example, a loop around $C_+$ brings the orbit back to the left of where is started out. A transition from oscillation around $C_+$ to oscillation around $C_-$ occurs when an orbit making a loop around $C_+$ comes back to the left of the dividing boundary.

The central band divides in half laterally at the bottom of this boundary, and each half, after having made a loop around the appropriate center, has become wide enough to cover almost the entire top of the band. Thus orbits are pulled apart laterally as they flow around the loops, and this accounts for the observed sensitive dependence on initial conditions.

A typical orbit on this structure wanders over the surface, coming arbitrarily close to each point infinitely often. There are, however, a great many nontypical orbits. We have already mentioned the orbits in the middle of the central band which simply converge down toward the origin. There are also many periodic orbits, all unstable, as well as orbits with more subtle kinds of atypical behavior.

We next take a closer look at the ribbons and argue that they cannot be simple surfaces but must rather have *infinitely many layers*. Start at the top of the central band where there are two approximately parallel ribbons, one on top of the other. As we have drawn the picture, the upper ribbon is made up of orbits returning to the central band after a loop around $C_+$; the lower, around $C_-$. As the orbits flow down the central band, the two ribbons are drawn together. At the bottom, they form a two-sheeted surface which proceeds to split laterally in two with half going left around $C_-$ and half right around $C_+$. Thus, the ribbon of orbits going around $C_+$ or $C_-$ has at least two sheets, the upper one made up of orbits whose previous circuit was around $C_+$, the lower of orbits whose previous circuit was around $C_-$. These sheets are carried closer together by the flow but the separation remains nonzero, so the upper ribbon at the top of the central band actually has two layers rather than just one. The same is true for the lower ribbon, and therefore the structure at the bottom of the central band actually has four layers rather than just two. Thus, the ribbons going around $C_+$ and $C_-$ are actually four-sheeted, and so on. Continuing to argue in this way we see that all the ribbons must have infinitely many sheets. This object is an instance of what has come to be called a *strange (Lorenz) attractor*.

Let us now investigate the Lorenz equations (42,9). The divergence of the phase flow (39,2) is negative for these equations:

$$\hat{\Lambda} = -(\sigma+b+1),$$

so that all trajectories migrate towards a certain set of zero volume. The quantity:

$$\mathcal{H} \equiv \left[ X^2 + Y^2 + (Z-r-\sigma)^2 \right]^{1/2}$$

satisfies the condition

(43,1)
$$\frac{d\mathcal{H}}{dt} < -C_1\mathcal{H} + C_2$$

with positive $C_1$ and $C_2$, so that all trajectories enter the sphere

$$\mathcal{H} \leq 2\frac{C_2}{C_1}.$$

The system does not change under the substitution

$$(X,Y,Z) \implies (-X,-Y,-Z).$$

For r<1, the unique fixed point is the stable vertex 0 at the coordinate origin. For r≥1 (the onset of convection) it loses its stability (becoming a saddle point with a two-dimensional stable manifold and two unstable one-dimensional ones- the separatrixes $\Gamma^+$ and $\Gamma^-$), and two new fixed points appear $C^+$, $C^-$, towards which the separatrixes move. For $\sigma<b+1$ they are stable,

while for σ>b+1 (following Lorenz, we will henceforth investigate the case σ=10, b=8/3) they are stable for

$$(43,2) \qquad 1 < r < r^* = \frac{\sigma(\sigma+b+3)}{\sigma-b-1} \cong 24,74 \ ,$$

while for r>r$^*$, they lose their stability.

For r=r$_1$≅13,92 we find a bifurcation for which the separatrixes return to saddle point. For r>r$_1$ out of the loops of the separatrixes there appear saddle-point periodic motion L$^+$, L$^-$ around the foci C$^+$, C$^-$ (at the same time, there appears an invariant set of curves Ω$_1$ which is not an attractor and which has Cantor structure, including a denumebrable set of saddle-point periodic motions); the separatrixes Γ$^+$, Γ$^-$ intersect and move towards the foci C$^-$, C$^+$.

For r=r$_2$≅24,06 the separatrixes Γ$^+$, Γ$^-$, in place of the foci, are curled around the cycles L$^-$, L$^+$, and instead of Ω$_1$, there appears an infinte Lorenz attractor Ω$_2$ whose region of attraction is limited by the stable manifolds of the cycles L$^-$, L$^+$ (so that excitation of randomness is " hard"). For r>r$_2$ it is stable, including 0, Γ$^+$, Γ$^-$ and therefore there is no strucural stability; on this attractor the periodic motions are everywhere dence (capable of undergoing a sequence of period-doubling bifurcations and of disappearing as r grows only by way of adhesion to the loops of the separatrixes). For r=r$^*$, the cycles L$^+$, L$^-$ contract to the points C$^+$, C$^-$ and the latter lose their stability.

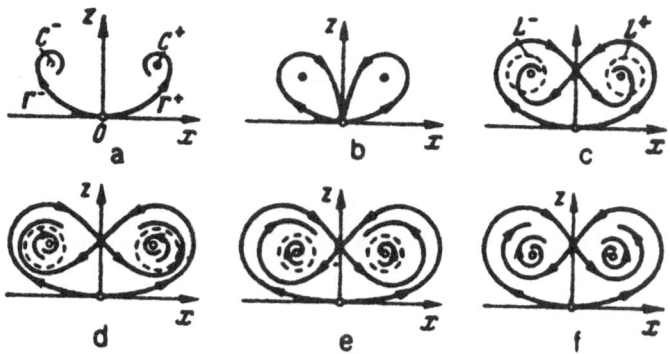

Fif.33: Bifurcations in the Lorenz system
(a) for 1 < r < r$_1$, (b) for r = r$_1$,
(c) for r$_1$< r < r$_2$, (d) for r = r$_2$,
(e) for r$_2$< r < r$_3$, (f) for r = r$_3$.

For $r^*<r<r^{**}\cong 220$ the Lorenz attractor is the unique stable limit set (we note that as r decrease from $r^{**}$ to $r_2$, the phase point M(t) stays within the attractor , while for $r<r_2$ it loses stability and M(t) moves toward $C^+$ or $C^-$). An example of a trajectory in the attractor (for r=28, intersecting the plane Z=27) is shown in Figure 34 below, which is the work of Oscar Lanford (Berkeley Univercity).

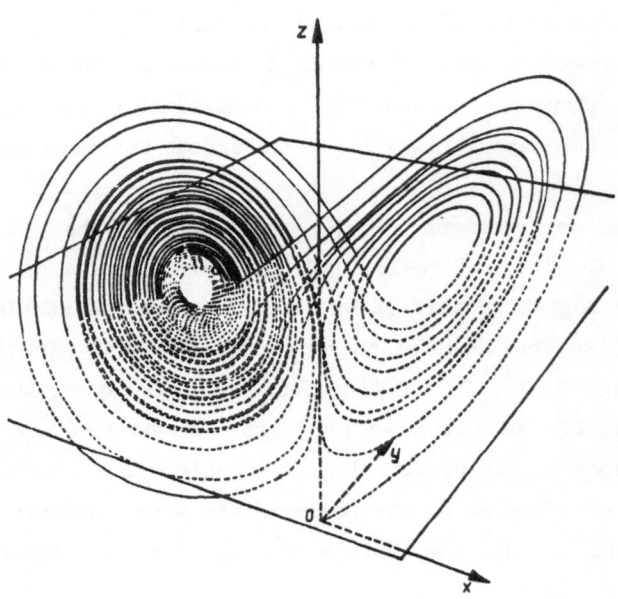

*Fig. 34: Lorenz strange attractor for* r = 28.

It start at the coordinate origin, circles $C^+$, and the unwinds and is drawn to $C^-$, leaves $C^-$ and spirals toward $C^+$, etc., while the period of rotation around $C^+$ or $C^-$ equals 0,62 and the radii of the spirals change by 6% per rotation.

As pointed out by Lorenz himself, for this example the Poincaré mapping $Z_{n+1}= \sqcap(Z_n)$ of successive maxima has the triangular form, while $|\sqcap'(Z)|>1$ everywhere; it is ergodic and has the mixing property.

Mori and Fujisaka (1980) have calculated the Lyapunov exponent $\lambda_1$ for the Lorenz attractor as a function of r (the second one $\lambda_2\equiv0$, the third $\lambda_3=\hat{\Lambda}-\lambda_1$ is negative and the dimension of the attractor $d=2+\lambda_1/|\lambda_3|$). For r<1 it is negative and for r=1 (the appearance of the convective "rollers") it reduces to zero; for $1<r<r^*$ it is again negative and for $r=r^*$ it reaches zero. However, for $r=r_2$ greater than zero there also appears a new branch of $\lambda_1$ which grows for $r>r_2$ and for very large r is multiply interrupted by "*lacunae*" of zero values, corresponding to periodic motion (see the example in Figure 35 below, for b=4, σ=16; here for r=40 the dimension of the Lorenz attractor is d=2,06).

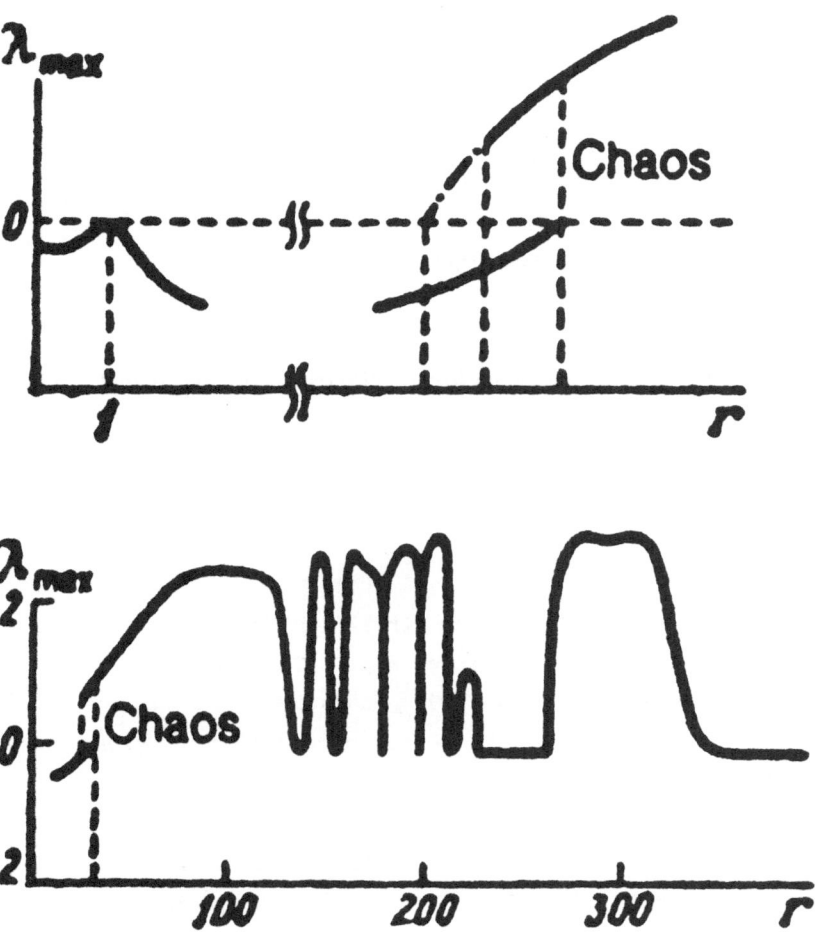

*Fig. 35: Graph of $\lambda_{max} \equiv \lambda_1$ as function of r ( b=4 and σ=16).*

278

*BACKGROUND READING*

For a good introduction to "Deterministic Chaos" the reader is referred to:

SCHUSTER, H.G. (1984) _ *Deterministic Chaos; An Introduction.*
Physik-Verlag, Weinheim (R.F.A.),

and also

BARENBLATT, G.I.
IOOSS, G.
and
JOSEPH, D.D.

(1983) _ *Nonlinear Dynamics and Turbulence.*
Pitman adv. Publ. Program
Boston-London-Melbourne.

Concerning the theory of the Lorenz system, see the book of:
SPARROW, C. (1982) _ *The Lorenz Equations: Bifurcations, chaos and strange*
*attractor.* Springer

*REFERENCES TO WORKS CITED*
*IN THE TEXT*

ARGOUL, F. and ARNEODO, A. (1984) _ in J.M.T.A, n°special 1984; pp.241-288.
BENARD, H. (1900) _ Rev. Gén. Sci. Pures Appl. 12, 1261-1271, 1309-1328.
COLLET, P. and ECKMANN, J.P. (1980) _ *Iterated maps on the interval as*
*dynamical system.* Progress in Physic,
Vol 1. Birkaüser, Basel.
COULLET, P.H. and SPIEGEL, E.A. (1983) _ S.I.A.M.J. Appl. Math., Vol.43,
n°4, 776-821.
COULLET, P.H. and TRESSER, C. (1984) _ in J.M.T.A, n°special 1984, pp.217-240.
DRAZIN, P.G. and REID, W.H. (1981) _ *Hydrodynamic stability.* Cambridge
University Press.
ECKMANN, J.P. (1981) _ Reviews of Modern Physics, vol.53, n°4, part 1,
pp.643-654.
ECKMANN, J.P. (1983) _ in *"Chaotic behavior in dissipative dynamical system".*
Les Houches Summer School (1981).
Eds. Iooss, Helleman and Stora, North-Holland.
ECKMANN, J.P. and RUELLE, D. (1985) _ Rev. Mod. Physics, 57; 617-656.
EPSTEIN, H. and LASCOUX, J. (1981) _ Comm. Math. Phys., vol.81; 437.
FEIGENBAUM, M.J. (1978) _ J. Stat. Phys., vol.19, 25-52.

FEIGENBAUM, M.J. (1980) _ Comm. Math. Phys ., 77; 65

                and also: Los Alamos Science, vol.1; 4-27.

GUCKENHEIMER, J. and HOLMES, Ph. (1986) _ *Nonlinear Oscillations, Dynamical systems, and Bifurcations of vectors-fields*. Springer-Verlag, New-York.

HOPF, E. (1948) _ Comm. Pure and Appl. Math., vol. 1; pp. 303-322.

IOOSS, G. (1984) _ in *"Turbulence and Chaotic Phenomena in Fluids"*; T.Tatsumi ed., p. 185. Elsevier Sci. Publ. B.V.(North-Holland).

JAKOBSON, M. (1980) _ in "Proceed. Intern. Conf. on Dynam. Syst". Northeastern University.

KIFER, J.I. (1974) _ Math. U.S.S.R. Izvestija, 8, 1083.

LANDAU, L.D. (1944) _ C.R.Acad.Sci. U.R.S.S., 44, 311-314.

LANFORD III (1982) _ Bull. Amer. Math. Soc., Vol.6; 427.

LORENZ, E.N. (1963) _ J. Atm. Sci., Vol.20; 130-141.

LOVEJOY, S. and SCHERTZER, D. (1986) _ Bull. of the American Meteo. Soc., Vol.67, n°1, (1986) pp.21-32.

MA, S.K. (1976) _ *Modern Theory of critical phenomena*. Benjamin, Reading, Mass.

MORI, H. and FUJISAKA, S. (1980) _ Lect. Notes in Physics, vol. 132, 181.

MONIN, A.S. (1986) _ Soviet Physics - USPEKHI, vol.29, n°9, pp.843-868.

PLATTEN, J.K. and LEGROS, J.C. (1984) _ *Convection in Liquids*. Springer-Verlag, Heidelberg.

POMEAU, Y. (1983) _ In *"Nonlinear Dynamics and Turbulence"*; p.295. Pitman Adv. Publ. Program.

POMEAU, Y. and MANNEVILLE, P. (1980) _ Comm. Math. Phys.; 77, 189.

RAYLEIGH, Lord (1916) _ Philos. Mag. (6) 32, 529-546.

RÖSSLER, O.E. (1976) _ Phys. Letter., ser.A, vol.57; p.397.

RUELLE, D. and TAKENS, F. (1971) _ Comm. in Math. Phys., 20, 167-192.

SREENIVASAN, K.R. and MENEVEAU, C. (1986) _ J. Fluid. Mech., vol.173; pp.357-386.

ZEYTOUNIAN, R.Kh. (1983) _ CRASc., Serie I, t.297; pp.271-274.

ZEYTOUNIAN, R.Kh. (1989) _ Int. J. Engng. Sci. Vol.27, N°11, pp. 1361-1366.

ZEYTOUNIAN, R.Kh. (1989) _ *An Essay on the Transition to Turbulence*. Archives of Mechanics, vol. 41, issue 2/3, pp. 383-418; Warszawa 1989.

CHAPTER  X

MISCELLANEA

**44 . INTERNAL SOLITARY WAVES**
   **IN AN ISOCHORIC FLOW**

We start here with the equation (14,21) for the function $\delta(x,z)$, which is the
displacement of a particle above its equilibrium height (in the steady case).
If $U_\infty$=constant then we obtain, instead of (14,21), the following equation:

(44,1)

$$\frac{\partial^2 \delta}{\partial x^2} + \frac{\partial^2 \delta}{\partial z^2} + \frac{1}{2}\left\{2\frac{\partial\delta}{\partial z} - \left[\left(\frac{\partial\delta}{\partial x}\right)^2 + \left(\frac{\partial\delta}{\partial z}\right)^2\right]\right\}\frac{1}{\rho_\infty}\frac{d\rho_\infty}{dz_\infty}$$

$$- \frac{g}{U_\infty^{02}}\frac{1}{\rho_\infty}\frac{d\rho_\infty}{dz_\infty}\delta = 0, \quad U_\infty^0 \equiv \text{constant},$$

where $z_\infty$ is the height, in the undisturbed region, of the fixed streamline
through the point $(x,z)$.

The mathematics is a bit simpler if we direct attention to isochoric flow with
an exponential basic density gradient

(44,2) $\qquad \rho_\infty(z_\infty) = \rho_\infty(0)\exp[-x_0 z_\infty]$.

We now non-dimensionalize according to the following scheme:

(44,3) $\qquad \xi=\frac{x}{\lambda}, \quad \eta=\frac{z}{h}, \quad \Delta=\frac{\delta}{\alpha h}, \quad \alpha=\frac{a}{h}, \quad \beta=x_0 h, \quad \gamma=\left(\frac{h}{\lambda}\right)^2,$

where we consider a flow in channel of height h; the characteristic horizontal
wave-length is $\lambda$ and the amplitude of the disturbance is of the order of a.
In terms of the quantities in (44,3), equation (44,1) with (44,2) becomes

(44,4) $\qquad \gamma\frac{\partial^2\Delta}{\partial\xi^2} + \frac{\partial^2\Delta}{\partial\eta^2} + \frac{\beta}{2}\left[\alpha\left(\frac{\partial\Delta}{\partial\eta}\right)^2 + \alpha\gamma\left(\frac{\partial\Delta}{\partial\xi}\right)^2 - 2\frac{\partial\Delta}{\partial\eta}\right] + \sigma^2\Delta = 0,$

where

(44,5) $\qquad \sigma^2 = \frac{gx_0 h^2}{U_\infty^{02}}.$

In non-dimensional form the boundary conditions are written

(44,6)    $\Delta=0$  on  $\eta=0$  and  $\eta=1$.

Let us now assume various expressions for $\gamma$ in terms of $\alpha$, $\beta$ and search for solutions of (44,4), (44,6) in the form[†]:

(44,7)
$$\begin{cases} \Delta = \Delta_{00} + \alpha\Delta_{10} + \beta\Delta_{01} + \alpha\beta\Delta_{11} + \ldots, \\ \sigma^2 = \sigma^2_{00} + \alpha\sigma^2_{10} + \beta\sigma^2_{01} + \alpha\beta\sigma^2_{11} + \ldots. \end{cases}$$

*THE CASE OF $\gamma=\beta$*

In this case we have: $\lambda^2 = h/x_0$. The first approximation is

(44,8)
$$\frac{\partial^2\Delta_{00}}{\partial\eta^2} + \sigma^2_{00}\Delta_{00} = 0$$

and since $\Delta_{00}$ must vanish at the top and bottom of the channel, the solution is

(44,9)    $\Delta_{00} = \mathcal{D}_0(\xi)\sin(n\pi\eta)$,    $\sigma^2_{00} = n^2\pi^2$,

where $\mathcal{D}_0(\xi)$ is arbitrary. A second approximation is:
$$\frac{\partial^2\Delta_{01}}{\partial\eta^2} + n^2\pi^2\Delta_{01} = -\sigma^2_{01}\Delta_{00} - \frac{\partial^2\Delta_{00}}{\partial\xi^2} + \frac{\partial\Delta_{00}}{\partial\eta}$$

or
$$\frac{\partial^2\Delta_{01}}{\partial\eta^2} + n^2\pi^2\Delta_{01} = -\left(\frac{d^2\mathcal{D}_0}{d\xi^2} + \sigma^2_{01}\mathcal{D}_0\right)\sin(n\pi\eta) + n\pi\mathcal{D}_0\cos(n\pi\eta)$$

with solution

(44,10)
$$\Delta_{01} = \mathcal{D}_1\sin(n\pi\eta) + \mathcal{D}_2\cos(n\pi\eta) + \frac{1}{2n\pi}\eta\cos(n\pi\eta)\left[\frac{d^2\mathcal{D}_0}{d\xi^2} + \sigma^2_{01}\mathcal{D}_0\right]$$
$$+ \frac{\mathcal{D}_0}{2}\eta\sin(n\pi\eta),$$

where $\mathcal{D}_1$ and $\mathcal{D}_2$ are arbitrary functions of $\xi$. But $\mathcal{D}_2 \equiv 0$, since $\Delta_{01}$ must vanish at $\eta=0$. Then the condition that $\Delta_{01}=0$ at $\eta=1$ yields

(44,11)
$$\frac{d^2\mathcal{D}_0}{d\xi^2} + \sigma^2_{01}\mathcal{D}_0 = 0.$$

This has *no* solution for $\mathcal{D}_0(\xi)$ that vanishes at either $\xi=\infty$ or $\xi=-\infty$. There is no steady-state disturbance for $\gamma=\beta$ that vanishes at infinity.

---

[†] According to R. Long (1965).

*THE CASE OF $\gamma=\alpha$*

Again the first approximation is

$$\Delta_{00} = \mathcal{D}_0(\xi)\sin(n\pi\eta), \quad \sigma^2_{00} = n^2\pi^2$$

and a second approximation is

(44,12)
$$\frac{\partial^2\Delta_{10}}{\partial\eta^2} + n^2\pi^2\Delta_{10} = -\frac{\partial^2\Delta_{00}}{\partial\xi^2} - \sigma^2_{10}\Delta_{00}$$

$$= -\sin(n\pi\eta)\left[\frac{d^2\mathcal{D}_0}{d\xi^2} + \sigma^2_{10}\mathcal{D}_0\right].$$

A solution satisfying $\Delta_{10}=0$ at $\eta=0$ is

$$\Delta_{10}(\xi,\eta) = \mathcal{G}_1(\xi)\sin(n\pi\eta) + \frac{1}{2n\pi}\eta\cos(n\pi\eta)\left[\frac{d^2\mathcal{D}_0}{d\xi^2} + \sigma^2_{10}\mathcal{D}_0\right]$$

and requirement that $\Delta_{10}=0$ at $\eta=1$ yields

$$\frac{d^2\mathcal{D}_0}{d\xi^2} + \sigma^2_{10}\mathcal{D}_0 = 0.$$

Again there is *no* steady-state disturbance for $\gamma=\alpha$ that vanishes at infinity.

*THE CASE OF $\gamma=\alpha\beta$*

We again have

$$\Delta_{00} = \mathcal{D}_0(\xi)\sin(n\pi\eta) \quad \text{and} \quad \sigma^2_{00} = n^2\pi^2$$

and the differential equation for $\Delta_{10}$ is the same as eq. (44,12), except that the first term on the right is missing.
Therefore,

(44,13)
$$\Delta_{10}(\xi,\eta) = \mathcal{G}_1(\xi)\sin(n\pi\eta) + \frac{\sigma^2_{10}}{2n\pi}\mathcal{D}_0(\xi)\eta\cos(n\pi\eta).$$

Since $\Delta_{10}=0$ at $\eta=1$, we see that:

(44,14)
$$\sigma^2_{10} = 0, \quad \Delta_{10} = \mathcal{G}_1(\xi)\sin(n\pi\eta).$$

The equation for $\Delta_{01}$ is

$$(44,14) \qquad \frac{\partial^2 \Delta_{01}}{\partial \eta^2} + n^2 \pi^2 \Delta_{01} = -\sigma_{01}^2 \Delta_{00} + \frac{\partial \Delta_{00}}{\partial \eta}$$

$$= -\sigma_{01}^2 \mathcal{D}_0(\xi) \sin(n\pi\eta) + n\pi \mathcal{D}_0(\xi) \cos(n\pi\eta).$$

The solution satisfying $\Delta_{01}=0$ at $\eta=0$ is

$$\Delta_{01}(\xi,\eta) = \mathcal{D}_1(\xi)\sin(n\pi\eta) + \frac{\sigma_{01}^2}{2n\pi}\mathcal{D}_0(\xi)\eta\cos(n\pi\eta) + \frac{\mathcal{D}_0}{2}\eta\sin(n\pi\eta).$$

Since $\Delta_{01}=0$ at $\eta=1$, we see that

$$(44,15) \qquad \sigma_{01}^2 = 0, \quad \Delta_{01} = \mathcal{D}_1 \sin(n\pi\eta) + \frac{\mathcal{D}_0}{2}\eta\sin(n\pi\eta).$$

Finally, the equation for $\Delta_{11}$ is

$$\frac{\partial^2 \Delta_{11}}{\partial \eta^2} + n^2 \pi^2 \Delta_{11} = -\frac{\partial^2 \Delta_{00}}{\partial \xi^2} - \frac{1}{2}\left[\frac{\partial \Delta_{00}}{\partial \eta}\right]^2 + \frac{\partial \Delta_{10}}{\partial \eta} - \sigma_{11}^2 \Delta_{00}$$

$$\equiv -\left(\frac{d^2 \mathcal{D}_0}{d\xi^2} + \sigma_{11}^2 \mathcal{D}_0\right)\sin(n\pi\eta) + n\pi \mathcal{G}_1(\xi)\cos(n\pi\eta)$$

$$- \frac{\mathcal{D}_0^2}{4} n^2 \pi^2 (1 + \cos(2n\pi\eta))$$

so that

$$(44,16) \qquad \Delta_{11} = \mathcal{D}_2(\xi)\sin(n\pi\eta) + \frac{\mathcal{D}_0^2(\xi)}{6}\cos(n\pi\eta) + \eta\frac{\cos(n\pi\eta)}{2n\pi}\left[\frac{d^2 \mathcal{D}_0}{d\xi^2} + \sigma_{11}^2 \mathcal{D}_0\right]$$

$$+ \mathcal{G}_1(\xi)\eta\frac{\sin(n\pi\eta)}{2} - \frac{1}{4}\mathcal{D}_0^2(\xi) + \frac{\mathcal{D}_0^2(\xi)}{12}\cos(2n\pi\eta).$$

Since $\Delta_{11}=0$ at $y=1$,

$$(44,17) \qquad [(-1)^n - 1]\frac{\mathcal{D}_0^2(\xi)}{6} + \frac{(-1)^n}{2n\pi}\left[\frac{d^2 \mathcal{D}_0}{d\xi^2} + \sigma_{11}^2 \mathcal{D}_0\right] = 0.$$

If $n$ is *even*, the first term vanishes completely, and the remaining differential equation has *no* solution vanishing at $\xi \rightarrow \infty$.

For *odd* $n$, however we get

$$(44,18) \qquad \frac{d^2 \mathcal{D}_0}{d\xi^2} + \sigma_{11}^2 \mathcal{D}_0 + \frac{2}{3}n\pi\mathcal{D}_0^2 = 0, \quad n=1,3,5,\ldots,$$

with solution

$$(44,19) \qquad \mathcal{D}_0(\xi) = -\frac{9\sigma_{11}^2}{4n\pi}\mathrm{sech}^2\left[\frac{\xi}{2}i\sigma_{11}\right], \quad n=1,3,5,\ldots$$

vanishing at $|\xi|=\infty$, if $\sigma_{11}^2$ is *negative*.

Notice that if we regard $\alpha$ in $(44,3)$ as the non-dimensional maximum amplitude of the disturbance, then

$$(44,20) \qquad -\frac{9\sigma_{11}^2}{4n\pi} = 1 \ , \quad n=1,3,5,\ldots$$

To the present order then,

$$(44,21) \qquad \begin{cases} \dfrac{\delta}{h} = \alpha\,\mathrm{sech}^2\left[\dfrac{x}{3h}\sqrt{n\pi\alpha\beta}\right]\sin(n\pi\frac{z}{h}), & n=1,3,5,\ldots, \\[2ex] \sigma^2 = n^2\pi^2 - \dfrac{4n\pi}{9}\alpha\beta \ , & n=1,3,5,\ldots \ . \end{cases}$$

## THE CASE OF $\gamma=\alpha^2\beta$

As in the preceding case, we have

$$\sigma_{00}^2 = n^2\pi^2, \qquad \Delta_{00} = \mathcal{D}_0(\xi)\sin(n\pi\eta);$$

$$\sigma_{10}^2 = 0 \ , \qquad \Delta_{10} = \mathcal{G}_1(\xi)\sin(n\pi\eta);$$

$$\sigma_{01}^2 = 0 \ , \qquad \Delta_{01} = \mathcal{D}_1(\xi)\sin(n\pi\eta) + \frac{\mathcal{D}_0}{2}\eta\sin(n\pi\eta).$$

In addition, the solution for $\Delta_{11}$ in $(44,16)$ and $(44,17)$ is applicable here if we omit the terms involving $\dfrac{d^2\mathcal{D}_0}{d\xi^2}$.

Equation $(44,17)$ then reveals that $\mathcal{D}_0=0$ if n *odd*, and that $\sigma_{11}^2=0$ is n if *even*. Hence

$$(44,22) \qquad \sigma_{11}^2 = 0 \quad \text{and} \quad \Delta_{11} = \mathcal{D}_2(\xi)\sin(n\pi\eta) + \mathcal{G}_1(\xi)\eta\frac{\sin(n\pi\eta)}{2}$$

$$+ \frac{\mathcal{D}_0^2(\xi)}{12}\left[2\cos(n\pi\eta) - 3 + \cos(2n\pi\eta)\right],$$

$$n=2,4,6,\ldots \ .$$

Continuing, we obtain

$$\sigma_{20}^2 = 0, \quad \Delta_{20} = \mathcal{G}_2(\xi)\sin(n\pi\eta), \quad n=2,4,6,\ldots,$$

and

$(44,23)$
$$\frac{\partial^2 \Delta_{21}}{\partial \eta^2} + n^2\pi^2\Delta_{21} = -\sigma_{21}^2\Delta_{00} - \frac{\partial^2\Delta_{00}}{\partial\xi^2} + \frac{\partial\Delta_{20}}{\partial\eta} - \frac{\partial\Delta_{00}}{\partial\eta}\frac{\partial\Delta_{10}}{\partial\eta}$$

$$\equiv -\left(\frac{d^2\mathcal{D}_0}{d\xi^2} + \sigma_{21}^2\mathcal{D}_0\right)\sin(n\pi\eta) + n\pi\mathcal{G}_2(\xi)\cos(n\pi\eta)$$

$$- \frac{1}{2}\mathcal{D}_0\mathcal{G}_1(\xi)n^2\pi^2(1 + \cos(2n\pi\eta)).$$

The solution satisfying $\Delta_{21}=0$ at $\eta=0$ is

$$\Delta_{21} = \mathcal{D}_3(\xi)\sin(n\pi\eta) + \frac{1}{3}\mathcal{D}_0(\xi)\mathcal{G}_1(\xi)\cos(n\pi\eta) + \eta\frac{\cos(n\pi\eta)}{2n\pi}\left(\frac{d^2\mathcal{D}_0}{d\xi^2} + \sigma_{21}^2\mathcal{D}_0\right)$$

$$+ \mathcal{G}_2(\xi)\eta\frac{\sin(n\pi\eta)}{2} - \frac{\mathcal{D}_0(\xi)}{2}\mathcal{G}_1(\xi) + \frac{\mathcal{D}_0(\xi)}{6}\mathcal{G}_1(\xi)\cos(2n\pi\eta). \quad n=2,4,6,\ldots$$

Applying the condition that $\Delta_{21}=0$ at $\eta=1$, we get
$$\frac{d^2\mathcal{D}_0}{d\xi^2} + \sigma_{21}^2\mathcal{D}_0 = 0.$$

There is *no* steady-state disturbance for $\gamma=\alpha^2\beta$ that vanishes at infinity.

*THE CASE OF $\gamma=\alpha\beta^2$*

As in the previous case ($\gamma=\alpha^2\beta$) we have:

$$\Delta_{00} = \mathcal{D}_0\sin(n\pi\eta), \quad \Delta_{10} = \mathcal{G}_1\sin(n\pi\eta), \quad \Delta_{20} = \mathcal{G}_2\sin(n\pi\eta);$$

$$\sigma_{00}^2 = n^2\pi^2, \quad \sigma_{10}^2 = \sigma_{20}^2 = 0, \quad \sigma_{01}^2 = 0, \quad \sigma_{11}^2 = 0,$$

and

$$\Delta_{01} = \mathcal{D}_1\sin(n\pi\eta) + \frac{\mathcal{D}_0}{2}\eta\sin(n\pi\eta),$$

$(44,24)$
$$\Delta_{11} = \left[\mathcal{D}_2 + \frac{1}{2}\mathcal{G}_1\frac{\eta}{2}\right]\sin(n\pi\eta) + \frac{\mathcal{D}_0^2}{12}[2\cos(n\pi\eta) - 3 + \cos(2n\pi\eta)].$$

As in the case: $\gamma=\alpha^2\beta$, the derivation of $(44,24)$ yields the requirement that n
be even. Also

$$\frac{\partial^2 \Delta_{02}}{\partial \eta^2} + n^2\pi^2\Delta_{02} = -\sigma_{02}^2\Delta_{00} + \frac{\partial \Delta_{01}}{\partial \eta}$$

$$\equiv -\sigma_{02}^2\mathcal{D}_0\sin(n\pi\eta) + \mathcal{D}_1 n\pi\cos(n\pi\eta)$$

$$+ \frac{\mathcal{D}_0}{2}\sin(n\pi\eta) + \frac{\mathcal{D}_0}{2}n\pi\eta\cos(n\pi\eta), \quad n=2,4,6,\ldots,$$

so that

(44,25)
$$\Delta_{02} = \mathcal{D}_4(\xi)\sin(n\pi\eta) - \frac{\mathcal{D}_0(\xi)}{4n\pi}\eta\cos(n\pi\eta) + \sigma_{02}^2\frac{\mathcal{D}_0(\xi)}{2n\pi}\eta\cos(n\pi\eta)$$

$$+ \frac{n\pi}{2}\mathcal{D}_0(\xi)\left[\frac{\eta}{4n^2\pi^2}\cos(n\pi\eta) + \frac{\eta^2}{4n\pi}\sin(n\pi\eta)\right]$$

$$+ \frac{\mathcal{D}_1(\xi)}{2}\eta\sin(n\pi\eta), \quad n=2,4,6,\ldots .$$

The condition at $\eta=1$ yields:

$$\mathcal{D}_0(\xi) = 4\sigma_{02}^2\mathcal{D}_0(\xi) \implies \sigma_{02}^2 = \frac{1}{4},$$

and

(44,26)
$$\Delta_{02} = \mathcal{D}_4(\xi)\sin(n\pi\eta) + \frac{\mathcal{D}_1(\xi)}{2}\eta\sin(n\pi\eta) + \frac{\mathcal{D}_0(\xi)}{8}\eta^2\sin(n\pi\eta).$$

Finally,

$$\frac{\partial^2 \Delta_{12}}{\partial \eta^2} + n^2\pi^2\Delta_{12} = -\frac{\partial^2 \Delta_{00}}{\partial \xi^2} - \frac{\partial \Delta_{01}}{\partial \eta}\frac{\partial \Delta_{00}}{\partial \eta} + \frac{\partial \Delta_{11}}{\partial \eta} - \sigma_{02}^2\Delta_{10} - \sigma_{12}^2\Delta_{00}$$

$$\equiv -\left(\frac{d^2\mathcal{D}_0}{d\xi^2} + \sigma_{12}^2\mathcal{D}_0\right)\sin(n\pi\eta)$$

$$- \mathcal{D}_0(\xi)n\pi\cos(n\pi\eta)\left[\mathcal{D}_1(\xi)n\pi\cos(n\pi\eta)\right.$$

$$\left. + \frac{\mathcal{D}_0(\xi)}{2}\sin(n\pi\eta)\right] + \mathcal{D}_2(\xi)n\pi\cos(n\pi\eta)$$

$$+ \frac{\mathcal{G}_1(\xi)}{4}\sin(n\pi\eta) + \frac{\mathcal{G}_1(\xi)}{2}n\pi\eta\cos(n\pi\eta)$$

$$- \frac{\mathcal{D}_0^2(\xi)}{6}n\pi\sin(n\pi\eta) - \frac{\mathcal{D}_0^2(\xi)}{6}n\pi\sin(2n\pi\eta).$$

The solution is:

(44,27)
$$\Delta_{12}(\xi,\eta) = -\left[\frac{1}{4}\mathscr{G}_1(\xi) - \frac{d^2\mathcal{D}_0}{d\xi^2} + \sigma_{12}^2\mathcal{D}_0(\xi) - \frac{n\pi}{6}\mathcal{D}_0^2(\xi)\right]\frac{\eta}{2n\pi}\cos(n\pi\eta)$$

$$+ \frac{n\pi}{2}\mathscr{G}_1(\xi)\left[\frac{\eta}{4n^2\pi^2}\cos(n\pi\eta) + \frac{\eta^2}{4n\pi}\sin(n\pi\eta)\right]$$

$$+ \frac{\mathcal{D}_2(\xi)}{2}\eta\sin(n\pi\eta) + \frac{5}{36}\mathcal{D}_0^2(\xi)\sin(2n\pi\eta)$$

$$- \mathcal{D}_0(\xi)\mathcal{D}_1(\xi)\left\{\frac{1}{2} - \frac{1}{6}\cos(2n\pi\eta)\right\} - \frac{\mathcal{D}_0^2(\xi)}{4}(\eta$$

$$+ n^2\pi^2)\left[-\frac{\eta}{3n^2\pi^2}\cos(2n\pi\eta) + \frac{4}{9n^3\pi^3}\sin(2n\pi\eta)\right]$$

$$+ \mathcal{D}_5(\xi)\sin(n\pi\eta) + \frac{1}{3}\mathcal{D}_0(\xi)\mathcal{D}_1(\xi)\cos(n\pi\eta).$$

The condition at $\eta=1$ yields

(44,28)
$$\frac{d^2\mathcal{D}_0}{d\xi^2} + \sigma_{12}^2\mathcal{D}_0 - \frac{n\pi}{6}\mathcal{D}_0^2 = 0.$$

This has a solution

(44,29)
$$\mathcal{D}_0(\xi) = \frac{9\sigma_{12}^2}{n\pi}\text{sech}^2\left[\frac{\xi}{2}i\sigma_{12}\right],$$

vanishing at $|\xi|=\infty$, if $\sigma_{12}^2$ is negative.
Thus

(44,30)
$$\frac{9\sigma_{12}^2}{n\pi} = -1$$

and, therefore, to the present order,

(44,31)
$$\begin{cases} \dfrac{\delta}{h} = -\alpha\,\text{sech}^2\left[\dfrac{x}{6h}\sqrt{n\pi\alpha\beta^2}\,\right]\sin(n\pi\frac{z}{h}), \quad n=2,4,6,\ldots, \\[2mm] \sigma^2 = n^2\pi^2 + \dfrac{1}{4}\beta^2 - \dfrac{n\pi}{9}\alpha\beta^2. \end{cases}$$

Notice that if we seek other solutions by putting $\gamma=\alpha^3\beta$, $\gamma=\alpha\beta^3$, etc... all subsequent cases will lead to equation (44,28) with the $d^2\mathcal{D}_0/d\xi^2$-term missing. It follows immediately that $\mathcal{D}_0(\xi)\equiv0$, and that there are not other solutions of the kind we are looking for.

The solutions (44,21) and (44,31) represent disturbances with maximum amplitude at $\xi=0$, dying off monotonically and symmetrically on both sides. The speeds of propagation are determined by values of $\sigma^2$ in (44,21) and (44,31) respectively. They are *internal solitary* disturbances similar to those discussed by Keulegan (1953).

Two examples, corresponding to n=1 and n=2, are schown in Figures 36 and 37 below. Since $\beta$ is always positive, we see from (44,21) and (44,31) that $\alpha>0$. The wave in the lowest portion of the channel is one of elevation if n odd, and one of depression if n even.

We recognize that if closed streamlines appear in the flow, we have a local situation in which density increases with height. Notice, finally, too that in some of the cases considered above, in which there is no solitary wave, it may be possible, nevertheless, to find solutions for an infinite train of finite-amplitude, internal waves.

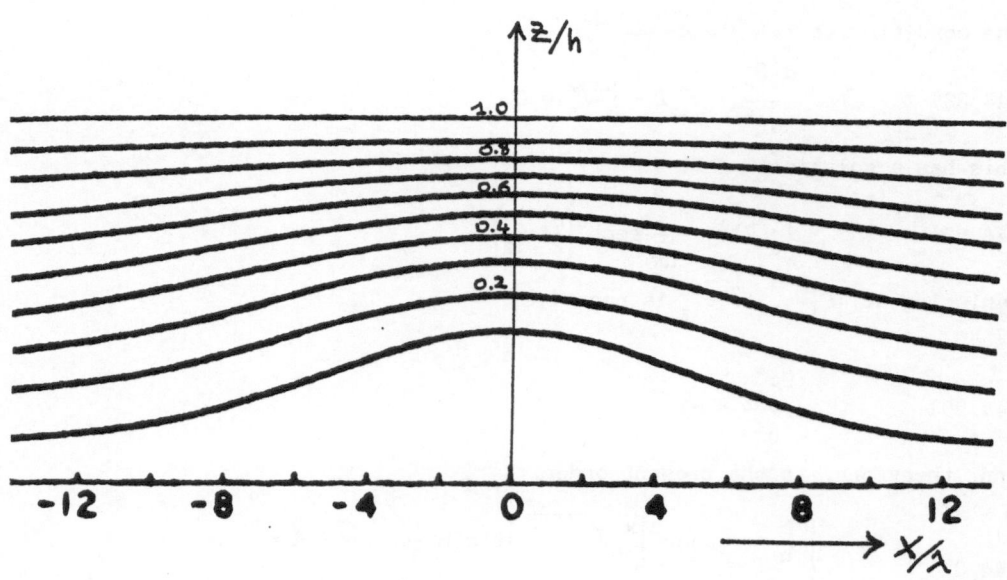

Fig. 36: Solitary wave in a isochoric flow
($\alpha=0,33$ and $\beta=0,10$).

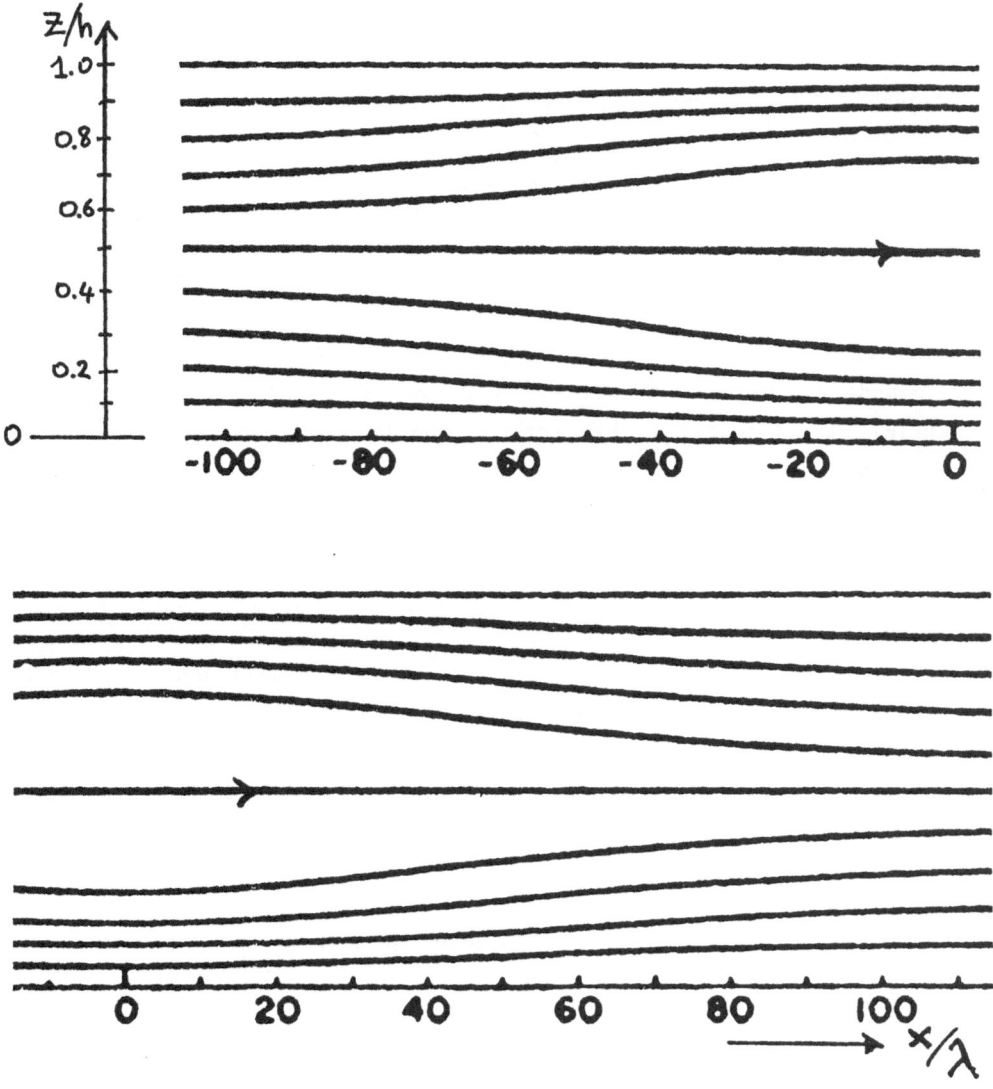

Fig. 37: Solitary wave in a isochoric flow
(α=0=17 and β=0, 10).

*FREE SURFACE SOLUTIONS*

If the upper boundary surface is *free*, then we have the following condition (instead of $\Delta=0$ on $\eta=1$)

$$(44,32) \qquad \sigma^2 \Delta = \beta \frac{\partial \Delta}{\partial \eta} - \frac{\alpha\beta}{2} \left[ \left( \frac{\partial \Delta}{\partial \eta} \right)^2 + \gamma \left( \frac{\partial \Delta}{\partial \xi} \right)^2 \right], \qquad \text{on } \eta=1+\alpha\Delta.$$

In the work of P.D. Weidman (1978) the mode shapes for internal solitary waves propagating in a *linearly* stratified fluid of finite depth are determined in the limit of "*weak*" stratification. The solutions obtained for both a fixed upper boundary and a free surface are compared with known results for exponential stratification. The profound changes which depending on whether the upper surface is fixed or free are derived from the strong influence of the free surface when the stratification is weak. Evidence is presented supporting the conjecture that the free surface can no longer effect qualitative differences in the modal structure when the fluid is heavily stratified.

Now we consider the case of $\gamma=\alpha\beta$. Since the Bernoulli equation must be evaluated at $\eta=1+\alpha\Delta$, the variable $\Delta$ and its derivatives in (44,32) are first expanded about the linearized position of the free surface, $\eta=1$. The ordered set of free surface conditions are then obtained by inserting the asymptotic developements (44,7) in the expanded form of (44,32). This gives:

$$(44,33a) \qquad \sigma_{00}^2 \Delta_{00} = 0 \quad \text{at } \eta=1;$$

$$(44,33b) \qquad \sigma_{00}^2 \left( \Delta_{10} + \Delta_{00} \frac{\partial \Delta_{00}}{\partial \eta} \right) + \sigma_{10}^2 \Delta_{00} = 0 \quad \text{at } \eta=1;$$

$$(44,33c) \qquad \sigma_{01}^2 \Delta_{00} + \sigma_{00}^2 \Delta_{01} = \frac{\partial^2 \Delta_{00}}{\partial \eta^2} \quad \text{at } \eta=1;$$

$$(44,33d) \qquad \sigma_{11}^2 \Delta_{00} + \sigma_{01}^2 \Delta_{10} + \sigma_{10}^2 \Delta_{01} + \left[ \Delta_{11} + \Delta_{00} \frac{\partial \Delta_{01}}{\partial \eta} + \Delta_{01} \frac{\partial \Delta_{00}}{\partial \eta} \right] = \frac{\partial \Delta_{10}}{\partial \eta}$$

$$+ \Delta_{00} \frac{\partial \Delta_{00}}{\partial \eta} (1 - \sigma_{01}^2) - \frac{1}{2} \left( \frac{\partial \Delta_{00}}{\partial \eta} \right)^2 \quad \text{at } \eta=1,$$

for the respective differential eqs. for $\Delta_{00}$, $\Delta_{10}$, $\Delta_{01}$ and $\Delta_{11}$. If one has solved the fixed boundary problem, it is a simple matter to make the necessary adjustements for the free surface.

As an example, we take Long's results (for exponential stratification) and solve the free surface problem. The application of (44,33) gives the free surface solutions:

$$\Delta_{00} = \mathcal{D}_0(\xi)\sin(n\pi\eta), \qquad \sigma_{00}^2 = n^2\pi^2;$$

$$\Delta_{10} = \mathcal{G}_1(\xi)\sin(n\pi\eta), \qquad \sigma_{10}^2 = 0 ;$$

$$\Delta_{01} = \mathcal{D}_1(\xi)\sin(n\pi\eta) + \frac{\mathcal{D}_{0}(\xi)}{2}\sin(n\pi\eta) + \frac{\mathcal{D}_0(\xi)}{n\pi}\eta\cos(n\pi\eta), \quad \sigma_{01}^2 = 2;$$

and the solution for $\Delta_{11}$ satisfying the lower boundary condition is

(44,34)
$$\Delta_{11} = \mathcal{D}_2(\xi)\sin(n\pi\eta) + \frac{\mathcal{D}_0^2(\xi)}{6}\cos(n\pi\eta) + \eta\frac{\cos(n\pi\eta)}{2n\pi}\left[ \frac{d^2\mathcal{D}_0}{d\xi^2} + \sigma_{11}^2\mathcal{D}_0 \right]$$

$$+ \mathcal{G}_1(\xi)\eta\frac{\sin(n\pi\eta)}{2} - \frac{1}{4}\mathcal{D}_0^2(\xi) + \frac{\mathcal{D}_0^2(\xi)}{12}\cos(2n\pi\eta)$$

$$+ \mathcal{G}_1(\xi)\frac{\eta}{n\pi}\cos(n\pi\eta).$$

The free surface condition (44,33d) is satisfied if:

(44,35)
$$\frac{d^2\mathcal{D}_0}{d\xi^2} + \sigma_{11}^2\mathcal{D}_0 - \frac{7n\pi}{3}\mathcal{D}_0^2 = 0 \qquad (n \; odd);$$

(44,36)
$$\frac{d^2\mathcal{D}_0}{d\xi^2} + \sigma_{11}^2\mathcal{D}_0 + 3n\pi\mathcal{D}_0^2 = 0 \qquad (n \; even),$$

showing that the even as well as the odd modes are determined when $\gamma=\alpha\beta$. Solutions of (44,35) and (44,36) vanishing at infinity agree with the results obtained by Benjamin (1966).

We now compare the results for a linearly stratified fluid with those obtained by previous Long's paper for an exponentialy stratified fluid.

For $\beta \ll 1$ the non-dimensional streamline deviations for both cases can be written

(44,37)
$$\Delta = sgn(\alpha)|\alpha|\sin(n\pi\eta)sech^2(\omega \xi)$$

and the associated non-linear phase speeds are then[†]

$$c^2 = c_n^2(1+4\omega^2), \quad c_n = \text{linear wave speed}.$$

We first observe that with a rigid upper boundary, both density profiles produce qualitatively similar wave shapes for both even and odd modes. Also,

---

[†] For example, if we consider the case of linear stratification with the free surface, then:

$$\omega^2 = \frac{n\pi}{18}\alpha\beta \text{ for the } odd \text{ modes and } \omega^2 = \frac{n\pi}{2}\alpha\beta \text{ for the } even \text{ modes.}$$

with the exception of the odd modes for linear stratification, all modes shapes for a free surface are essentially inverted from what they were with a rigid top plate. It is remarkable that such profound changes can be brought about by simply removing the rigid boundary. Although there is a distinct character change for the odd mode internal waves with a free surface in going from an exponential to a linear density profile, this is apparently not a manifestation of a basic difference between constant an variable Brunt frequency fluid. Rather, all the above sensitive changes in mode shape seem to be due to the strong influence of the free surface when the fluid is weakly stratified.

## 45 . THE DEEP CONVECTION
EQUATIONS

The atmospheric flows are *low* Mach number flows

$$Mo^2 \equiv \frac{U_0^2}{\gamma RT_\infty(0)} \ll 1$$

and if we wish to avoid the constraint (8,9) imposed in the Boussinesq approximation (see, the section 8) it becomes necessary to analyze flows with very low Froude number since, according to (3,7),

(45,1)
$$Bo = \varepsilon_0 \frac{\gamma Mo^2}{Fr^2} \implies Fr \approx Mo \ll 1,$$

if it is assumed tha $Bo = gH_0/RT_\infty(0)$, $\gamma = c_p/c_v$ and $\varepsilon_0 = H_0/L_0$ are of the order unity.

However, the limiting process $Fr \to 0$, i.e., in fact $Mo \to 0$ in the atmospheric equations (for instance the equations (8,4) where the expression $BoFr^2$ must be substituted for $\gamma Mo^2$, when $\varepsilon_0 \equiv 1$ and $Fr^2 \equiv U_0^2/gH_0$) leads to a very strong degeneracy of these equations to order zero[†].

IT soon become clear that it should be assumed, according to Zeytounian (1974), that when $Fr \to 0$, the term (with the dimensionless quantities)[††]

$$\frac{Bo}{T_\infty(z_\infty)}\left\{\frac{\gamma-1}{\gamma} + \frac{dT_\infty}{dz_\infty}\right\} \equiv \alpha_\infty^0 N_\infty^2(z_\infty) \to 0,$$

---

[†] This degeneracy is related to the so-called *quasi-solenoidal* model which is obtained in the section 46.

[††] See the formula (3,11) or the system of equations (10,2).

which means that we must consider the following double limiting process:

(45,2) $\qquad$ $Fr \rightarrow 0$ , $\alpha_\infty^0 \rightarrow 0$ , with $\dfrac{\alpha_\infty^0}{Fr^2} = O(1)$.

In order to confirm (45,2) it is sufficient to understand that when $Fr \rightarrow 0$, the first equation of (8,4) implies that $\theta \sim Fr^2$. Because of this the third equation of (8,4), in fact, implies the constraint written in (45,2). If the latter is not satisfied, then (when $\alpha_\infty^0 = O(1)$) $\vec{u}.\vec{K} \equiv w \rightarrow 0$ with $Fr \rightarrow 0$. This is precisely what bring on a very strong degeneracy of the basic exact equations (8,4) at order zero.

Hence it is necessary that the characteristic value of the Väisälä internal frequency (with dimensions and for this we use a superscript asterisk) satisfy the relation:

(45,3) $\qquad$ $N_\infty^*(0) \cong \dfrac{U_0}{H_0} \approx 10^{-3} 1/s$

since $Bo \cong 1$ implies that $H_0 \approx RT_\infty(0)/g \approx 10^4 m$.

The constraint (45,3) is, in fact, the one imposed by Ogura and Phillips (1962) in order to obtain the so called "anelastic" equations via the system of equations (which follows of Euler's $f_0$-plane equations):

(45,4) $\qquad \begin{cases} (\gamma-1)M_\infty^2 \dfrac{D\vec{u}}{Dt} + \theta\vec{\nabla}\sqcap + \dfrac{\gamma-1}{\gamma}Bo\vec{K} = 0; \\[2ex] \dfrac{D\theta}{Dt} = 0; \\[2ex] (\gamma-1)\vec{\nabla}.\vec{u} + \dfrac{DLog\sqcap}{Dt} = 0, \end{cases}$

where

(45,5) $\qquad$ $\sqcap = p^{-(\gamma-1)/\gamma}$ , $\theta = T\sqcap$ , $\dfrac{D}{Dt} \equiv \dfrac{\partial}{\partial t} + \vec{u}.\vec{\nabla}$,

with $S \equiv 1$, $\varepsilon_0 \equiv 1$ and $Ro \equiv \infty$.

*THE "ANELASTIC" EQUATIONS*
*OF OGURA and PHILLIPS*

Let us return to (45,4) and replace $M_\infty^2$ by $BoFr^2/\gamma$. When $Bo/\gamma = O(1)$ and $Fr \rightarrow 0$, the limiting form of these equations (45,4) can be sought by postulating the following asymptotic representation:

$$(45,6) \quad \begin{cases} \vec{u} = \vec{u}_0 + \ldots ; \\ \Pi = \Pi_0 + Fr^2 \Pi_2 + \ldots ; \\ \theta = \theta_0 + Fr^2 \theta_2 + \ldots . \end{cases}$$

To order zero, we have the following from the first equation of (45,4)

$$(45,7) \quad \theta_0 \vec{\nabla} \Pi_0 + \frac{\gamma-1}{\gamma} Bo \, \vec{k} = 0 \implies \Pi_0 = \Pi_0(t,z), \quad \theta_0 = \theta_0(t,z).$$

But, the second equation in this system (45,4) shows that

$$\frac{\partial \theta_0}{\partial t} + \vec{u}_0 \cdot \vec{k} \frac{\partial \theta_0}{\partial z} = 0,$$

and if at the initial instant $\theta_0 \equiv \theta_{00}(z)$, then[†]:

$$(45,8) \quad \theta_0 \equiv \theta_{00}(z) \text{ and } \Pi_0 \equiv \Pi_{00}(z).$$

Again from the first equation in (45,4), to order $Fr^2$, we obtain:

$$(45,9) \quad \frac{\gamma-1}{\gamma} Bo \frac{D\vec{u}_0}{Dt} + \theta_{00}(z) \vec{\nabla} \Pi_2 = \frac{\gamma-1}{\gamma} Bo \frac{\theta_2}{\theta_{00}(z)} \vec{k},$$

given the fact that

$$\theta_2 \frac{d\Pi_{00}}{dz} \equiv -\frac{\gamma-1}{\gamma} Bo \frac{\theta_2}{\theta_{00}}.$$

To order zero, the last equation of systeme (45,4) yields:

$$\vec{\nabla} \cdot \vec{u}_0 = (\vec{u}_0 \cdot \vec{k}) \frac{d}{dz} \left[ Log \frac{\theta_{00}}{\Pi_{00}^{1/\gamma-1}} \right] = -w_0 \frac{1}{\rho_{00}(z)} \frac{d\rho_{00}}{dz},$$

since $\rho = \Pi^{1/\gamma-1}/\theta$. Therefore, to order zero, the following continuity equation results:

$$(45,10) \quad \vec{\nabla} \cdot (\rho_{00}(z) \vec{u}_0) = 0,$$

where

$$\rho_{00} = \Pi_{00}^{1/\gamma-1}/\theta_{00}.$$

Let us consider the second equation of system (45,4).

---

[†] It might easily be thought that, in fact, $\theta_{00}(z) \equiv 1$ since $\dfrac{d\theta_{00}}{dz} = 0$, if $w_0 \equiv \vec{u}_0 \cdot \vec{k} \neq 0$. It will be seen further on, however, that because of the similarity relation (45,11), we have
$$\theta_{00}(z) = 1 + Fr^2 \theta_{02}(z).$$

It gives:

$$w_0 \frac{d\theta_{00}}{dz} + Fr^2 \frac{D\theta_2}{Dt} + \ldots = 0,$$

and it is observed that the following similarity relation must be imposed

(45,11)
$$\frac{d\theta_{00}}{dz} = \lambda_0 \Gamma_{02}(z) Fr^2,$$

where $\lambda_0$=constant and $\Gamma_{02}(z)$ is function of order unity which takes into account the influence of the reference stratification. As a matter of fact (45,11) necessarily implies that $\partial\theta_{00}/\partial t=0$. On the hypothesis (45,11), for $\theta_2$ it follows that:

(45,12)
$$\frac{D\theta_2}{Dt} + \lambda_0 \Gamma_{02}(z) w_0 = 0.$$

Let us now return to the zeroth order relation (see (45,7))

$$\theta_{00} \frac{d\Pi_{00}}{dz} + \frac{\gamma-1}{\gamma} Bo = 0$$

and derive it with respect to z. There results:

$$\theta_{00} \frac{d^2\Pi_{00}}{dz^2} + \frac{d\theta_{00}}{dz} \frac{d\Pi_{00}}{dz} \equiv \theta_{00} \frac{d^2\Pi_{00}}{dz^2} + \lambda_0 \Gamma_{02}(z) Fr^2 \frac{d\Pi_{00}}{dz} = 0,$$

i.e., to order zero, when Fr$\to$0, we have

$$\theta_{00} \frac{d^2\Pi_{00}}{dz^2} = 0 \implies \Pi_{00} = 1 + c_0 z$$

or even

$$c_0 \theta_{00} + \frac{\gamma-1}{\gamma} Bo = 0 \implies c_0 = -\frac{\gamma-1}{\gamma} Bo,$$

since $\underset{Fr\to0}{\text{Lim }} \theta_{00} = 0$.

Therefore

(45,13)
$$\begin{cases} \theta_{00}(z) = 1 + \underbrace{(\lambda_0 \int \Gamma_{02}(z)dz)\, Fr^2}_{\theta_{02}(z)} \\[2mm] \Pi_{00}(z) = 1 - \frac{\gamma-1}{\gamma} Bo\, z + P_{02}(z) Fr^2, \end{cases}$$

with

(45,14)
$$\frac{dP_{02}}{dz} = \frac{\gamma-1}{\gamma} Bo\lambda_0 \int \Gamma_{02}(z)dz.$$

Finally, we arrive at the conclusion that the following asymptotic representation must be postulated:

(45,15)
$$\begin{cases} \vec{u} = \vec{u}_0 + Fr^2\vec{u}_2 + \ldots ; \\ \Pi = 1 - \frac{\gamma-1}{\gamma}Boz + Fr^2\{P_{02}(z) + \Pi_2\} + \ldots ; \\ \theta = 1 + Fr^2\Xi_2 + \ldots , \end{cases}$$

to obtain the following Ogura and Phillips (1962) type, anelastic, equations for $\vec{u}_0$, $\Pi_2$ and $\Xi_2$:

(45,16)
$$\frac{\gamma-1}{\gamma} Bo \frac{D\vec{u}_0}{Dt} + \vec{\nabla}\Pi_2 = \frac{\gamma-1}{\gamma} Bo \Xi_2\vec{k};$$

(45,17)
$$\vec{\nabla}.(\rho_{00}(z)\vec{u}_0) = 0;$$

(45,18)
$$\frac{D}{Dt}(\Xi_2) = 0;$$

where

(45,19)
$$\begin{cases} \Xi_2 \equiv \theta_2 + \lambda_0 \int \Gamma_{02}(z)dz \\ \rho_{00}(z) = \left[1 - \frac{\gamma-1}{\gamma}Boz\right]^{1/\gamma-1} \end{cases}$$

and
$$\frac{D}{Dt} \equiv \frac{\partial}{\partial t} + \vec{u}_0.\vec{\nabla}.$$

*THE DEEP CONVECTION EQUATIONS*
*ACCORDING TO ZEYTOUNIAN*
*(Case of the adiabatic atmosphere)*

Here, our starting point is the system of Euler equations (8,4) for $\vec{u}$, $\pi$, $\omega$ and $\theta$, written without dimensions. We will consider the limiting process $M_0 \rightarrow 0$ with Bo fixed at the order unity. The variables t, x, y, z, as well as all the other parameters $(S, \gamma)$ remain fixed at the order unity when $M_o \to 0$. When $M_o \to 0$ we suppose satisfied the following asymptotic representation:

$$(45,20) \quad \begin{cases} \vec{u} = \vec{u}_a + \ldots \; ; \\[2mm] \omega = M_o^2 \omega_a + \ldots \; ; \\[2mm] \theta = M_o^2 \theta_a + \ldots \; ; \\[2mm] \pi = M_o^2 \pi_a + \ldots \; . \end{cases}$$

In this case, we have the following adiabatic *deep* convection equations for the functions $\vec{u}_a$, $\omega_a$, $\theta_a$ and $\pi_a$:

$$(45,21) \quad \begin{cases} S \dfrac{D\vec{u}_a}{Dt} + \dfrac{T_\infty^0(z_\infty)}{\gamma} \vec{\nabla}\pi_a = \dfrac{Bo}{\gamma} \theta_a \vec{k}; \\[4mm] \vec{\nabla}.\vec{u}_a = \dfrac{Bo}{\gamma T_\infty^0(z_\infty)} w_a, \quad w_a \equiv \vec{u}_a.\vec{k}; \\[4mm] S \dfrac{D\theta_a}{Dt} - \dfrac{\gamma-1}{\gamma} S \dfrac{D\pi_a}{Dt} + \dfrac{Bo}{T_\infty^0(z_\infty)} \chi_\infty(z_\infty)w_a = 0; \\[4mm] \pi_a = \omega_a + \theta_a . \end{cases}$$

where $S\dfrac{D}{Dt} \equiv S\dfrac{\partial}{\partial t} + \vec{u}_a.\vec{\nabla}$, and once the following hypothesis is made:

$$(45,22) \quad \dfrac{dT_\infty}{dz_\infty} = -\dfrac{\gamma-1}{\gamma} + \chi_\infty(z_\infty)M_o^2,$$

$\chi_\infty(z_\infty)$ being a function which takes into account a weak stratification with the altitude of the standard atmosphere and which is assumed of the order unity in absolute values. In the limiting system (45,21) we have:

$$(45,23) \quad T_\infty^0(z_\infty) \equiv 1 - \dfrac{\gamma-1}{\gamma} z_\infty; \quad z_\infty \equiv Bo z.$$

It is again pointed out that if (45,21) is to remain asymptotically valid, then according (45,22), the temperature gradient $-dT_\infty/dz_\infty$ must be very close to $(\gamma-1)/\gamma$.

*STEADY TWO-DIMENSIONAL CASE*

We consider now the following steady two-dimensional adiabatic deep convection equations, according to (45,21),

$$\begin{cases} u_a \dfrac{\partial u_a}{\partial x} + w_a \dfrac{\partial u_a}{\partial z} + \dfrac{1-\frac{\gamma-1}{\gamma}Boz}{\gamma}\dfrac{\partial \pi_a}{\partial x} = 0; \\[3mm] u_a \dfrac{\partial w_a}{\partial x} + w_a \dfrac{\partial w_a}{\partial z} + \dfrac{1-\frac{\gamma-1}{\gamma}Boz}{\gamma}\dfrac{\partial \pi_a}{\partial z} = \dfrac{Bo}{\gamma}\theta_a; \\[3mm] \dfrac{\partial u_a}{\partial x} + \dfrac{\partial w_a}{\partial z} = \dfrac{Bo}{\gamma}\dfrac{w_a}{1-\frac{\gamma-1}{\gamma}Boz}; \\[3mm] u_a \dfrac{\partial \theta_a}{\partial x} + w_a \dfrac{\partial \theta_a}{\partial z} - \dfrac{\gamma-1}{\gamma}\left[ u_a \dfrac{\partial \pi_a}{\partial x} + w_a \dfrac{\partial \pi_a}{\partial z}\right] + \dfrac{Bo}{1-\frac{\gamma-1}{\gamma}Boz}\chi_\infty(Boz)w_a = 0; \end{cases}$$

(45,24)

an we note that $\dfrac{dLog\rho_\infty^0(Boz)}{dz} \equiv -\dfrac{Bo}{\gamma}\dfrac{1}{1-\frac{\gamma-1}{\gamma}Boz}.$

The continuity equation (thied equation of system (45,24)) is integrated if we introduce the following generalized stream function $\psi_a(x,z)$:

(45,25)
$$\begin{cases} u_a = -\exp\left[\dfrac{Bo}{\gamma}\displaystyle\int\dfrac{dz}{1-\frac{\gamma-1}{\gamma}Boz}\right]\dfrac{d\psi_a}{dz}; \\[4mm] w_a = +\exp\left[\dfrac{Bo}{\gamma}\displaystyle\int\dfrac{dz}{1-\frac{\gamma-1}{\gamma}Boz}\right]\dfrac{d\psi_a}{dx}. \end{cases}$$

The fourth equation of (45,24) then leads to the following first integral:

(45,26)
$$\theta_a - \dfrac{\gamma-1}{\gamma}\pi_a + Bo\int\dfrac{\chi_\infty(Boz)}{1-\frac{\gamma-1}{\gamma}Boz}dz = \Xi(\psi_a).$$

Furthermore, the following vorticity equation is obtained from the first two equations of (45,24) when the third equation is also put to use:

(45,27)
$$u_a \dfrac{\partial \Omega_a}{\partial x} + w_a \dfrac{\partial \Omega_a}{\partial z} + \dfrac{1}{1-\frac{\gamma-1}{\gamma}Boz}w_a\Omega_a = \dfrac{Bo}{\gamma}\dfrac{\partial}{\partial x}\left[\theta_a - \dfrac{\gamma-1}{\gamma}\pi_a\right],$$

where $\Omega_a = \dfrac{\partial w_a}{\partial x} - \dfrac{\partial u_a}{\partial z}$. By taking use of (45,25) and (45,26), we obtain from

(45,27) , a second first integral:

(45,28)
$$\exp\left[\dfrac{Bo}{\gamma}\int\dfrac{dz}{1-\frac{\gamma-1}{\gamma}Boz}\right]\Omega_a - \dfrac{Bo}{\gamma}\dfrac{d\Xi}{d\psi_a}z = \omega(\psi_a),$$

since

$$\left[\frac{\partial \psi_a}{\partial z}\frac{\partial}{\partial x} - \frac{\partial \psi_a}{\partial x}\frac{\partial}{\partial z}\right]\frac{d\Xi}{d\psi_a} \equiv 0.$$

But

$$\exp\left[\frac{Bo}{\gamma}\int\frac{dz}{1-\frac{\gamma-1}{\gamma}Boz}\right] = \left[1 - \frac{\gamma-1}{\gamma}Boz\right]^{-\frac{1}{\gamma-1}},$$

and hence

$$\frac{\partial w_a}{\partial x} = \left[1 - \frac{\gamma-1}{\gamma}Boz\right]^{-\frac{1}{\gamma-1}}\frac{\partial^2\psi_a}{\partial x^2};$$

$$\frac{\partial u_a}{\partial z} = -\left[1 - \frac{\gamma-1}{\gamma}Boz\right]^{-\frac{1}{\gamma-1}}\frac{\partial^2\psi_a}{\partial z^2} - \frac{Bo}{\gamma}\left[1 - \frac{\gamma-1}{\gamma}Boz\right]^{-\frac{\gamma}{\gamma-1}}\frac{\partial\psi_a}{\partial z}.$$

Therefore, from (45,28), the following equation in $\psi_a$ is derived:

(45,29)
$$\frac{\partial^2\psi_a}{\partial x^2} + \frac{\partial^2\psi_a}{\partial z^2} + \frac{Bo}{\gamma}\left[1 - \frac{\gamma-1}{\gamma}Boz\right]^{-1}\frac{\partial\psi_a}{\partial z}$$

$$= \left[1 - \frac{\gamma-1}{\gamma}Boz\right]^{\frac{2}{\gamma-1}}\left\{\omega(\psi_a) + \frac{Bo}{\gamma}\frac{d\Xi}{d\psi_a}z\right\}.$$

The arbitrary functions of $\psi_a$, $\omega(\psi_a)$ and $\Xi(\psi_a)$, must be determined from the boundary conditions.

## 46 . MODEL EQUATIONS FOR LOW
## MACH NUMBER ATMOSPHERIC FLOWS

It soon become apparent that the quasi-geostrophic model equation considered in the section 9 was too approximate for many synoptic situations. It was thus necessary to devise a new approximation which would also lead to a simple limiting model, but a more efficient one.

It thus seemed reasonable to preserve the Mach number $M_0$ as small parameter (atmospheric motions always low Mach number flows) but, on the other hand, to drop the idea of a low Kibel number. The latter would be assumed of the order unity.

The starting point once again consists of the primitive equations which we write in the following form in accordance with the results of section 7 (see the equations (7,9)):

(46,1)
$$S \frac{\partial \vec{V}}{\partial t} + (\vec{V}.\vec{D})\vec{V} + \omega \frac{\partial \vec{V}}{\partial p} + \left[\frac{1}{Ro} + \beta y\right](\vec{k} \wedge \vec{V}) + \frac{Bo}{\gamma M_0^2} \vec{D}\mathcal{H} = 0;$$

(46,2)
$$\left[S \frac{\partial}{\partial t} + \vec{V}.\vec{D}\right]\frac{\partial \mathcal{H}}{\partial p} + \frac{\omega}{p}\left\{\frac{\partial}{\partial p}\left[p\frac{\partial \mathcal{H}}{\partial p}\right] - \frac{\gamma-1}{\gamma}\frac{\partial \mathcal{H}}{\partial p}\right\} = 0;$$

(46,3)
$$\vec{D}.\vec{V} + \frac{\partial \omega}{\partial p} = 0.$$

The system of equations (46,1)-(46,3) forms a closed system for: $\vec{V}=u\vec{i}+v\vec{j}$, $\omega$ and $\mathcal{H}$, which functions of t,x,y and p.

For system (46,1)-(46,3), the following slip condition is prescribed:

(46,4)
$$\left[S \frac{\partial}{\partial t} + \vec{V}.\vec{D} + \omega \frac{\partial}{\partial p}\right]\mathcal{H} = 0, \quad on \ \mathcal{H}=0.$$

We introduce in the equations (46,1)-(46,3) the horizontal divergence

(46,5)
$$\mathcal{D} \equiv \vec{D}.\vec{V} = \frac{\partial u}{\partial x} + \frac{\partial v}{\partial y}$$

and the vertical component of the eddy

(46,6)
$$\Omega \equiv \vec{k}.(\vec{D}\wedge\vec{V}) = \frac{\partial v}{\partial x} - \frac{\partial u}{\partial y}.$$

It now becomes possible, after a rather long but simple calculation to replace the vectorial equation (46,1) with two scalar equations for $\mathcal{D}$ and $\Omega$:

(46,7)
$$\left[S \frac{\partial}{\partial t} + \vec{V}.\vec{D} + \omega \frac{\partial}{\partial p}\right]\mathcal{D} + \mathcal{D}^2 + 2\left[\frac{\partial v}{\partial x}\frac{\partial u}{\partial y} - \frac{\partial v}{\partial y}\frac{\partial u}{\partial x}\right] + \frac{\partial u}{\partial p}\frac{\partial \omega}{\partial x} + \frac{\partial v}{\partial p}\frac{\partial \omega}{\partial y}$$
$$- \left[\frac{1}{Ro} + \beta y\right]\Omega + \beta u + \frac{Bo}{\gamma M_0^2} \vec{D}^2\mathcal{H} = 0;$$

(46,8)
$$\left[S \frac{\partial}{\partial t} + \vec{V}.\vec{D} + \omega \frac{\partial}{\partial p}\right]\Omega + \mathcal{D}\Omega + \frac{\partial v}{\partial p}\frac{\partial \omega}{\partial x} - \frac{\partial u}{\partial p}\frac{\partial \omega}{\partial y} + \left[\frac{1}{Ro} + \beta y\right]\mathcal{D} + \beta v = 0.$$

It will be remarked that the slip condition (46,4) can be written for

(46,9)
$$p = p_s(t,x,y) \text{ provided that } \mathcal{H}(t,x,y,p_s)=0.$$

According to the definition of $\omega$ (see (7,2)), it is seen that $p_s(t,x,y)$ must satisfy the equation:

$$(46,10) \qquad S \frac{\partial p_s}{\partial t} + \vec{V}.\vec{D}p_s = \omega(t,x,y,p_s).$$

## THE SO-CALLED "CLASSICAL" QUASI-SOLENOIDAL (QUASI-NONDIVERGENT) MODELS EQUATIONS

Here we consider the classiacal Monin-Charney case which is based on the limiting process:

$$(46,11) \qquad M_0 \longrightarrow 0, \quad \text{with } t,x,y \text{ and } p \text{ fixed}.$$

In conjunction with the limiting process (46,11) let us consider the following *main* asymptotic expansion, for the system of equations (46,1)-(46,3)

$$(46,12) \qquad \begin{pmatrix} \vec{V} \\ \omega \\ \mathcal{H} \\ T \\ p_s \end{pmatrix} = \begin{pmatrix} \vec{V}_0 \\ \omega_0 \\ \mathcal{H}_0 \\ T_0 \\ p_{s0} \end{pmatrix} + M_0^2 \begin{pmatrix} \vec{V}_2 \\ \omega_2 \\ \mathcal{H}_2 \\ T_2 \\ p_{s2} \end{pmatrix} + \ldots .$$

To order zero (46,1) leads to

$$\frac{\partial \mathcal{H}_0}{\partial x} = \frac{\partial \mathcal{H}_0}{\partial y} = 0$$

and from (46,2) it is noticed that

$$(46,13) \qquad S \frac{\partial^2 \mathcal{H}_0}{\partial t \partial p} + \omega_0 \left[ \frac{\partial^2 \mathcal{H}_0}{\partial p^2} + \frac{1}{\gamma p} \frac{\partial \mathcal{H}_0}{\partial p} \right] = 0,$$

which means it can be assumed that $\mathcal{H}_0 \equiv \mathcal{H}_0(t,p)$. In this case, we also have $\omega_0 = \omega_0(t,p)$ and from (46,3) we find that $\mathcal{D}_0 \equiv \vec{D}.\vec{V}_0 = \mathcal{D}_0(t,p)$. Finally, according to (46,10), $p_{s0} \equiv p_{s0}(t)$ satisfies the following relation:

$$(46,14) \qquad S \frac{\partial p_{s0}}{\partial t} = \omega_0(t,p_{s0}).$$

It is obvious that the order is consistent only with a flat ground. In addition, the components $u_0$ and $v_0$ of $\vec{V}_0$ can be represented in the following form:

$$(46,15) \quad \begin{cases} u_0 = \frac{1}{2}\mathcal{D}_0 x + \dfrac{\partial \varphi_0^h}{\partial x} - \dfrac{\partial \psi_0}{\partial y}; \\[3mm] v_0 = \frac{1}{2}\mathcal{D}_0 y + \dfrac{\partial \varphi_0^h}{\partial y} + \dfrac{\partial \psi_0}{\partial x}, \end{cases}$$

with

$$(46,16) \quad \psi_0(t,x,y,p) = \frac{1}{2\pi} \iint\limits_{-\infty}^{+\infty} \Omega_0(t,x,y,p) \mathrm{Log}\left\{(x-x')^2 + (y-y')^2\right\}^{1/2} dx'dy',$$

where $\Omega_0 \equiv \dfrac{\partial v_0}{\partial x} - \dfrac{\partial u_0}{\partial y}$ and $\varphi_0^h(x,y)$ is an arbitrary function which is harmonic with respect to x and y ($\partial^2\varphi_0^h/\partial x^2 + \partial^2\varphi_0^h/\partial y^2 = 0$).

The problem which now arises is to make sure that to order zero, our functions are independent of the time t, i.e., that they characterize the standard atmosphere as a function only of the altitude p. To this end the full non adiabatic equation for the temperature T must be used, instead of (46,2), and the expansion (46,12) must be extended to the term proportional to $M_0^\alpha$, if we take into account the following similarity relation

$$(46,17) \quad \frac{1/\sqrt{\mathrm{Re}_\perp}}{M_0^\alpha} = \hat{\mathscr{S}}_0 = O(1), \quad \alpha > 0$$

between $\mathrm{Re}_\perp$ and $M_0$ (for the definition of $\mathrm{Re}_\perp$ see (6,1)). In this case, so that secularities do not appear during a sufficiently long period of time (since the fluctuation of temperature $T_\alpha$ must remain bounded), it is mandatory that the right hand member, of the approximate non adiabatic equation for $T_\alpha$, be zero[†]:

$$(46,18) \quad \frac{d}{dp}\left[\rho_0(p)\frac{dT_0}{dp}\right] = \sigma_0\frac{d\hat{R}_\infty}{dp}.$$

We then recover the standard atmosphere

$$(46,19) \quad \mathcal{H}_0 \equiv \mathcal{H}_0(p), \quad T_0 \equiv T_0(p), \quad \omega_0 = \frac{dp_{s0}}{dt} \equiv 0,$$

associated with the thermal balance equation (46,18).

---

[†] See, for the definition of $\sigma_0$ and $\hat{R}_\infty$ the section 3 and the equation (4,33).

For this case, the following system is derived from the system of equations (46,1)-(46,3) for $\vec{V}_0$ and $\pi_2 \equiv \dfrac{Bo}{\gamma} \mathcal{H}_2$:

(46,20)

$$\begin{cases} S\, \dfrac{\partial \vec{V}_0}{\partial t} + (\vec{V}_0 \cdot \vec{D})\vec{V}_0 + \left[\dfrac{1}{Ro} + \beta y\right](\hat{\vec{k}} \wedge \vec{V}_0) + \vec{D}\pi_2 = 0, \\[3mm] \vec{D} \cdot \vec{V}_0 = 0. \end{cases}$$

System (46,20) describes the flow of an incompressible atmosphere along isobaric surfaces p=constant;this plane flow was projected onto the β-plane. This flow is strongly uncoupled with respect to the altitude. The only means of obtaining a coupling with respect to p is to impose on this flow initial and lateral conditions (in x and y).

From the equations (46,7) and (46,8) we find to order zero the following system for $\Omega_0$ and $\mathcal{H}_2$, which is equivalent to (46,20):

(46,21)

$$\begin{cases} \vec{D}^2 \mathcal{H}_2 = \dfrac{\gamma}{Bo}\left\{\left[\dfrac{1}{Ro} + \beta y\right]\vec{D}^2\psi_0 + \beta\, \dfrac{\partial \psi_0}{\partial y} + 2\hat{J}\left(\dfrac{\partial \psi_0}{\partial x}, \dfrac{\partial \psi_0}{\partial y}\right)\right\}; \\[4mm] \left[S\, \dfrac{\partial}{\partial t} + \hat{J}(\psi_0, \cdot)\right]\vec{D}^2\psi_0 + \beta\, \dfrac{\partial \psi_0}{\partial x} = 0, \end{cases}$$

if we take into account that $\mathcal{D}_0 \equiv 0$ and $\omega_0 \equiv 0$.
In (46,21) we have

$$\hat{J}(a,b) = \dfrac{\partial a}{\partial x}\dfrac{\partial b}{\partial y} - \dfrac{\partial a}{\partial y}\dfrac{\partial b}{\partial x}.$$

System (46,21) forms the so-called classical quasi-solenoidal (or quasi-nondivergent) model. The first equation in (46,21) is the so-called "*balance*" equation whereas the second one of (46,21) is an *evolution* equation for $\psi_0$.

Certain remarks can now be made.

Firstly, system (46,21) is of *first* order in t and necessitates an initial condition : $\psi_0|_{t=0} = \psi_0^0$. How can $\psi_0^0$ be found?

Solution resides in posing the problem of the unsteady adjustment by introducing a short time $\tilde{t}=t/M_0$ in order to take into account the internal gravity waves (which exist at the level of the primitive equations (46,1)-(46,3)) which were filtered out during the limiting process (46,11).

Secondly in the quasi-solenoidal model (46,21), why is there no derivation with respect to the altitude p. As has already been pointed out, this quasi-solenoidal model discribes the motion of an incompressible (barotropic) atmosphere stratified in horizontal layers in the planes p=constant and the latter being totally independent of each other. Hence, to order zero, the cancelling out of $\omega_0$ leads to the undesirable consequence of forcing the motion into the horizontal planes p=constant. The problem might be expected to be remedied if the expansion (46,12) is carried out to the next following order. If $\alpha > 1$ in (46,17) then from the (46,2) it follows that:

$$(46,22) \qquad \omega_2 = \frac{1}{Bo} \frac{p}{K_0(p)} \left[ S \frac{\partial T_2}{\partial t} + \vec{V}_0 \cdot \vec{D} T_2 \right],$$

where

$$K_0(p) \equiv - \frac{p}{Bo} \left( \frac{dT_0}{dp} - \frac{\gamma-1}{\gamma} \frac{T_0}{p} \right),$$

and $T_2$ is directly related to $\pi_2$, satisfying with $\vec{V}_0$ to (46,20), by the relation:

$$(46,23) \qquad T_2 = -\gamma p \frac{\partial \pi_2}{\partial p}.$$

Next, we have

$$(46,24) \qquad \mathcal{D}_2 \equiv \vec{D} \cdot \vec{V}_2 = - \frac{\partial \omega_2}{\partial p},$$

and then

$$(46,25) \qquad S \frac{\partial \vec{V}_2}{\partial t} + (\vec{V}_0 \cdot \vec{D}) \vec{V}_2 + (\vec{V}_2 \cdot \vec{D}) \vec{V}_0 + \omega_2 \frac{\partial \vec{V}_0}{\partial p} + \vec{D}\pi_4 = 0,$$

where $\pi_4 \equiv \frac{Bo}{\gamma} \mathcal{H}_4$ $(\mathcal{H} = \mathcal{H}_0(p) + M_0^2 \mathcal{H}_2 + M_0^4 \mathcal{H}_4 + \ldots)$.

It will be remarked that a coupling with the altitude, p, exists thanks to the terms $\partial \omega_2 / \partial p$ and $\omega_2 \partial \vec{V}_0 / \partial p$. However this is not enough and this main description remains highly degenerated.

Let us now take a look at what happens to the slip condition on the flat ground. First of all, we have the following

$$(46,26) \qquad \mathcal{H}_0(p_{s0}) = 0 \implies p_{s0} = 1.$$

We denote by $f_s \equiv f(t, x, y, p_{s0})$ and we then have:

$$(46,27) \qquad \frac{d\mathcal{H}_0}{dp}\bigg|_s p_{s2} + \mathcal{H}_{2s} = 0 \implies p_{s2} = -\frac{\mathcal{H}_2}{\dfrac{d\mathcal{H}_0}{dp}\bigg|_s}.$$

The above relation (46,27) is indeed compatible with the one which results from (46,10):

$$(46,28) \qquad S\,\frac{\partial p_{s2}}{\partial t} + \vec{V}_0 . \vec{D}\, p_{s2} = \omega_{2s},$$

given the fact that

$$\left[ S\,\frac{\partial \mathcal{H}}{\partial t} + \vec{V}.\vec{D}\mathcal{H} + \omega\,\frac{\partial \mathcal{H}}{\partial p} \right]_{p=p_s} = 0.$$

## THE GUIRAUD and ZEYTOUNIAN'S[†]
## RECENT RESULTS

The equations (46,20) corresponds to a kind of Froude blocking within isobaric surfaces as discovered by Drazin (1961). Such a blocked flow is unable to ride over any relief and must turn over it. This a serious drawback of such a kind of approximation as is the fact that flows within two isobaric surfaces, close from each other, are apparently disconnected. Some connection is uncovered at higher approximations but only in a parametric way (see, for instance (46,25)). A much stronger correction is got in the far field. We put:

$$(46,29) \qquad x = \frac{1}{M_0}\,\hat{x}, \quad y = \frac{1}{M_0}\,\hat{y}$$

and we let: $M_0 \to 0$ while $\hat{x}$ and $\hat{y}$ remains $O(1)$. We find that:

$$(46,30) \qquad \begin{cases} \vec{V} = \vec{V}_\infty(p) + M_0\,\hat{\vec{V}}, \quad \omega = M_0^2\,\hat{\omega}, \\[2mm] \mathcal{H} = \mathcal{H}_0(p) + M_0^2\,\hat{\mathcal{H}}. \end{cases}$$

Again $\mathcal{H}_0(p)$ is the standard distribution of altitude, as a function of pressure, but $\vec{V}_\infty(p)$ is an horizontal wind with an arbitrary dependence on pressure or, which is the same through $\mathcal{H}_0(p)$, on altitude. Two situations must be considered separately. The first one deals with the case when there is no relief at all, even at finite horizontal distances. We set up an expansion:

---

† See, Zeytounian and Guiraud (1984).

$$\begin{cases} \hat{\vec{v}} = \hat{\vec{v}}_0 + M_0 \hat{\vec{v}}_1 + M_0^2\, \hat{\vec{v}}_2 + \ldots; \\ \hat{\omega} = \qquad\qquad M_0^2\, \hat{\omega}_2 + \ldots; \\ \hat{\mathcal{H}} = \qquad\qquad M_0^2\, \hat{\mathcal{H}}_2 + \ldots; \end{cases}$$

(46,31)

and the following results emerge. The horizontal velocity $\hat{\vec{v}}_0$ is a potential vortex which depends on the distribution of vertical vorticity associated with the field $\vec{v}_0$. As a matter of fact, Drazin's model is unable to explain how vorticity is created but it may describe how such a vorticity evolves one created. If one assumes that this distribution of vorticity is localized it generates the potential vortex $\hat{\vec{v}}_0$ in the far field. One finds that $\hat{\vec{v}}_1$ is a doublet of potential vortices and both $\hat{\vec{v}}_0$ and $\hat{\vec{v}}_1$ do not afford any new information which is not contained in the so-called quasi-solenoidal field $\vec{v}_0$. At the level of $\hat{\vec{v}}_2$ and $\hat{\mathcal{H}}_2$ new features occur. One finds that $\hat{\mathcal{H}}_2$ has to be a solution of[t]:

(46,32)
$$\left\{ \frac{\partial^2}{\partial\hat{x}^2} + \frac{\partial^2}{\partial\hat{y}^2} + \gamma S^2 \frac{\partial^2}{\partial t^2}\left[\frac{\partial}{\partial p}\left(\frac{p^2}{K_0(p)}\frac{\partial}{\partial p}\right)\right] \right\} \hat{\mathcal{H}}_2 + \frac{2\gamma/\text{Bo}}{(\hat{x}^2+\hat{y}^2)^2}\left[\frac{\bar{\bar{\Omega}}_0(p)}{2\pi}\right]^2 = 0.$$

This equation plays with respect to the adiabatic primitive equations (46,1)-(46,3) a role analogous to the one played by the acoustic equations with respect to the Euler ones (see, for instance, Viviand (1970)). As far as aerodynamics is concerned, acoustics rules the phenomenon of adaptation to incompressible flow, here (46,32) rules the phenomenon of adaptation to the quasi-solenoidal approximation. We observe that $\bar{\bar{\Omega}}_0(p)$ is the total amount of vertical vorticity which is contained in the isobaric surface p=constant and which drives the potential vortex $\hat{\vec{v}}_0$.

We have been dealing with one of two situations, the second corresponds to the case when there is some relief at finite distance. Then one must put:

(46,33)
$$\hat{\vec{v}} = \hat{\vec{v}}_0 + M_0\hat{\vec{v}}_1 + \ldots, \quad \hat{\omega} = M_0\hat{\omega}_1 + \ldots, \quad \hat{\mathcal{H}} = M_0\hat{\mathcal{H}}_1 + \ldots$$

and one finds, again, that $\hat{\vec{v}}_0$ is a potential vortex but $\hat{\vec{v}}_1$ is no longer a doublet of potential vortices. Rather $\hat{\vec{v}}_1$ and $\hat{\mathcal{H}}_1$ play now the role that was played $\hat{\vec{v}}_2$ and $\hat{\mathcal{H}}_2$ when there was no relief.

---

[t] We note that: $\bar{\bar{\Omega}}_0(p) = \displaystyle\iint_{-\infty}^{+\infty} \Omega_0(t,x,y,p)\,dx\,dy.$

A number of problems arise from this low Mach number approximation. First blocking remains a mystery and we have only explained how blocking is released in the far field. Related to that one would like to understand how waves are generated in the lee of mountains in the context of low Mach number atmospheric flows. Another point concerns the three-dimensional nature of the low Mach number approximation near the top of a relief (see Hunt and Snyder (1980) for some considerations about that topic). Finally one would like to understand how the low Mach number approximation works at high altitude when p→0. Clearly further researchs are necessary.

## 47 . FRACTALS IN ATMOSPHERIC TURBULENCE

Scaling notions are associated with power law spectra, lack of characteristic scales over wide ranges, and the appearance of fractal dimensions and structures. More precisely, a system is said to be scaling (or scale invariant) over a range if the small - and large - scale structures are related by a scale-changing operation that involves only the scale ratio.

In Meteorological Fluid Dynamics, the existence of scaling regimes can often be argued directly from the dynamical equations themselves: The only scales associated with the Navier-Stokes atmospheric equations are a large scale defined by the largest scale of energy injection and a small viscous scale where the dissipation occurs.

In the atmosphere, these scales are roughly of the order of thousands of kilometers and several millimeters, respectively, allowing the possibility of a scaling regime spanning over 9 orders of magnitude in scale. Furthermore, the notion of scaling regimes in the atmosphere can be traced back to Richardson (1965), the father of numerical weather prediction, who in the 1920s also suggested a model of atmospheric dynamics involving a self-similar cascade of energy from large to small scales. Since then, scaling ideas have been central to studies of turbulence, a fact that is most notably expressed by ubiquity of the scaling $k^{-5/3}$ Kolmogorov energy spectrum of velocity fluctuations in atmospherical flows.

The turbulent velocity field $\vec{V}$ affords a prototypical example with which to develop the basic scaling ideas. The first scaling of interest might be called "simple scaling", since it occurs when only one parameter is sufficient to specify the scaling of all the statistical properties. Assuming statistical translation invariance and isotropy (including reflection symmetry), the fluctuations of the velocity depend only on the distance

$\ell=|\vec{\ell}|$ between the points $\vec{x}$ and $\vec{x}+\vec{\ell}$:

(47,1)         $\Delta V(\ell) = |\vec{V}(\vec{x}+\vec{\ell}) - \vec{V}(\vec{x})|.$

In this case, in dividing the scale $\ell$ by $\lambda$ we reduce fluctuation by the factor $\lambda^H$:

(47,2)         $\Delta V(\ell/\lambda) =^d \Delta V(\ell)/\lambda^H;$

H is the (single) scaling parameter. The equality "$=^d$" is understood in the sense of probability distributions; hence the scaling of the various high-order statistical moments follows:

(47,3)         $\langle\Delta V(\ell/\lambda)^h\rangle = \langle\Delta V(\ell)^h\rangle/\lambda^{\xi(h)}$

where $\xi(h)=hH$ and $\langle\rangle$ indicates ensemble average. Since the energy spectrum is the Fourier transform of the covariance, we have a spectrum

$k^{-\beta}$ with $\beta=2H+1.$

If one assumes that there exists a nonfluctuating density of energy flux to small scales ($\varepsilon$) that is scale-invariant (the nonlinear terms in the N-S Eqs. conserve this flux, while breaking up large eddies into smaller and smaller subeddies), then dimensional analysis gives

(47,4)         $\Delta V \propto \varepsilon^{1/3}\ell^{1/3}$, hence $H=\frac{1}{3}$, $\beta=\frac{5}{3}$.

Note that since $\frac{\partial V}{\partial x}\approx\ell^{-2/3}$ (which diverges as $\ell\rightarrow0$), such behavior is already associated with velocity fields with singular shears. The problem of such singular behavior was first discussed by Leray (1934).

In the 1960s, Kolmogorov (1962) and Obukhov (1962) (see Monin and Yaglom (1975) for summary and development) pointed out that scaling generally involves an infinite number of parameters: $\xi(h)$ is no longer linear in h. This is a richer behavior called "multiple scalling".The simplest way of expressing this is to now consider $\varepsilon$ as a fluctuating quantity. Its ensemble spatial average is independent of scale (as before) but is highly vriable in each realization of the cascade process. This variability (intermittency) is built up step by step; large eddies *multiplicatively* modulate the flux to smaller and smaller scales.

Multiplicative processes are associated with a number of interesting phenomena, including a hierarchy of singularities of different orders (not just one, as in the simple scaling described above), the divergence of high-order statistical moments, and the related phenomenon of statistical "outliers" in the data. These processes (which are easy to simulate numerically), create fields that are extremely intermittent. Because a whole hierarchy of dimensions is now involved (there are in fact two dual-dimension functions: one for the singularities, the other for their statistical moments), the statistical properties of such multiplicative processes depend not only on the scale but also on the dimension (for example, line, plane, or fractal set) over which they are averaged. We must therefore speak of "multifractal" rather than "fractal" properties.

The recent book by Schertzer and Lovejoy (1988) on NVAG1 gives ample development of these topics. The second NVAG workshop will be at the (former) Ecole Polytechnique, Paris, June 27 - July 1, 1988.

## DO DISSIPATIVE STRUCTURES IN FULLY DEVELOPED TURBULENCE FORM A FRACTAL SET?

What is the fractal dimension of this set? Answers to these questions are interesting because they bring the theory of fractals closer to application to turbulence and shed new hight on some classical problems in  turbulence -for example, the growth of material lines in a turbulent environment. The overwhelming conclusion of recent works[†] is that several aspects of turbulence can be described roughly by fractals and that their fractal dimensions can be measured. However, it is not clear how (or whether), given the dimension for several of its facets, on can solve (up to a useful accurary) the inverse problem of reconstructing the original set (that is, the turbulent flow it-self). Speculations abound that several *facets* of fully turbulent flows are *fractals* ! Although the earlier leading work of Mandelbrot (1974,1975) suggests that these speculations, initiated largely by himself, are plausible, no effort has yet been made to put them on firmer ground by ressorting to actual measurements in turbulent shear flows. The paper of Sreenivasan and Meneveau (1986) is a first attempt at filling this gap.

---

† See, for example, Sreenivasan and Meneveau (1986).

starting with Richardson (1965), it has been thought that fully developed turbulence consists of a hierarchy of eddies, or scales of various orders. The mechanism responsible of this situation is assumed to be that eddies of a given order (or size) arise as a result of the loss of stability of larger eddies of the preceding order; these in turn are assumed to lose their stability and generate eddies of a smaller order to which they transmit their energy. This reccuring scheme is expected to terminate at scales small enough to be stable - that is, scale whose characteristic Reynolds number is unity. It is well known that this lower bound on the scale size is of the order of the Kolmogorov scale. This theory of cascade, verbalized in a memorable rhyme by Richardson, in the 1920s, and cultivated by Kolmogorov (1941,1962), Obukhov (1941,1962), Onsager (1945) and Weizsäcker (1948), has made remarkable strides in advancing our undestanding of turbulent flows.

It is this description of turbulent flows -namely that they are "objects" consisting of a hierarchy of scales- that leads to the expectation that the theory of fractals[†] must be applicable to turbulence. In the most basic sense, *fractals are objects that display self-similarity over a wide range of scales.*

Mandelbrot, for example, has remarked that " turbulence involves many fractal facets" and claimed that a proper investigation of the geometric aspects of turbulence-which has been ignored all along in the vast literature on turbulence-must necessarily involve fractals; concepts from Euclidean geometry are totally inadequate. He has also led the way by his own investigations but, in his own words, "they involve suggestions with few hard results as yet".

Analogous to the Euclidean dimension of classical objects, each fractal object is associated with a characteristic dimension called the *fractral dimension* which form a basic measure of it fragmentation or *roughness*; it has the property that it is strictly greater than the object's topological dimension. It appears as a certain exponent $D=LogN/Log(1/\varepsilon)$, characteristic of a selfsimilar object which is made of N part, each of which is obtained from the whole by a reduction of ratio $\varepsilon$.

Of course, a complete description of fractal sets demands a specification of other quantities such as *lacunarity* -which, loosely speaking, is a measure of how far the fractral object is from being dust-like- or the entire spectrum of scaling functions only one of which is the fractal dimension.

---

[†] See, Mandelbrot (1982), to which reference must be made for an enjoyable and original account of fractals.

It is known (Batchelor and Townsend (1949)) that the small structure of turbulence is *intermittent*, and that scale-similarity arguments are very helpful in discribing it. The essence of scale-similarity arguments in this context is the following. Within a given field of (fully developed) turbulence, consider a cube with sides of length $L_0$, where $L_0$ is an integral scale of turbulence. If we divide this cube into a number (n>>1) of smaller cubes of $L_1 = L_0/n^{1/3}$, the density of dissipation rate in each of these smaller cubes is distributed according to a certain probability law. Further subdivision of these cubes into second-order cubes of length $L_2 = L_1/n^{1/3}$ leaves the probability distribution unaltered. This similarity extends to all scales of motion until ones reaches sizes directly affected by viscosity. The simplest distribution is the binary one according to which a given high-order box either contains dissipation or does not. It is this simple picture that we shall pursue.

## AN UPDATE OF MANDELBROT'S WORK

Let $\mathcal{D}$ be the fractral dimension of the dissipative field. When we have resolved the smallest scales $\eta$, and determined the number N of boxes of size $\eta$ required to cover the entire dissipation regions, $\mathcal{D}$ can be calculated according to its definition:

$$(47,5) \qquad \mathcal{D} = \frac{\text{LogN}}{\text{Log}(L_0/\eta)} \quad \text{or}, \quad N = \left[L_0/\eta\right]^{\mathcal{D}}.$$

Since each cube has a volume of the order $(L_0/\eta)^3$, the total volume occupied by the cubes of active dissipation is

$$(47,6) \qquad \wedge \!\! s = \left(\frac{L_0}{\eta}\right)^{\mathcal{D}-3}.$$

Since all dissipation is contained in these cubes, the level of dissipation in them is $(L_0/\eta)^{3-\mathcal{D}}$ times the global average value. Assuming local isotropy, this means that $(du/dx)^2$ in the dissipating cubes is $(L_0/\eta)^{3-\mathcal{D}}$ times the global mean. Consequently, the Kurtosis (or the flatness factor) of $du/dx$, defined as

$$(47,7) \qquad K = \overline{\left[\frac{\partial u}{\partial x}\right]^4} \Big/ \overline{\left[\frac{\partial u}{\partial x}\right]^2}^2$$

will be given by $(L_0/\eta)^{2(3-\mathcal{D})}$ times the volumes occupied by the dissipating cubes.

From[†] $\mathcal{D}=2+D_c$ we have:

(47,8)
$$K \propto (L_0/\eta)^{3-\mathcal{D}} \propto Re_\lambda^{(3/2)(3-\mathcal{D})},$$

where $Re_\lambda = \dfrac{u'\lambda}{\nu}$; $\lambda$ being the Taylor microscale and $u'$ the root-mean-square streamwise velocity. If we invoke Taylor's frozen-field hypothesis, the flatness factor of $(du/dx)$ is the same as that of $(du/dt)$. A plot of LogK, where now K is the Kurtosis of $(du/dt)$, vs. $LogRe_\lambda$ will yield the co-dimension $(3-\mathcal{D})$.

Mandelbrot used this argument and, from an examination of the Kurtosis data from Kuo and Corrsin (1971), estimated $\mathcal{D}$ to be 2,6. More data have become available since then, and are plotted in Figure 16 of the paper of Sreenivasan and Meneveau (1986; p.376). The data may be considered to collapse on a line with a slope of 0.4, yielding a $\mathcal{D}$ of 2,73, a version from Mandelbrot's earlier estimate. This means that the fractional volume $v=(L_0/\eta)^{3-\mathcal{D}}$ occupied by dissipation field is given by

$$\left(\frac{L_0}{\eta}\right)^{-0,27}.$$

For $Re_\lambda < 150$, the slope in this Figure 16 is decidely smaller ($\approx 0,15$), which yields a $\mathcal{D}$ of 2,9. This indicates either that the dissipation regions at low Reynolds numbers are less spotty or that local isotropy does not obtain. Both are likely.

As a final remark we note that it seems that turbulence genuinely loses its fractalike behaviour when viewed on very long times scales. Turbulence is perhaps a collection of a number of fractals each of which is slightly different. We think that this view can be reconciled roughly with the view of turbulence now in vogue as an ensemble of semi-organized motions.

---

[†] Let $\mathcal{F}$ be an object in three-dimensional space with a fractal interface of dimension $\mathcal{D}$. Let $\ell$ a line element, intersecting the object, gives a set of isolated point-akin to the Cantor discontinuum- whose dimension $D_c$ can be measured.

Fig. 38: A smoke photograph of a turbulent boundary layer developing on a flat plate. The thickness of the intersecting light sheet is of the order of the Kolmogorov thickness.

*REFERENCES TO WORKS CITED IN THE TEXT*

BATCHELOR, G.K. and TOWNSEND, A.A. (1949) _ Proc. Roy. Soc. London A199,239.

BENJAMIN, T.B. (1966) _ J. Fluid Mech. 25, 241-270.

DRAZIN, P.G. (1961) _ Tellus, 13, 239-251.

HUNT, J.C.R. and SNYDER, W.H. (1980) _ J. Fluid Mech. 96, 4, 671.

KEULEGAN, G.H. (1953) _ NBSJ. of Research 51, 133-140.

KOLMOGOROV, A.N. (1941) _ C.R. Acad. Sci. URSS, 30,299.

KOLMOGOROV, A.N. (1962) _ J. Fluid Mech. 13, 82.

KUO, A.-Y. and CORRSIN, S. (1971) _ J. Fluid Mech. 50, 285.

LERAY, J. (1934) _ Acta Math., 63, 193.

LONG, R.R. (1965) _ Tellus 17, 1, 46-52.

MANDELBROT, B.B. (1974) _ J. Fluid Mech. 62, 331.

MANDELBROT, B.B. (1975) _ J. Fluid Mech. 72, 401.

MANDELBROT, B.B. (1982) _ *The Fractal Geometry of Nature*. W.H.Freeman and Company, New-York.

MONIN, A.S. and YAGLOM, A.M. (1975) _ *Statistical Fluid Mechanics: Mechanics of Turbulence*, vol. 2. MIT. Press, Cambridge, Mass.

OBUKHOV, A.M. (1941) _ C. R. Acad. Sci. URSS ,32, 22.

OBUKHOV, A.M. (1962) _ J. Fluid Mech. 13,77.

OGURA, Y. and PHILLIPS, N.A. (1962) _ J. Atm. Sci. vol. 19, 173-179.

ONSAGER, L. (1945) _ Phys. Rev. 68, 286 (abstract only).

RICHARDSON, L.F. (1965) _ *Weather Prediction by Numerical Process*, Dover, New-York.

SCHERTZER, D. and LOVEJOY, S. (1988) _ *Scaling, Fractals, and Non-linear Variability in Geophysics*. 1. D. Reidel, Hingham, Mass.

SREENIVASAN, K.R. and MENEVEAU, C. (1986)_J. Fluid Mech.,vol.173, pp. 357-386.

VIVIAND, H. (1970) _ J. de Mécanique, 9, 573.

WEIDMAN ,P.D. (1978) _ Tellus, 30, 177-184.

WEIZSÄCKER, C.F.Von (1948) _ Z. Phys. 124, 614.

ZEYTOUNIAN, R.Kh. (1974) _ Archiwum Mechaniki Stosowanej, 26, 3, 499-509 (Warszawa).

ZEYTOUNIAN ,R.Kh. and GUIRAUD, J.P. (1984) _ in:"*Adv. in Computational methods for boundary and interior layers*". Boole Press, ed. J.J.H.Miller. Dublin.

# APPENDIX 1

# BOUNDARY LAYER TECHNIQUES
# FOR THE STUDY OF SINGULAR
# PERTURBATION PROBLEMS


Throughout the discussion of Meteorological Fluid Dynamics we have made use of boundary layer techniques to approximate the solution of atmospheric equations with small (or high) parameters.

Such methods are characteristic for the tools used to study certain singular perturbation problems in Theoretical Fluid Mechanics. For the reader who is not familar with these methods we will give a short introduction to the topic of boundary-layer theory. A much more extensive treatement can be found in the book by Van Dyke (1975).

The original idea, that a fluid of small viscosity could be approximated by an inviscid (perfect, non viscous) fluid in almost every spatial region except for narrow boundary layers, was first presented by Prandtl (see, Prandtl (1905)). This treatement has been give the name, "the method of inner and outer expansions", or in more recent literature "the method of matched asymptotic expansions[†]".

Altghough these methods have not yet been shown to be completely rigorous in all situations, boundary layer theory have proved to be extremely important in the analysis of the Navier-Stokes equations (see, for example, Zeytounian (1986 and 1987)).

In general, it may be stated, *boundary-layer theory is asymptotically correct.*

In this Appendix 1 we discuss perturbative method for solving a differential equation whose highest derivative is multiplied by the perturbing parameter $\varepsilon \ll 1$ this method (the most elementary) is called boundary-layer theory.

A boundary layer is a narrow region where the solution of a differential equation changes rapidly. By definition, the thickness of a boundary layer must approch 0 as $\varepsilon \to 0$. In this Appendix 1 we will concerned with differential equations whose solutions exhibit only isolated (well-separated) narrow regions of rapid variation. It is possible for a solution to a perturbation problem to undergo rapid variation over a *thick* region (one whose thickeness does not vanish with $\varepsilon \to 0$). However, such region *is not* a boundary layer. We will consider such problems in Appendix 2.

---

† MMAE.

There are two standard approximations that one makes in boundary-layer theory. In the *outer* region (away from a boundary layer) the solution y(x) is slowly varying, so it is valid to neglect any derivatives of y(x) which are multiplied by ε. Inside a boundary layer the derivatives of y(x) are large, but the boundary layer is so narrow that we may approximate the coefficient functions of the differential equation by constants.

Thus we can replace a single differential equation by a sequence of much simpler approximate equations in each of several inner and outer regions. In every region the solution of the approximate equation will contain are or more unknown constants of integration. These constants are then determined from the boundary or initial conditions using the technique of asymptotic *matching*.


*LARGE* O *and SMALL* o.
*ASYMPTOTIC SEQUENCES*
*AND EXPANSIONS*


Consider a family of boundary value problems $\mathcal{P}_\varepsilon$ depending on a small parameter ε (<<1). Under many conditions, a solution $y_\varepsilon(x)$ of $\mathcal{P}_\varepsilon$ can be constructed by the well-known "method of perturbation" i.e., as a power series in ε with first term $y_0$ being the solution of the problem $\mathcal{P}_0$.

When such an expansion converges as ε→0 *uniformly in* x, we have a regular perturbation problem. When $y_\varepsilon(x)$ *does not* have a uniform limit in x as ε→0, this regular perturbation method will fail and we have a *singular* perturbation problem.

Below, it will be convenient to use the Landau order symbols O and o which are defined as follows:

Given two functions f(ε) and g(ε), we write

(A1,1)
$$\begin{cases} f = O(g) & \text{as } \varepsilon \to 0 \\[2mm] \text{if } \left| \dfrac{f(\varepsilon)}{g(\varepsilon)} \right| \text{ is } \textit{bounded as } \varepsilon \to 0. \end{cases}$$

We write[†]

(A1,2)
$$\begin{cases} f = o(g) & \text{as } \varepsilon \to 0 \\[2mm] \text{if } \dfrac{f(\varepsilon)}{g(\varepsilon)} \to 0 & \text{as } \varepsilon \to 0. \end{cases}$$

---

[†] Often, f<<g is used as an equivalent notation.

Consider now a sequence $\{\delta_n(\varepsilon)\}$ n=1,2,... of functions of $\varepsilon$. Such a sequence is called an *asymptotic sequence* if

(A1,3) $\qquad\qquad \delta_{n+1}(\varepsilon) = o(\delta_n(\varepsilon))$ as $\varepsilon \to 0$

for each n=1,2,... .
A sum of terms of the form

$$\sum_{n=1}^{N} a_n(x)\delta_n(\varepsilon)$$

is called an *asymptotic expansion* of the function $f(x;\varepsilon)$ to N terms (N may be infinite) as $\varepsilon \to 0$ with respect to the sequence $\{\delta_n(\varepsilon)\}$ if

(A1,4) $\qquad\qquad f(x;\varepsilon) - \sum_{n=1}^{M} a_n(x)\delta_n(\varepsilon) = o(\delta_M)$ as $\varepsilon \to 0$

for each M=1,2,...,N.
if N=$\infty$, the following notation is generally used

(A1,5) $\qquad\qquad f(x;\varepsilon) \sim \sum_{n=1}^{\infty} a_n(x)\delta_n(\varepsilon)$ as $\varepsilon \to 0$.

Clearly, an equivalent definition for an asymptotic expansion is that

$$f(x;\varepsilon) - \sum_{n=1}^{M-1} a_n(x)\delta_n(\varepsilon) = O(\delta_M)$$ as $\varepsilon \to 0$

for each M=2,...,N.

*INTUITIVE APPROACH*
*OF THE MMAE*[†]

In any perturbation problem involving a small positive parameter $\varepsilon$ it is natural to seek a solution of the form

(A1,6) $\qquad\qquad y_\varepsilon^0(x) \sim \sum_{j=0}^{\infty} a_j(x)\delta_j(\varepsilon)$ as $\varepsilon \to 0$,

where x ranges over some (usually bounded) domain $\mathcal{D}$ and $\{\delta_j(\varepsilon)\}$ is an asymptotic sequence (often the power sequence $\{\varepsilon^j\}$) as $\varepsilon \to 0$. In a singular perturbation problem, such an expansion cannot be valid uniformly in x; for example, this solution may fail to satisfy all boundary conditions (moreover, in applications, physical considerations will often indicate which boundary conditions are so omitted).

[†] The following discussion is taken from the book of O'Malley (1974).

Instead, the expansion $y_\varepsilon^0$ will be generally satisfactory in the "outer region" away from (part of) the boundary of $\mathcal{D}$. It will be called an *outer* asymptotic expansion (or outer solution). In order to investigate regions of nonuniform convergence, one introduces one or more stretching transformations

$$(A1,7) \qquad \xi = \psi(x;\varepsilon)$$

which "blow up" a region of nonuniformity (near a part of the boundary with neglected boundary conditions, for example).
Thus

$$\xi = \frac{x}{\varepsilon^\alpha}, \quad \alpha > 0,$$

might be used for nonuniform convergence at x=0. Then if $\xi$ is fixed and $\varepsilon \rightarrow 0$, $x \rightarrow 0$, while if x>0 is fixed and $\varepsilon \rightarrow 0$, $\xi \rightarrow \infty$.
Selection of correct stretching transformations is an art sometimes motivated by physical considerations, or mathematically (as in Kaplun's concept of principal limits).
In terms of the stretched variable $\xi$, one might seek an asymptotic solution of the form

$$(A1,8) \qquad y_\varepsilon^1(\varepsilon) \sim \sum_{j=0}^{\infty} b_j(\varepsilon)\lambda_j(\varepsilon) \quad \text{as } \varepsilon \rightarrow 0$$

where the sequence $\{\lambda_j(\varepsilon)\}$ is asymptotic as $\varepsilon \rightarrow 0$, valid for values of $\xi$ in some "inner region". This will be called an inner asymtotic expansion and this inner expansion often accounts for boundary conditions neglected by the outer expansion. The inner region will generally shrink completely as $\varepsilon \rightarrow 0$ when expressed in terms of the outer variable x. Hence, the inner expansion is *local*.

In most problems, it is impossble to determine both the outer and the inner expansions $y_\varepsilon^0$ and $y_\varepsilon^1$ completely by straightforward expansion procedures. Since both expansions should represent the solution of the original problem asymptotically in different regions, one might attempt to *match* them, i.e., to formally relate the outer expansion $y_\varepsilon^0$ in the inner region: $(y_\varepsilon^0)^1$ and the inner expansion $y_\varepsilon^1$ in the outer region: $(y_\varepsilon^1)^0$ through use of the stretching $\xi=\psi(x;\varepsilon)$.
The rules for even formally accomplishing this, in all generality, can be very complicated (see, Fraenkel (1969)). Justification in particular examples through use of an *overlap* domain an *intermediate limits* is difficult (see, Lagerstrom and Casten (1972) and also the book of Kevorkian and Cole (1981))

and, in general, there is no apriori reason to believe that an overlap domain (where matching is possible) exists.

Once matching is accomplished, however, the asymptotic solution to well-posed problems becomes completely known in both the inner and outer regions. Frequently, it is convenient to obtain a *composite expansion* $y_\varepsilon^c$ *uniformly valid* in $\mathcal{D}$. One method of doing so is to let[†]

(A1,9)
$$y_\varepsilon^c = y_\varepsilon^0 + y_\varepsilon^1 - (y_\varepsilon^1)^0,$$

making the appropriate modification if several regions of nonuniform convergence (several inner regions) are necessary.

We refer the reader to the cited references for more details of this important technique and many examples of its application.

*ASYMPTOTIC MATCHING*
*PRINCIPLE*

One general asymptotic matching which involves the intermediate limits is that due to Kaplun (1967). Another widely used asymptotic matching principle is that due to Van Dyke (1964). The amount of work involved in using Van Dyke's matching principle is considerable. Fraenkel (1969), who probed into the distinction between the idea of overlapping and the asymptotic matching principle *cannot* derive justification for itself from Kaplun's principle of intermediate limits. Further, there is a possibility, as Fraenkel demonstrated, that Van Dyke's matching principle can fail when the asymptotic sequence contains logarithms.

In order to relate the outer solution (A1,6) and the inner solution (A1,8) to each other, one presupposes the existence of a region of overlapping, where the two expansions are valid. Kaplun and Lagerstrom (1957) admitted that there is no apriori reason for the regions of validity of the inner and outer expansions to overlap. Therefore, the results should be assumed to be apriori asymptotically correct.

First we note that the *simplest* asymptotic matching principle due to Prandtl (1928) states

(A1,10)
$$\lim_{x \to 0} y_\varepsilon^0 = \lim_{\xi \to \infty} y_\varepsilon^1, \quad \text{where } \xi = \frac{x}{\varepsilon^\alpha}.$$

[†] Note that $(y_\varepsilon^1)^1 \equiv y_\varepsilon^1$ so that $(y_\varepsilon^c)^1 = y_\varepsilon^1$. Likewise, in the outer region, $(y_\varepsilon^c)^0 = y_\varepsilon^0$.

Another widely used asymptotic matching principle is that due to Van Dyke (1964), which states:

⇒ [The m-terms inner expansion of (the n-terms outer expansion)] = [The n-terms outer expansion of (the m-terms inner expansion)],

where m and n are any two integers. Thus in order to find the left hand side, on writes n terms of the outer expansion in terms of the inner variable, expands it for small $\varepsilon$ keeping the inner variable $\xi$ fixed, and truncates the resulting expansion after m terms; and similarly for the right hand side.

We proceed to give here[†] a variant form of asymptotic matching principle which is more expeditious in this context. One still writes here (A1,10). But more precisely, the inner limit of the outer expansion, which is what the left hand side in (A1,10) essentially implies, is represented by a formal Laurent's series about the inner boundary. That is, in the neighborhood of x=0, one writes

$$a_j(x) = a_j(0) + x \left.\frac{da_j}{dx}\right|_{x=0} + O(x^2), \quad j=1,2,\ldots,$$

or

$$y_\varepsilon^0(x) \sim a_0(0) + \varepsilon\left[\frac{x}{\varepsilon} \left.\frac{da_0}{dx}\right|_{x=0} + a_1(0)\right] + O(\varepsilon^2),$$

with $\delta_j(\varepsilon) \equiv \varepsilon^j$, and written in terms of the inner variable $\xi = \frac{x}{\varepsilon}$ ($\alpha \equiv 1$),

(A1,11) $$y_\varepsilon^0(x) \sim a_0(0) + \varepsilon\left[\xi \left.\frac{da_0}{dx}\right|_{x=0} + a_1(0)\right] + O(\varepsilon^2).$$

It is clear that this principle is simply a rational generalisation of the basic principle due to Prandtl. Note that this principle puts a less stringent restriction on the domain of validity of the outer solution in that the latter is required to extend merely to the neighborhood of the inner boundary whereas the basic principle due to Prandtl requires the domain of validity of the outer solution to extend right up to the inner boundary

-a probable source of the difficulties the latter method develops at the higher order problems. This may also be the reason why the present principle succeeds where Prandtl's principle fails.

Formally one may enunciate the present principle as follows:

⇒ [The n-terms formal Laurent's series expansion of the outer expansion about the inner boundary written in terms of the inner variable] = [ The n-terms formal outer limit of the inner expansion].

---

† According to Shivamoggi (1978).

*A SIMPLE BOUNDARY-LAYER*

*PROBLEM*

$$(A1,12) \qquad \begin{cases} \varepsilon y'' + y' + y = 0, \quad 0 \le x \le 1, \ \varepsilon \ll 1, \\[2mm] y(0) = a, \quad y(1) = b. \end{cases}$$

The exact solution of the two-point problem (A1,12) is

$$(A1,13) \qquad y_\varepsilon = (a\,e^{s_2} - b)e^{s_1 x} + (b - a\,e^{s_1})e^{s_2 x},$$

where

$$s_{1,2} = \frac{-1 \pm (1 - 4\varepsilon)^{1/2}}{2\varepsilon}.$$

One approximates (A1,13) by

$$(A1,14) \qquad y_\varepsilon = b e^{1-x} + (a-be)\,e^x\,e^{-x/\varepsilon} + O(\varepsilon)$$

-a form which is uniformly valid throughout $0 \le x \le 1$ as $\varepsilon \to 0$. Note that this expansion cannot be obtained keeping either $x$ or $x/\varepsilon$ fixed. In the former case one obtains

$$y_\varepsilon^0 = b e^{1-x} + O(\varepsilon) \quad \text{for } x > 0$$

which is not valid in the boundary layer near $x=0$ since

$$y_\varepsilon^0(0) = be \ne a.$$

In the latter case one obtains

$$y_\varepsilon^1 = be + (a-b)e^{-x/\varepsilon} + O(\varepsilon),$$

which is not valid as $x \to 1$.

This suggests that we represent the solution by two different asymptotic expansion using the variables $x$ and $\frac{x}{\varepsilon} \equiv \xi$ -this is the MMAE. Since they are different asymptotic representations of the *same* function, they should be related to each other in a rational manner -this leads to the asymptotic matching principle (that besides makes the two different problems determinate).

Thus seek an outer expansion

$$(A1,15) \qquad y_\varepsilon^0(x;\varepsilon) \sim \sum_{n=0}^{N-1} a_n(x)\varepsilon^n + O(\varepsilon^N)$$

where in accordance with the outer limit process one has

(A1,16)
$$a_m(x) = \lim_{\substack{\varepsilon \to 0 \\ x \text{ fixed}}} \frac{y_\varepsilon^0 - \sum_{n=0}^{m-1} a_n(x)\varepsilon^n}{\varepsilon^m}$$

Substituting (A1,15) in equation (A1,12), and equating the coefficients of equal powers of $\varepsilon$, one obtains

(A1,17)
$$a_0' + a_0 = 0, \quad a_1' + a_1 = -a_0'', \quad \text{etc.}$$

The outer solution is valid everywhere except in the region $x = O(\varepsilon)$[†], so that one has:

(A1,18)
$$a_0(1) = b, \quad a_1(1) = 0, \quad \text{etc.}$$

Thus one obtains

(A1,19)
$$a_0(x) = be^{1-x}, \quad a_1(x) = b(1-x)e^{1-x}, \quad \text{etc.,}$$

so that

(A1,20)
$$y_\varepsilon^0 = b[1+\varepsilon(1-x)]e^{1-x} + O(\varepsilon^2).$$

For small $\varepsilon$, this solution is close to the exact solution (A1,14) everywhere, except in a small interval at $x=0$, where the exact solution (A1,14) changes rapidly in order to retrieve the boundary condition there which is about to be lost.

In order to determine an expansion valid in the boundary layer, $x = O(\varepsilon)$, one magnifies the independent variable as

$$\xi = \frac{x}{\varepsilon}$$

so that equation (A1,12) becomes

(A1,21)
$$\frac{d^2y}{d\xi^2} + \frac{dy}{d\xi} + \varepsilon y = 0.$$

Seek now an inner expansion

(A1,22)
$$y_\varepsilon^i(\varepsilon\xi;\varepsilon) = \sum_{n=0}^{N-1} b_n(\xi)\varepsilon^n + O(\varepsilon^N)$$

---

[†] Since the system (A1,17) cannot take on both of the boundary conditions, in general, and one of these boundary conditions, $y(0)=a$, should be dropped.

where in accordance with the inner limit process

(A1,23)
$$b_m(\xi) = \lim_{\substack{\varepsilon \to 0 \\ \xi \text{ fixed}}} \frac{y_\varepsilon^i - \sum_{n=0}^{m-1} b_n(\xi)\varepsilon^n}{\varepsilon^m} .$$

Substituting (A1,22) in equation (A1,12), and equating the coefficients of equal powers of $\varepsilon$, one obtains

(A1,24)
$$\frac{d^2 b_0}{d\xi^2} + \frac{db_0}{d\xi} = 0, \qquad \frac{d^2 b_1}{d\xi^2} + \frac{db_1}{d\xi} = -b_0, \quad \text{etc.}$$

Noting that the inner solution is valid only in the region $x=O(\varepsilon)$, one has

(A1,25)
$$b_0(0) = a, \quad b_1(0) = 0, \quad \text{etc.}$$

Thus one finds

(A1,26)
$$\begin{cases} b_0(\xi) = a - A_0(1-e^{-\xi}); \\ b_1(\xi) = A_1(1-e^{-\xi}) - [a - A_0(1+e^{-\xi})]\xi; \\ \text{etc.}, \end{cases}$$

so that

(A1,27)
$$y_\varepsilon^i = a - A_0(1-e^{-\xi}) + \varepsilon\left\{A_1(1-e^{-\xi}) - [a - A_0(1+e^{-\xi})]\xi\right\} + O(\varepsilon^2).$$

Asymptotic matching principle, according to (A1,11), states

$$be + \varepsilon[be-be\xi] + O(\varepsilon^2) = (a - A_0) + \varepsilon[A_1 - (a-A_0)\xi] + O(\varepsilon^2),$$

from which one has immediately

(A1,28)
$$A_0 = a - be, \quad A_1 = be$$

so that

(A1,29)
$$y_\varepsilon^i = be + (a-be)e^{-\xi} + \varepsilon\left\{be(1-e^{-\xi}) - [be-(a-be)e^{-\xi}]\xi\right\} + O(\varepsilon^2).$$

Note that writing the outer expansion (A1,20) in terms of the inner variable $\xi=x/\varepsilon$, then, we have

$$y_\varepsilon^0 = bee^{-\varepsilon\xi} + \varepsilon(1-\varepsilon\xi)bee^{-\varepsilon\xi} + O(\varepsilon^2)$$

which leads to the $O(\varepsilon^2)$ approximation

(A1,30)      $(y_\varepsilon^0)^1 \cong be[1 + \varepsilon(1-\xi)].$

Analogously, writing the inner expansion (A1,27) in terms of the outer variable x for x>0, the $e^{-\xi}$ terms are asymptotically negligible and we have the $O(\varepsilon^2)$ approximation

(A1,31)      $(y_\varepsilon^1)^0 \cong (a-A_0)(1-x) + \varepsilon A_1.$

Since $\xi=\frac{x}{\varepsilon}$, matching (to this order) will be accomplished by selecting

$$a-A_0 = be \implies A_0 = a-be, \quad A_1 = be,$$

which agree with (A1,28).

Exressing all approximations in terms of the outer variable x, then we have the composite approximation

(A1,32)      $y_\varepsilon^c(x) \approx [be^{1-x} + (a-be)(1+x)e^{-x/\varepsilon}] + \varepsilon[(1-x)be^{1-x} - bee^{-x/\varepsilon}],$

which should be compared with the exact solution previously given.

The inner and outer expansions are depicted in Fig.1. The asymptotic solution (see, Fig.2) follows the inner solution near x=0 and the outer solution for x>0.

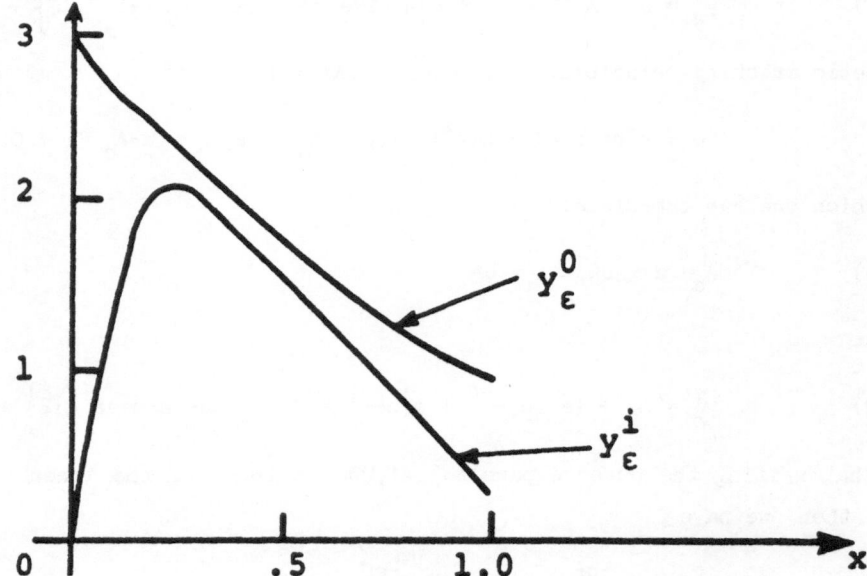

Figure 1. The inner expansion $y_\varepsilon^i$ and the outer expansion $y_\varepsilon^0$ for $\varepsilon y'' + y' + y = 0$, $y(0)=0$, $y(1)=1$; $\varepsilon=0,1$.

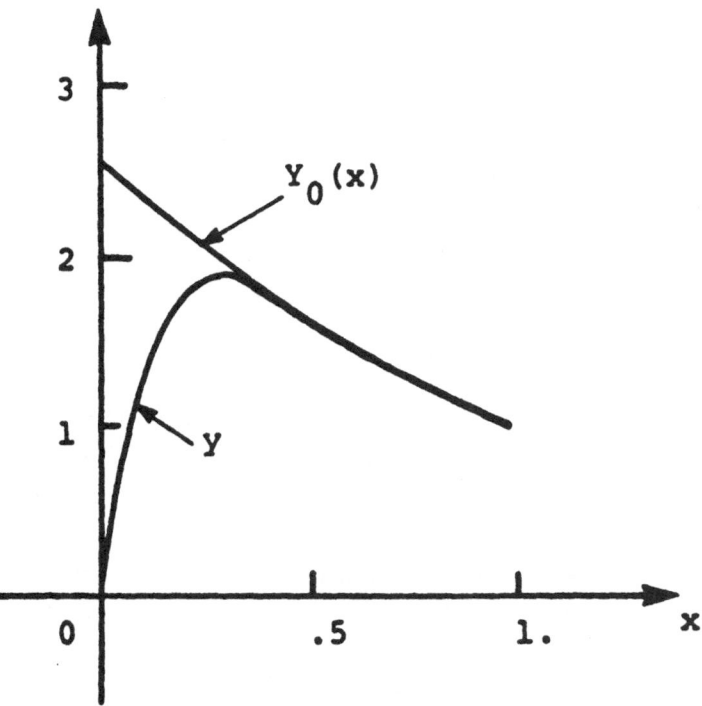

Figure 2. Nonuniform convergence of the solution $y_\varepsilon$ of $\varepsilon y''+y'+y=0$, $y(0)=0$, $y(1)=1$ to the solution $e^{1-x}$ of the reduced problem.

*REFERENCES TO WORK CITED*
*IN THE TEXT*

FRAENKEL, L.E. (1969) _ Proc. Cambridge Phil. Soc. 65, 209-284.

KAPLUN, S. (1967) _ In *Fluid Mechanics and Singular Perturbations* (P. A. Lagerstrom, L. N. Howard, and C. S. Liu, eds.) Academic Press, New York.

KAPLUN, S. and LAGERSTROM, P.A. (1957) _ J. Math. Mech. 6, 585.

KEVORKIAN, J. and COLE, J.D. (1981) _ *Perturbation Method in Applied Mathematics.*
Springer-Verlag, New York.

LAGERSTROM, P.A. and CASTEN, R.G. (1972) _ S.I.A.M. Review 14, 63-120.

O'MALLEY, R.E.,Jr (1974) _ *Introduction to Singular Perturbations.* Academic Press, New York.

PRANDTL, L. (1905) _ Verh. Int. Math. Kongr.3rd, pp.484-491. Tuebner, Leipzig.

PRANDTL, L. (1928) _ NACA TM-452.

SHIVAMOGGI, B.K. (1978) _ ZAMM, 58, 354-356.

VAN DYKE, M.D. (1964) _ *Perturbation Methods in Fluid Mechanics.* Accademic Press, New york.

VAN DYKE, M.D. (1975) _ *Perturbation Methods in Fluid Mechanics.* Parabolic Press, Stanford.

ZEYTOUNIAN, R.Kh. (1986) _ *Les Modèles Asymptotiques de la Mécanique des Fluides I*. Springer-Verlag, Heidelberg. Lecture Notes in Physics, vol. 245.

ZEYTOUNIAN, R.Kh. (1987) _ *Les Modèles Asymptotiques de la Mécanique des Fluides II*. Springer-Verlag, Heidelberg. Lecture Notes in Physics, vol. 276.

## TWO-VARIABLE EXPANSIONS

Various physical problems are characterized by the presence of a small disturbance which, because of being active over a long time, has a non-negligible *cumulative* effect.

In solving a given problem, the main is to combine appropriate techniques to construct an expansion which is uniformly valid over long time intervals. A central feature of the method is the *non existence*, for long times, of a limit process expansion of the type used in the previous Appendix 1. Since limit process expansions are not applicable, successive terms in the solution cannot be calculated by the repeated application of limits, and more importantly, rules must be established for the calculation of these terms.

A typical elementary example illustrating the kinds of problems which arise is the effect of a small linear damping on a linear oscillator.

We have the following problem:

$$(A2,1) \qquad \frac{d^2y}{dt^2} + 2\varepsilon \frac{dy}{dt} + y = 0, \quad y(0)=0, \qquad \frac{dy}{dt}\bigg|_{t=0} =1,$$

where $\varepsilon \ll 1$.

The physical phenomena described by equation (A2,1) occur over two-time scales as can be seen clearly if the exact solution is written:

$$(A2,2) \qquad y = \frac{e^{-\varepsilon t}}{\sqrt{1-\varepsilon^2}} \sin(\sqrt{1-\varepsilon^2}\, t).$$

For this example it is evident that nontrivial correction terms can be obtained only if $y(t;\varepsilon)$ is expanted in powers of $\varepsilon$ as follows

$$(A2,3) \qquad y(t;\varepsilon) = \sin t - \varepsilon t \sin t + O(\varepsilon^2 t^2)$$

and in this approximation, the amplitude decays linearly with time. If we consider the exact solution (A2,2) we see that the amplitude $(1-\varepsilon t)$ is the expansion of $e^{-\varepsilon t}$ to this order. Clearly, this expansion is only valid for t *fixed and finite*. Thus, this regular perturbation is identical to an inner limit process expansion as discussed in Appendix 1. For any finite time interval $0<t<T$, $\varepsilon$ can be chosen sufficiently small so that solution (A2,3) is

a good approximation (uniformly valid to $O(\varepsilon)$ of the exact result of (A2,2).

Because in this limit $e^{-\varepsilon t}$ and $\sin(\sqrt{1-\varepsilon^2}\, t)$ are approximated by power series in $\varepsilon t$ and $\varepsilon^2 t$, the results are not uniformly valid over the entire time interval.

Therefore, this expansion is associated with the limit process

$$\varepsilon \to 0, \quad t \text{ fixed}$$

and is only initially valid due to the presence of the $\varepsilon t \sin t$ term. In this example, the first mixed *secular* term we encounter is due to the nonuniform representation for large times of the $e^{-\varepsilon t}$ term in the exact solution. Mixed secular terms of the form $\varepsilon^2 t \cos t$ will occur next from the nonuniform representation of the $\sin(\sqrt{1-\varepsilon^2}\, t)$ term in (A2,2).

It is also evident that mutually contradictory requirements will arise if we wish to represent $e^{-\varepsilon t}$ and $\sin(\sqrt{1-\varepsilon^2}\, t)$ uniformly for long times. In fact, the only uniformly valid representation of $e^{-\varepsilon t}$ for long times is $e^{-\varepsilon t}$ itself; therefore we need the limit process $\varepsilon \to 0$, $\tilde{t} = \varepsilon t$ fixed. However, this limit process *does not* exist for $\sin(\sqrt{1-\varepsilon^2}\, t) = \sin\left[\dfrac{\sqrt{1-\varepsilon^2}}{\varepsilon}\, \tilde{t}\right]$ as the argument of the sine tends to infinity. Another way of saying this is that the function defined by (A2,2) does not have an outer expansion.

Any general method that is proposed must, be able to cope simultaneously with these two difficulties. The essence of the two-variable method is to represent the solution as a general asymptotic expansion where each term can be uniquely expressed as a function of the two time variables

(A2,4)
$$\tau = \left[1 + \sum_{n=2}^{\infty} \omega_n \varepsilon^n\right] t = \text{"fast time"};$$
$$\tilde{t} = \varepsilon t = \text{"slow time"},$$

where the $\omega_n$ are unknown constants.

*TWO-VARIABLE ANALYSIS*

In addition to the ("slow-time") variable $x$, one introduces another ("fast-time") variable $\eta$ ranging over an unbounded interval. If, for example, nonuniform convergence occurred at $x = x_0$, the fast variable

(A2,5)
$$\eta = \frac{1}{\varepsilon} \int_{x_0}^{x} g(s)ds$$

for some positive g might be appropriate.

Selection of a proper fast variable may be based on different physical times scales occurring, may be motivated by simple model equations, or may be left somewath arbitrary initially.

One seeks a solution $y(x,\eta;\varepsilon)$ which is a function of *both* the slow and fast variables. The original differential equation becomes a *partial* differential equation in the variables x *and* $\eta$ and an asymptotic solution of it is sough having a power series solution

(A2,6)
$$y(x,\eta;\varepsilon) \sim \sum_{j=0}^{\infty} y_j(x,\eta)\varepsilon^j \quad \text{as } \varepsilon \to 0,$$

where the coefficients are bounded for $0 \leq x \leq 1$ and for all $\eta \geq 0$.

Substituting this expansion into the differential equation and equating coefficients of likes powers of $\varepsilon$, we obtain partial differential equations for the coefficients $y_j(x,\eta)$. Applying the boundary conditions will generally not suffice to determine the term $y_j$ successively. In addition to the *boundedness* condition on $y_j$'s, additional conditions must be imposed[†]. Following Poincaré, one usually asks that certain "*secular terms*" (like, e.g., $\eta^k e^{-\eta}$, k>0) be *eliminated*. These somewhat arbirary requirements can be motivated mathematically. We note that it was necessary to determine the form of the second-order coefficient $y_1$ (by boundedness and *secularity* conditions) in order to completely obtain the first coefficient $y_0$. This is typically the case of for two-time methods.

It is important to also observe that equations with slowly varying coefficients arise in a variety of physical applications and engeneering approximations. Two-time techniques are well suited to such problems.

As an illustration of the two-variable method, consider the two-point problem

(A2,7)
$$\begin{cases} \varepsilon\frac{d^2y}{dx^2} + a(x)\frac{dy}{dx} + b(x)y = 0, \\ \\ y(0) = \alpha, \quad y(1) = \beta, \end{cases}$$

where a(x) and b(x) are infinitely differentiable functions on $0 \leq x \leq 1$ and

---

[†] We do not expect to determine the $y_j$'s uniquely, since the expansion sought is a generalized asymptotic expansion, but we need to eliminate some arbitrariness.

$a(x)>0$ there. By analogy to the constant coefficient problem treated previously, one might expect nonuniform convergence as $\varepsilon\to 0$ near x=0 and that the fast variable[†]

(A2,8)
$$\eta = \frac{1}{\varepsilon} \int_0^x a(s)ds$$

might be appropriate.

Proceeding with this $\eta$ requires the solution $y(x,\eta;\varepsilon)$ to satisfy the *partial* differential equation

$$\frac{a^2(x)}{\varepsilon}\left[ \frac{\partial^2 y}{\partial \eta^2} + \frac{\partial y}{\partial \eta} \right] + 2a(x)\frac{\partial^2 y}{\partial x \partial \eta} + \frac{da}{dx}\frac{\partial y}{\partial \eta} + a(x)\frac{\partial y}{\partial x}$$

$$+ b(x)y + \varepsilon\frac{\partial^2 y}{\partial x^2} = 0$$

since

$$\frac{d}{dx} = \frac{\partial}{\partial x} + \frac{a(x)}{\varepsilon}\frac{\partial}{\partial \eta}$$

and

$$\frac{d^2}{dx^2} = \frac{\partial^2}{\partial x^2} + \frac{2a(x)}{\varepsilon}\frac{\partial^2}{\partial x \partial \eta} + \frac{da}{dx}\frac{1}{\varepsilon}\frac{\partial}{\partial \eta} + \frac{a^2(x)}{\varepsilon^2}\frac{\partial^2}{\partial \eta^2}.$$

---

[†] The condition under which boundary layer theory (Appendix 1) can be applied are easily deduced from a canonical form of equation (A2,7).
Under the transformation
$$y(x) = \exp\left[-\frac{1}{2\varepsilon}\int_0^x a(s)ds\right]Y(x)$$
the equation (A2,7) takes the form
$$\varepsilon\frac{d^2Y}{dx^2} - \left\{\frac{a^2(x)}{4\varepsilon} + \frac{1}{2}\frac{da}{dx} - b(x)\right\}Y = 0.$$

A sufficient condition for the existence of *simple boundary layers* as $\varepsilon\to 0$ is merely that
$$a(x)\neq 0, \quad \left|\frac{da}{dx}\right|<\infty, \quad b(x)<\infty, \quad 0\leq x\leq 1.$$

Then it always is possible to find sufficiently small $\varepsilon$ so that the inequality
$$\frac{a^2(x)}{4\varepsilon} + \frac{1}{2}\frac{da}{dx} - b(x) > 0, \quad 0\leq x\leq 1,$$
is satisfied and in this case $d^2Y/dx^2 > 0$ when $Y>0$.

Substituting the expansion

$$y(x, \eta, \varepsilon) \sim \sum_{j=0}^{\infty} y_j(x, \eta)\varepsilon^j$$

into this differential equation, we formally equate coefficients of each power of $\varepsilon$ separately to zero. Thus, from the coefficient of $\frac{1}{\varepsilon}$, we have

$$a^2(x)\left[ \frac{\partial^2 y_0}{\partial \eta^2} + \frac{\partial y_0}{\partial \eta} \right] = 0$$

and, integrating, we obtain

(A2,9)             $y_0(x, \eta) = A_0(x) + C_0(x)e^{-\eta},$

where $A_0$ and $C_0$ are undetermined. Likewise, from the coefficient of $\varepsilon^0$, we have

$$a^2(x)\left[ \frac{\partial^2 y_1}{\partial \eta^2} + \frac{\partial y_1}{\partial \eta} \right] = -2a(x)\frac{\partial^2 y_0}{\partial x \partial \eta} - \frac{da}{dx}\frac{\partial y_0}{\partial \eta} - a(x)\frac{\partial y_0}{\partial x} - b(x)y_0.$$

Integrating with respect to $\eta$, then,

$$a^2(x)\left[ \frac{\partial y_1}{\partial \eta} + y_1 \right] = -2a(x)\frac{\partial y_0}{\partial x} - \frac{da}{dx} y_0 - \int^{\eta}\left[ a(x)\frac{\partial y_0}{\partial x} - b(x)y_0 \right]d\eta$$

and, substituting for $y_0$, we have

$$a^2(x)\left[ \frac{\partial y_1}{\partial \eta} + y_1 \right] + \left[ a(x)\frac{dA_0}{dx} + b(x)A_0 \right]\eta$$

$$+ \left[ a(x)\frac{dC_0}{dx} + \frac{da}{dx} C_0 - b(x)C_0 \right]e^{-\eta} = a^2(x)A_1(x)$$

for $A_1$ arbitrary. We then integrate with respect to $\eta$. Since $y_1$ must remain bounded as $\eta \rightarrow \infty$ (i.e., as $\varepsilon \rightarrow 0$ for $x>0$), we must have

(A2,10)             $a(x)\frac{dA_0}{dx} + b(x)A_0(x) = 0.$

Likewise, to avoid a secular term in $y_1$ which is a multiple of $\eta e^{-\eta}$, we must have

(A2,11)             $a(x)\frac{dC_0}{dx} + \frac{da}{dx} C_0 - b(x)C_0 = 0.$

Thus, integration implies that $y_1$ has the form

(A2,12)             $y_1(x, \eta) = A_1(x) + C_1(x)e^{-\eta}$

where $A_1$ and $C_1$ are so far arbitrary. Applying the boundary conditions to

$$y = A_0(x) + C_0(x)e^{-\eta} + \varepsilon(\ldots),$$

noting that $e^{-\eta}$ is asymptotically negligible at $x=0$, implies that we must select

(A2,13) $\qquad A_0(1) = \beta \quad$ and $\quad C_0(0) = \alpha - A_0(0)$.

Summarizing, then, we have begun to develop an asymptotic solution of the form

(A2,14) $\qquad y(x,\eta;\varepsilon) = A_0(x) + C_0(x)e^{-\eta} + \varepsilon[A_1(x) + C_1(x)e^{-\eta}] + O(\varepsilon^2)$

where

(A2,15)
$$\begin{cases} A_0(x) = \beta \, \exp\left[-\int_1^x \frac{b(s)}{a(s)} \, ds\right], \\[2mm] C_0(x) = \frac{a(0)}{a(x)}\left[\alpha - A_0(0)\right]\exp\left[\int_0^x \frac{b(s)}{a(s)} \, ds\right], \end{cases}$$

and $A_1$ and $C_1$ are, as yet, undetermined.

*LINEAR OSCILLATOR*
*WITH SMALL DAMPING*[†]

In finish discussion of two-variable method by studying the damped linear oscillator. We consider the problem (A2,1) and we assume that the solution has a general asymptotic expansion of the form

(A2,16) $\qquad y = F_0(\tau,\tilde{t}) + \varepsilon F_1(\tau,\tilde{t}) + \varepsilon^2 F_2(\tau,\tilde{t}) + \ldots \, .$

involving the fast and slow times

$$\tau = (1 + \varepsilon^2\omega_2 + \varepsilon^3\omega_3 + \ldots)t, \quad \tilde{t} = \varepsilon t.$$

We then use the chain rule to calculate

---

[†] According to Kevorkian and Cole (1981).

$$\frac{dy}{dt} = \frac{\partial F_0}{\partial \tau}(1 + \varepsilon^2\omega_2) + \varepsilon\frac{\partial F_0}{\partial \tilde{t}} + \varepsilon\frac{\partial F_1}{\partial \tau} + \varepsilon^2\frac{\partial F_1}{\partial \tilde{t}} + \varepsilon^2\frac{\partial F_2}{\partial \tau} + O(\varepsilon^3),$$

(A2,17)

$$\frac{d^2y}{dt^2} = \frac{\partial^2 F_0}{\partial \tau^2}(1 + 2\varepsilon^2\omega_2) + 2\varepsilon\frac{\partial^2 F_0}{\partial \tau\partial \tilde{t}} + \varepsilon^2\frac{\partial^2 F_0}{\partial \tilde{t}^2} + \varepsilon\frac{\partial^2 F_1}{\partial \tau^2} + 2\varepsilon^2\frac{\partial^2 F_1}{\partial \tau\partial \tilde{t}}$$

$$+ \varepsilon^2\frac{\partial^2 F_2}{\partial \tau^2} + O(\varepsilon^3).$$

Thus, the sequence of equation for the $F_n$ are

(A2,18a) $$\mathcal{L}(F_0) \equiv \frac{\partial^2 F_0}{\partial \tau^2} + F_0 = 0,$$

(A2,18b) $$\mathcal{L}(F_1) = -2\frac{\partial^2 F_0}{\partial \tau\partial \tilde{t}} - 2\frac{\partial F_0}{\partial \tau},$$

(A2,18c) $$\mathcal{L}(F_2) = -2\omega_2\frac{\partial^2 F_0}{\partial \tau^2} - \frac{\partial^2 F_0}{\partial \tilde{t}^2} - 2\frac{\partial^2 F_1}{\partial \tau\partial \tilde{t}} - 2\frac{\partial F_0}{\partial \tilde{t}} - 2\frac{\partial F_1}{\partial \tau},$$

etc.

The first of these is the equation for the free oscillations, while the remainder have the appearance of forced linear oscillations. Howevere, since $F_0 = F_0(\tau, \tilde{t})$, the free linear oscillations which are the solutions to (A2,18a) have the possibility of being slowly modulated. Thus, we have

(A2,19) $$F_0(\tau, \tilde{t}) = A_0(\tilde{t})\cos\tau + B_0(\tilde{t})\sin\tau.$$

According to expansions (A2,16) and (A2,17), the initial conditions of the problem (A2,1) become

$$F_0(0,0) = 0, \quad \frac{\partial F_0}{\partial \tau}(0,0) = 1,$$

(A2,20) $$F_1(0,0) = 0, \quad \frac{\partial F_1}{\partial \tau}(0,0) = -\frac{\partial F_0}{\partial \tilde{t}}(0,0),$$

$$F_2(0,0) = 0, \quad \frac{\partial F_2}{\partial \tau}(0,0) = -\frac{\partial F_1}{\partial \tilde{t}}(0,0) - \omega_2\frac{\partial F_0}{\partial \tau}(0,0).$$

Conditions for $F_0$ yields initial conditions for $A_0$ and $B_0$:

(A2,21) $$A_0(0) = 0, \quad B_0(0) = 1.$$

Nothing more can be found out about $A_0(\tilde{t})$ and $B_0(\tilde{t})$ without considering $F_1$.

334

Substituting for $F_0$ into the right-hand side of equation (A2,18b) gives:

(A2,22)
$$\mathcal{L}(F_1) = 2\left[\frac{dA_0}{d\tilde{t}} + A_0\right]\sin\tau - 2\left[\frac{dB_0}{d\tilde{t}} + B_0\right]\cos\tau.$$

The bracketed terms on the right-hand side of equation (A2,22) are functions of $\tilde{t}$ *only*. Therefore the particular solutions corresponding to these terms would be functions of $\tilde{t}$ multiplied by the mixed secular terms $\tau\sin\tau$ or $\tau\cos\tau$. Such terms cannot be permitted to occur in the solution because they lead to unboundedness with respect to $\tau$. Alternately, we can use the consistency argument mentioned earlier: A term like $\varepsilon\tau\sin\tau$ is inconsistent with a unique $F_1$ since it could equally well be relabelled $\tilde{t}\sin\tau + O(\varepsilon^2)$ and would become $O(1)$ instead of being $O(\varepsilon)$ as assumed.

Therefore, we must eliminate all homogeneous solutions of $\mathcal{L}(F_1)=0$ and this give the two first order ordinary differential equations for $A_0$ and $B_0$:

$$\frac{dA_0}{d\tilde{t}} + A_0 = 0, \qquad A_0(0) = 0$$

$$\frac{dB_0}{d\tilde{t}} + B_0 = 0, \qquad B_0(0) = 1$$

and we find that

(A2,23)
$$A_0(\tilde{t}) = 0, \quad B_0(\tilde{t}) = e^{-\tilde{t}}.$$

The uniformly valid expansion thus far is

(A2,24)
$$y(t,\varepsilon) = e^{-\tilde{t}}\sin\tau + \varepsilon\left\{A_1(\tilde{t})\cos\tau + B_1(\tilde{t})\sin\tau\right\} + O(\varepsilon^2).$$

Comparing this with the exact solution (A2,2), we see that the first term $e^{-\varepsilon t}\sin t$ is indeed the correct uniformly valid approximation of the exact result to $O(1)$.

Now, $A_1(\tilde{t})$, $B_1(\tilde{t})$ and the frequency shift $\omega_2$ are to be found from similar considerations applied to (A2,18c) for $F_2$. Thus far, we have

$$F_0 = e^{-\tilde{t}}\sin\tau, \quad F_1 = A_1(\tilde{t})\cos\tau + B_1(\tilde{t})\sin\tau,$$

and

(A2,25)
$$\mathcal{L}(F_2) = \left\{2\left[\frac{dA_1}{d\tilde{t}} + A_1\right] + (2\omega_2 + 1)e^{-\tilde{t}}\right\}\sin\tau - 2\left\{\frac{dB_1}{d\tilde{t}} + B_1\right\}\cos\tau.$$

First, repeating the argument that homogeneous solutions of $\mathcal{L}(F_2)=0$ cannot be permitted, we must set the bracketed terms in (A2,25) equal to zero. Solving the resulting equations for $A_1$ and $B_1$ subject to the initial conditions $A_1(0)=B_1(0)=0$ (which follow from (A2,20), for $F_2$) we find

(A2,26)
$$A_1(\tilde{t}) = -\left[\omega_2 + \frac{1}{2}\right]\tilde{t}e^{-\tilde{t}}, \qquad B_1(\tilde{t}) = 0.$$

This means that $\varepsilon F_1$ would be proportional to $\varepsilon \tilde{t} e^{-\tilde{t}}\cos\tau$. Again, such a term cannot be consistent because it can also be written as $\varepsilon^2 e^{-\tilde{t}}\tau\cos\tau + O(\varepsilon^3)$ and shift to $O(\varepsilon^2)$ in the expansion. One could also have required that $|F_2/F_1|$ be bounded to disallow such a term. Therefore, we must set

(A2,27)
$$\omega_2 = -\frac{1}{2}.$$

All the necessary reasoning has now been explained to carry out the solution to any order and, in fact, to solve a wide variety of weakly nonlinear problems having the form

$$\frac{d^2y}{dt^2} + y + \varepsilon f\left[y, \frac{dy}{dt}\right] = 0.$$

For the problem (A2,1), the result

(A2,28)
$$y(t,\varepsilon) = \exp(-\varepsilon t)\sin\left[\left[1 - \frac{\varepsilon^2}{2} + O(\varepsilon^2)\right]t\right] + O(\varepsilon^2)$$

is seen to be the uniformly valid general asymptotic expansion of the exact solution to $O(\varepsilon)$ for times t of order $1/\varepsilon$.

For a general discussion of multiple scale methods see also Nayfeh (1973; Chapter 6).

*REFERENCES TO WORK CITED IN THE TEXT*

KEVORKIAN, J. and COLE, J.D. (1981) _ *Perturbation Methods in Applied Mathematics*. Springer-Verlag, New York.

NAYFEH, A.H. (1973) _ *Perturbation Methods*. Wiley, New York.

# BIBLIOGRAPHY

BEER, T. 1974 _ *Atmospheric waves*
Adam Hilger, London.


DRAZIN, P.G. and REID, W.H. 1981 _ *Hydrodynamic stability*.
Cambridge University Press.


ESKINAZI ,S. 1975 _ *Fluid Mechanics and Thermodynamics of our environment*.
Academic Press Inc., New York.


FRIEDLANDER, S. 1980 _ *An introduction to the Mathematical Theory of Geophysical Fluid Dynamics*.
North-Holland Maths. Studies, 41. Amsterdam.


HOUGHTON, J.T. 1977 _ *The Physics of Atmospheres*.
Cambridge University Press.


KIBEL, I.A. 1963 _ *An Introduction to the Hydrodynamical Method of short period weather Forecasting*.
The Mac Milan Company.


LEBLOND, P.H. and MYSAK, L.A. 1978 _ *Waves in the Ocean*. Elsevier Scientific Publishing Company, Amsterdam.


LIGHTHILL, J. 1978 _ *Waves in Fluids*.
Cambridge University Press.


MANDELBROT, B.B. 1982 _ *The Fractal Geometry of Nature*.
W.H. Freeman and Company, New York.


MEYER, R.E. 1971 _ *Introduction to Mathematical Fluid Dynamics*.
Wiley-Interscience, New York.


MONIN, A.S. 1972 _ *Weather Forecasting as a Problem in Physics*.
MIT Press. Cambridge Mass. U.S.A.

338

NAYFEH, A.H. 1973 _ *Perturbation Methods.*
      Wiley, New York.


PEDLOSKY, J. 1979 _ *Geophysiscal Fluid Dynamics.*
      Springer-Verlag, New York.


SCHUSTER, H.G. 1984 _ *Deterministic chaos; An Introduction.*
      Physik-Verlag, Weinheim, R.F.A.


SCORER, R.S. 1978 _ *Environmental Aerodynamics.*
      Wiley, England.


VAN DYKE, M. 1975 _ *Perturbations Methods in Fluid Mechanics.*
      Parabolic Press, Stanford.


WHITHAM, G.B. 1974 _ *Linear and nonlinear Waves.*
      J. Wiley et sons.


YIH, C-S. 1980 _ *Stratified Flows.*
      Academic Press, London.


ZEYTOUNIAN, R.Kh. 1990 _ *Asymptotic modeling of Atmospheric flows.*
      Springer-Verlag, Heidelberg.


*Suggestions for further reading:*


BENGTSSON, L., GHIL, M. and KÄLLÉN, E. (eds). 1981 _ *Dynamic Meteorology: Data*
                    *Assimilation Methods.*
                  Springer-Verlag, N-Y.


BERGER, A. (ed.). 1981 _ *Climatic Variations and Variability : Facts and*
          *Theories.* D. Reidel, Dordrecht.


BLACKMON, M. (ed.). 1978 _ *The general Circulation : Theory, Modeling and*
          *Observations.* Notes from a summer Colloquium.
          NCAR/CQ _6+1978_ ASP, NCAR, Boulder, Co.

BURRIDGE, D.M. and KÄLLÉN, E. (eds.). 1984 _ Problems and Prospects in Long and Medium Range Weather Forecasting. Springer-Verlag, Berlin.

CHAMBERLAIN, J.W. 1978 _ Theory of Planetary atmospheres: An Introduction to Their Physics and Chemistry. Academic Press, N-Y.

CHANG, J. (ed.). 1977 _ General Circulation Models of the Atmosphere. Methods Comput. Phys. 17, Academic Press, N-Y.

CHARNEY, J.G. 1973 _ Planetary Fluid Dynamics. In MOREL (1973), pp. 97-351.

COANTIC, M.F. 1978 _ An Introduction to Turbulence in Geophysics, and Air-Sea Interactions. AGARDOGRAPH N°232, London.

COLBECK, S.C. (ed.). 1980 _ Dynamics of Snow and Ice masses. Academic Press, N-Y.

CORBY, G.A. (ed.). 1969 _ The Global Circulation of the Atmosphere. Royal Meteo. Soc., London.

CRAIG, R.A. 1965 _ The upper Atmosphere. Meteorology and Physics. Academic Press, N-Y.

GHIL, M., BENZI, R. and PARISI, G. (eds.). 1985 _ Turbulence and Predictability in Geophysical Fluid Dynamics and Climate Physics. North-Holland, Amsterdam.

GILL, A.E. 1982 _ Atmosphere-Ocean Dynamics. Academic Press.

HALTINER, G.J. and WILLIAMS, R.T. 1980 _ Numerical Prediction and Dynamic Meteorology, 2nd. ed. J. Wiley, N-Y.

HOLLOWAY, G. and WEST, B.J. (eds.). 1984 _ Predictability of Fluid Motions. American Institute of Physics, N-Y.

HOLTON, J.R. 1979 _ *An Introduction to Dynamical Meteorology*, 2nd ed. Academic Press, N-Y.

HORTON, C.W., REICHL, L.E. and SZEBEHELY (eds.).1983 _ *Long-Time Prediction in Dynamics*. J.Wiley, N-Y.

HOSKINS, B.J. and PEARCE, R.P. (eds). 1983 _ *Large-scale Dynamic Processes in the atmosphere*. Academic Press, London.

LORENZ, E.N. 1967 _ *The Nature and Theory of the General Circulation of the Atmosphere*. WMO, Geneva.

MOREL, P. (ed.). 1973 _ *Dynamic Meteorology*. D. Reidel Publ. Co., Dordrecht Holland.

NICOLIS, G. (ed.). 1987 _ *Irreversible Phenomena and Dynamical Systems Analysis in the Geosciences*. D.Reidel Dordrecht Holland.

ROBERTS, P.H. and SOWARD A.M. (eds.). 1978 _ *Rotating Fluid in Geophysics*. Academic Press, London.

SWINNEY, H.L. and GOLLUB, J.P. (eds.). 1981 _ *Hydrodynamic Instabilities and the Transition to Turbulence*. Springer-Verlag, Berlin.

ZEYTOUNIAN, R.Kh. (ed.). 1988 _ *Atmospheric Flows: Asymptotic Modelling and Numerical Simulations*. Special issue J.M.T.A, supplement n°2 to vol. 7, Paris, Gauthier-Villars.

# AUTHOR INDEX[†]

---

† See also the Bibliography pp. 337-340.

# SUBJECT INDEX

Absolute acceleration 2
— Velocity 1,2
Acoustic waves 40,85
Ackerblom's problem 159
Adjustment 90,91,97,101,105,117
Airy functions 51,52
Anelastic equations 296
Angular velocity 1
Antibreeze 165
Aperiodic regime 239
Asymptotic expansions 317
— sequences 317
Average value 58
Balance equation 303
Baroclinic 191
Baroclinic instability 197,200
Barotropic 40,57,58,190
Barotropic instability 203
Bénard problem 222,225
Bessel functions 126
$\beta$-parameter 23
$\beta$-plane approximation 20-22
Bifurcations 238,275
Bjerkne's theorem 4
Boundary layer problem 155,
172-173,315,321
Boussinesq approximation 68,214
— equations 30,264
— filtering 86
— gravity waves 44
— number 7,213,258,292
Breezes 161
Brunt-Väisälä frequency 8

Burger number 193
Cantor set 236,312
Chaos 234-279
Characteristic rays 132
Conditions for instability 202
Conservation law 58
Consistency 118
Contraction of volume 244
Convective instability 212-231
Coriolis acceleration 3
— parameter 3
Correlation function 235,237
Curvilinear coordinates 16,21
Deep convection equations 263,296-299
Depth parameter 264
Dimension 246
Dispersion relation 190,218
Dispersive wave 57
Dissipation function 15
Dissipative flow 234
Domain of attraction 235
Dominant equations 130,262
Dorodnitsyn method 140
Doublings 248-249
Dynamical system 234
Eady problem 191
Eigenvalue problem 216
— relation 196
Ekman number 9-10,198
— region 156
Energy equation 12
Entropy 245

## M. Hotine

### *Differential Geodesy*

Edited with commentaries by J.D.Zund,
New Mexico State University, Las Cruses, NM

With contributions by J.Nolten, B.Chovitz, C.Whitten

1991. Approx. 200 pp. 6 figs. 2 tabs. Hardcover
ISBN 3-540-53799-6

## R. P. Gupta

### *Remote Sensing Geology*

1991. Approx. 375 pp. 284 figs. 34 tabs. Hardcover
ISBN 3-540-52805-9

## R. J. Huggett

### *Climate, Earth Processes and Earth History*

1991. XIV, 281 pp. 71 figs. Softcover
ISBN 3-540-53419-9

## R. Zeytounian

### *Asymptotic Modeling of Atmospheric Flows*

Translated from the French by L. Bry

1990. XII, 396 pp. 6 figs.
Hardcover ISBN 3-540-19404-5

Springer-Verlag
Berlin
Heidelberg
New York
London
Paris
Tokyo
Hong Kong
Barcelona
Budapest

# Lecture Notes in Physics

For information about Vols. 1–365
please contact your bookseller or Springer-Verlag

## New Series m: Monographs